Impulsive differential equations:
periodic solutions and applications

π Pitman Monographs and
Surveys in Pure and Applied Mathematics 66

Impulsive differential equations: periodic solutions and applications

Drumi Bainov

Academy of Medicine, Sofia

and

Pavel Simeonov

Plovdiv University

CRC Press
Taylor & Francis Group
Boca Raton London New York

CRC Press is an imprint of the
Taylor & Francis Group, an **informa** business

First published 1993 by Longman Scientific & Technical

Published 2019 by CRC Press
Taylor & Francis Group
6000 Broken Sound Parkway NW, Suite 300
Boca Raton, FL 33487-2742

First issued in paperback 2019

No claim to original U.S. Government works

ISBN 13: 978-0-367-44984-1 (pbk)
ISBN 13: 978-0-582-09639-4 (hbk)

Visit the Taylor & Francis Web site at
http://www.taylorandfrancis.com

and the CRC Press Web site at
http://www.crcpress.com

AMS Subject Classification: 34C25

ISSN 0269-3666

British Library Cataloguing in Publication Data

A catalogue record for this book is
available from the British Library

Library of Congress Cataloging-in-Publication Data

Bainov, D. (Dimitur)
    Impulsive differential equations : periodic solutions and
applications / D. Bainov and P. Simeonov.
    p. cm. -- (Pitman monographs and surveys in pure and applied
mathematics ; 66)
    Includes bibliographical references and index.
    1. Differential equations. I. Simeonov, P. S. (Pavel S.)
II. Title. III. Series.
QA372.B273 1993
515'.352--dc20
                                    93-19957
                                      CIP

Contents

Preface

Impulsive differential equations are suitable for the mathematical simulation of evolutionary processes in which the parameters undergo relatively long periods of smooth variation followed by a short-term rapid change (that is, jumps) in their values. Processes of this type are often investigated in various fields of science and technology.

Particular impulsive differential equations simulating the work of concrete systems have been considered by many authors. As an example of this we shall point out the investigations of the shock model of a clock mechanism (Andronov et al. [6], Bautin [17], Kalitin [47–49], and Krylov and Bogolyubov [53]) and the investigations of the relaxation oscillations of electromechanical systems (Andronov et al. [6]).

The mathematical theory of systems that change by jumps is developing in two main directions.

One of these directions is delineated in the works on impulsive systems by Halanay and Wexler [34], Pandit and Deo [61], and Zavališčin et al. [91]. In these publications the apparatus of generalized functions is used and differential equations with fixed moments of the impulse effect are investigated regardless of the fact that the impulse effects can occur at moments when certain spatial or space–time relations are valid.

Investigations in the other direction were begun by Mil'man and Myshkis [57]. They give some general concepts of systems with an impulse effect and obtain the first results on the stability of their solutions. Their results make it possible to investigate impulsive systems for which the jumps take place when certain space–time relations are valid. The well-known classical apparatus of ordinary differential equations is used. The authors of the present book adhere to this direction.

The possibility of wide practical application of impulsive differential equations explains the still growing interest of many authors in the investigation of these equations in recent years and the publication of monographs on this subject by A. M. Samoilenko and N. A. Perestyuk [73], V. Lakshmikantham, D. D. Bainov, and P. S. Simeonov [55], and D. D. Bainov and P. S. Simeonov [15].

One of the important trends in the investigation of impulsive differential equations is related to the periodic solutions of these equations. To this subject many of the results obtained by M. U. Ahmetov, D. D. Bainov, V. I. Guţu, S. G. Hristova, V. Lakshmikantham, Xinzhi Liu, N. A. Perestyuk, A. M. Samoilenko, and P. S. Simeonov are devoted.

In the present book a systematic exposition of the results related to periodic solutions of impulsive differential equations is given and the potential for their application is illustrated.

The book consists of six chapters. Chapter I is introductory and offers a description of impulsive differential equations, as well as a brief exposition of auxiliary assertions about these equations which are used in subsequent chapters.

In Chapter II linear impulsive differential equations with fixed moments of an impulse effect are considered. The Floquet theory for these equations is expounded, the problem of the existence of periodic solutions of non-homogeneous linear equations in the non-critical and the critical cases is solved, and Hamiltonian periodic impulsive equations are considered.

In Chapters III and IV the method of the small parameter is elaborated for the three main classes of impulsive equations considered in the book. In Chapter III the non-critical case is considered, and Chapter IV deals with the critical case.

In Chapter V the problem of the existence of periodic solutions of non-linear impulsive equations is discussed, applying the method of the monodromy operator along trajectories (the Poincaré operator) and the method of upper and lower solutions.

In Chapter VI various approximate methods for finding periodic solutions of impulsive differential equations are considered, such as the monotone-iterative technique, a numerical-analytical method, the method of bilateral approximations, a projection-iterative method, and the method of boundary layer functions for singularly perturbed impulsive equations.

Possible practical applications of the results obtained are illustrated by a large number of concrete mathematical models, and an extensive bibliography on the subjects treated in the book is given.

The authors wish to express their gratitude to the staff at Longman and to Dr Alan Jeffrey for their support in publishing the book. They also express sincere thanks to Dr V. Covachev who helped in the preparation of the manuscript of the book, and to the Bulgarian Ministry of Education and Science for their partial support under Grant MM-7.

D. D. Bainov
P. S. Simeonov

Pleven, August 1992

Notation

$\mathbb{N}$	= set of all positive integers
$\mathbb{Z}$	= set of all integers
$\mathbb{R}$	= set of all real numbers
$\mathbb{C}$	= set of all complex numbers
$\mathbb{R}^n$ ($\mathbb{C}^n$)	= real (complex) n-dimensional Euclidean space
$[a, b]$	= closed interval $a \leq x \leq b$
(a, b)	= open interval $a < x < b$
$\langle a, b \rangle$	= arbitrary interval
$[a; b]$	= interval $[a, b]$ if $a < b$ or interval $[b, a]$ if $b \leq a$
$\mathbb{R}_+ = [0, \infty)$	= set of all non-negative real numbers
$\Re(z) = x, \Im(z) = y$	= real and imaginary parts of the complex number $z = x + iy$ $(i = \sqrt{-1})$
$z^* = x - iy$	= complex conjugate of $z = x + iy$
$x = \text{col}(x_1, \ldots, x_n)$	= column matrix, element of the space $\mathbb{R}^n$ ($\mathbb{C}^n$)
$(x \mid y) = \sum_{j=1}^n x_j y_j^*$	= scalar product in $\mathbb{R}^n$ (or $\mathbb{C}^n$)
$\mid x \mid$	= norm of the element $x \in \mathbb{R}^n$ ($x \in \mathbb{C}^n$)
$B_r(x_0) = \{x \in \mathbb{R}^n : \mid x - x_0 \mid < r\}$	
$d(x, M) = \inf_{y \in M} \mid x - y \mid$	= distance from $x \in \mathbb{R}^n$ to $M \subset \mathbb{R}^n$
$\bar{G}$	= closure of the set G
$A = (a_{ij})$	= matrix with entries a_{ij}
$A^* = (a_{ji}^*)$	= conjugate matrix of the matrix $A = (a_{ij})$
$\mid A \mid = \sup_{\mid x \mid = 1} \mid Ax \mid$	= norm of the matrix A
$\det A$	= determinant of the matrix A
$\text{Tr } A$	= trace of the matrix A
$\lambda_j(A)$	= eigenvalue of the matrix A
$\text{diag}(A_1, \ldots, A_k)$	= block diagonal matrix with blocks $A_1, \ldots, A_k$
$[X_1, \ldots, X_m]$	= matrix with columns $X_1, \ldots, X_m$
E	= identity operator or the unit $n \times n$ matrix
E_m	= unit $m \times m$ matrix
O_{mn}	= zero $m \times n$ matrix
O_m	= zero $m \times m$ matrix
$\mathbb{R}^{n \times m}$ ($\mathbb{C}^{n \times m}$)	= set of all matrices with real (complex) entries consisting of n rows and m columns

$$\frac{\partial f}{\partial x} = \left(\frac{\partial f_i}{\partial x_j}\right), \quad i = 1,\ldots,m, \quad j = 1,\ldots,n$$

$$= \text{Jacobi matrix of the function } f : \mathbb{R}^n \to \mathbb{R}^m,$$
$$x \to f(x)$$

$$\dot{x} = x' = \frac{dx}{dt} \qquad\qquad = \text{time derivative of the function } x(t)$$

$$x(a^+) = \lim_{t \to a+} x(t)$$
$$x(a^-) = \lim_{t \to a-} x(t)$$

If $-\infty < a \le b < +\infty$ and the sequence $\{\tau_k\}$ has no finite points of accumulation, then we denote by $i\langle a, b\rangle$ the number of members of the sequence $\{\tau_k\}$ lying in the interval $\langle a, b\rangle$.

If $\{p_k\}$ is a sequence, then $\sum_{\tau_k \in \langle a,b\rangle} p_k$ and $\prod_{\tau_k \in \langle a,b\rangle} p_k$ stand for the sum and the product of the numbers p_k such that $\tau_k \in \langle a, b\rangle$.

In the case $i\langle a, b\rangle = 0$, $\sum p_k = 0$ and $\prod p_k = 1$.

For:

$J^+, J^-, J,$	see pp. 15 and 16
$\mathrm{PC}(D, F)$, $\mathrm{PC}^r(D, F)$,	see pp. 16 and 17
$x_j(t) \overset{B}{\to} x_0(t)$ as $j \to \infty$,	see p. 25.

Chapter I

Preliminary Notes

In this chapter we shall formulate some auxiliary assertions about impulsive differential equations; we shall use these ideas in the subsequent chapters.

1. Impulsive differential equations

Before we describe impulsive differential equations we shall consider two examples in order to show that the impulses appear in a natural way in the investigation of concrete processes.

Example 1.1   Consider the scalar differential equation of Verhulst:

$$\frac{\mathrm{d}x}{\mathrm{d}t} = rx(1 - \frac{x}{K}). \tag{1.1}$$

Equation (1.1) describes the change of the population number $x(t)$ of an isolated population of some biological species with incessant reproduction inhabiting a stationary environment with limited resources. The constants $r > 0$ and $K > 0$ are called the *rate coefficient* and the *capacity of the environment*, respectively.

The forms of the graphs of the positive solutions of equation (1.1) are given in Fig. 1.1. For the sake of definiteness we shall assume that $x(t)$ is the quantity of biomass of a certain species of micro-organism cultivated in a bioreactor. External effects on the development of the species can cause jumps in the quantity of biomass $x(t)$. For instance, this is possible with a single removal of part of the biomass or with the introduction of a supplementary quantity of biomass into the bioreactor.

As a result of an external effect at the moment $t = \tau_k$ let the quantity of biomass $x(t)$ suffer an increment δ_k, that is,

$$\Delta x(\tau_k) = x(\tau_k^+) - x(\tau_k^-) = \delta_k$$

where $x(\tau_k^-)$ and $x(\tau_k^+)$ are respectively the quantities of biomass before and after the impulse effect. If there is an addition of biomass the increment δ_k is positive; if biomass is removed from the bioreactor the increment is negative. A natural constraint in this case is

$$x(\tau_k^+) = x(\tau_k^-) + \delta_k > 0$$

which means that the population is not destroyed as a result of the impulse effect. We shall also assume that the increment δ_k depends on $x(\tau_k^-)$, that is,

$$\Delta x = \delta_k(x) \quad \text{for } t = \tau_k. \tag{1.2}$$

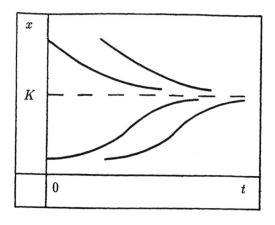

Fig. 1.1

For instance, when part of the biomass is removed we can assume that

$$\Delta x = \gamma x \quad \text{for } t = \tau_k$$

where $\gamma \in (-1, 0)$ is a constant.

The moments τ_k of the impulse effect can be chosen in various ways (such as randomly, for instance). However, the following two ways of choosing are of practical interest.

(i) The moments τ_k are fixed beforehand, that is, the moments of the impulse effect are determined by a previously given schedule. Of interest is the case when the moments τ_k are T-periodic:

$$\tau_{k+q} = \tau_k + T \quad (k \in \mathbb{Z})$$

where $q \geq 1$ is an integer. If $q = 1$ then the impulse effect takes place after equal intervals of time: $\tau_k = \tau_0 + kT \;\; (k \in \mathbb{Z})$.

(ii) The moments of the impulse effect occur when the quantity $x(t)$ satisfies some relation $\phi(x) = 0$. This means that the impulse effect takes place when $x(t)$ acquires some of the values $x_1, \ldots, x_m$ which are the roots of the equation $\phi(x) = 0$. If this equation has just one root, $x = N$, and the increments δ_k do not depend on k, then

$$x(\tau_k^-) = N, \qquad \delta_k(x(\tau_k^-)) = \delta(N) = C.$$

Then the jump condition (1.2) takes the form

$$\Delta x = C \quad \text{for } x = N \tag{1.3}$$

where $N + C > 0$.

It is easily verified that in this case the impulsive equation (1.1) with (1.3) has a T-periodic solution

$$
x(t) = \begin{cases} \dfrac{K(N+C)}{N+C+(K-N-C)e^{-rt}} = x_0(t) & \text{for } 0 < t \leq T, \\ x_0(t - kT) & \text{for } kT < t \leq kT + T \end{cases}
$$

where

$$
T = \frac{1}{r}\ln \frac{(K-N-C)N}{(N+C)(K-N)}.
$$

The graph of this solution is given in Fig. 1.2 (for $0 < N+C < N < K$) and in Fig. 1.3 (for $K < N < N+C$).

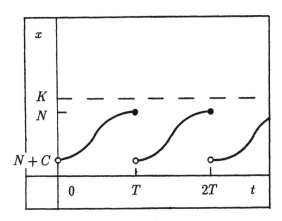

Fig. 1.2

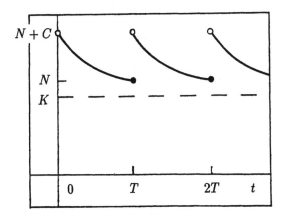

Fig. 1.3

We note that the change by jumps in the quantity of biomass $x(t)$ of Example 1.1 is caused by an external effect. However the system considered in the next example is characterized by jumps in the values of its parameters that are intrinsic to the system itself.

Example 1.2   Consider the electronic circuit shown in Fig. 1.4.

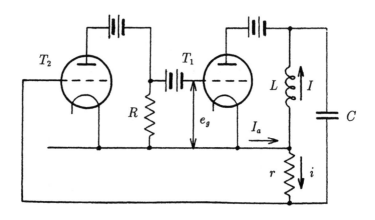

Fig. 1.4

Let the tube T_2 operate in the rectilinear section of its characteristic and let the amplification factor of T_2 be equal to κ. Then the current intensities I and i satisfy the equations

$$I = \varphi(\kappa r i) - i, \qquad L\frac{dI}{dt} - ri - \frac{1}{C}\int i\, dt = 0 \qquad (1.4)$$

where I is the current through the inductance L,

$$V = \frac{1}{C}\int i\, dt = L\frac{dI}{dt} - ri$$

is the voltage of the capacitance C, $e_g = \kappa r i$ is the potential of the control grid of tube T_1, and $\varphi : \mathbb{R} \to \mathbb{R}, e_g \to I_a(e_g)$ is the characteristic of tube T_1 (Fig. 1.5).

Differentiating equations (1.4) with respect to t and setting $e_g = \kappa r i = x, dI/dt = y$, we obtain the system

$$\frac{dx}{dt} = \frac{y}{\varphi'(x) - \frac{1}{\kappa r}}, \qquad \frac{dy}{dt} = \frac{x}{\kappa r L C} + \frac{1}{\kappa L}\frac{y}{\varphi'(x) - \frac{1}{\kappa r}}. \qquad (1.5)$$

Assume that the function $\varphi(x)$ is odd, that the derivative $\varphi'(x)$ is positive and has a maximum for $x = 0$, and that the relations

$$\lim_{x\to\infty} \varphi(x) = \frac{I_s}{2},$$

$$\lim_{x\to\infty} \varphi'(x) = 0, \qquad \varphi'(0) > \frac{1}{\kappa r} \qquad (1.6)$$

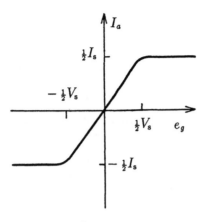

Fig. 1.5

hold, where I_s is the saturation current of tube T_1 and $\varphi'(0)$ is the steepness of the characteristic of tube T_1 at the operating point.

From (1.6) it follows that the equation

$$\varphi'(x) - \frac{1}{\kappa r} = 0$$

has just two roots, $x = \pm x_1$. Hence for each $y \in \mathbb{R}$ the derivatives dx/dt and dy/dt are infinitely large at the points $x = x_1$ and $x = -x_1$. Hence the motion of the point (x, y) in a neighbourhood of the straight lines $x = x_1$ and $x = -x_1$ can be simulated by a system with an impulse effect. The jump conditions of the point (x, y) are determined by the requirement of continuity at the moments of a jump in the voltage $V = Ly - x/\kappa$ of the capacitance C and the current $I = \varphi(x) - x/(\kappa r)$ through the inductance L.

Hence, if the point (x, y) at the moment before the jump was in the position (x_j, y_j) and after the jump it moves to the position (x_j^+, y_j^+), then the following conditions have to be fulfilled:

$$\varphi(x_j^+) - \frac{x_j^+}{\kappa r} = \varphi(x_j) - \frac{x_j}{\kappa r}, \qquad Ly_j^+ - \frac{x_j^+}{\kappa} = Ly_j - \frac{x_j}{\kappa}. \qquad (1.7)$$

In practice, the characteristic $I_a = \varphi(x)$ is a linear function outside small neighbourhoods $(\frac{1}{2}V_s - \delta, \frac{1}{2}V_s)$, $(-\frac{1}{2}V_s, -\frac{1}{2}V_s + \delta)$ of the points x_1 and $-x_1$, respectively (Fig. 1.5). Therefore, we can assume that $\varphi(x)$ is piecewise linear, that is,

$$\varphi(x) = \begin{cases} -\frac{1}{2}I_s, & x < -\frac{1}{2}V_s \\ Sx, & \text{for } |x| \le \frac{1}{2}V_s \\ \frac{1}{2}I_s, & x > \frac{1}{2}V_s \end{cases} \qquad (1.8)$$

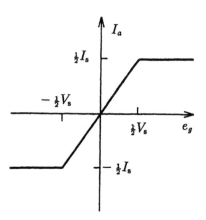

Fig. 1.6

where V_s is the saturation voltage and $S = I_s/V_s$ is the steepness of the characteristic of tube T_1 (Fig. 1.6).

In view of (1.5), (1.7) and (1.8) we obtain the following linear autonomous system with impulse effect:

$$
\begin{aligned}
\frac{dx}{dt} &= \begin{cases} \dfrac{\kappa r y}{\kappa r S - 1}, & \text{for } |x| < \tfrac{1}{2}V_s \\[2mm] -\kappa r y, & \text{for } |x| > \tfrac{1}{2}V_s \end{cases} \\[4mm]
\frac{dy}{dt} &= \begin{cases} \dfrac{x}{\kappa r L C} + \dfrac{ry}{L(\kappa r S - 1)}, & \text{for } |x| < \tfrac{1}{2}V_s \\[2mm] \dfrac{x}{\kappa r L C} - \dfrac{ry}{L}, & \text{for } |x| > \tfrac{1}{2}V_s \end{cases}
\end{aligned}
\tag{1.9}
$$

with a jump when the point (x, y) meets one of the straight lines $x = \tfrac{1}{2}V_s$ or $x = -\tfrac{1}{2}V_s$:

$$
\begin{aligned}
\Delta x &= x_j^+ - x_j = -2\kappa r S x, \\[2mm]
\Delta y &= y_j^+ - y_j = \frac{1}{\kappa L}\Delta x = -\frac{2rS}{L}x.
\end{aligned}
\tag{1.10}
$$

Thus, when investigating the operation of the above electronic circuit, instead of the non-linear system (1.5) we can consider the linear impulsive system (1.9), (1.10).

In these two examples the impulsive systems are given by an ordinary differential equation, together with relations defining the jump condition. We shall give a brief description of impulsive differential equations.

Let Ω be the phase space of some evolutionary process, i.e., the set of its states. Denote by P_t the point representing the state of the process at time t, and assume that the state of the process is determined by n parameters. Then the mapping point P_t can be interpreted as a point (t, x) of the $(n+1)$-dimensional space $\mathbb{R}^{n+1}$ and Ω as a set in $\mathbb{R}^n$. The set $\mathbb{R} \times \Omega$ will be called an extended phase space of the evolutionary process under consideration. Assume that the law of evolution of the process is described by:

(a) a differential equation

$$\frac{dx}{dt} = f(t, x) \tag{1.11}$$

where $t \in \mathbb{R}$, $x = \operatorname{col}(x_1, \ldots, x_n) \in \Omega$, $f : \mathbb{R} \times \Omega \to \mathbb{R}^n$;

(b) the sets $M_t, N_t \subset \Omega$ for each $t \in \mathbb{R}$; and

(c) the operator $A_t : M_t \to N_t$ for each $t \in \mathbb{R}$.

The motion of the point P_t in the extended phase space is performed in the following way: the point P_t begins its motion from the initial point (t_0, x_0) and moves along the curve $(t, x(t))$ described by the solution $x(t)$ of equation (1.11) with initial condition $x(t_0) = x_0$ until the instant $\tau_1 > t_0$ when P_t meets the set M_t. At time τ_1 the operator A_{τ_1} 'instantly' transfers the point P_t from the position $P_{\tau_1} = (\tau_1, x(\tau_1))$ to the position $(\tau_1, x_1^+) \in N_{\tau_1}$, $x_1^+ = A_{\tau_1} x(\tau_1)$. The point P_t continues its motion along the curve $(t, x(t))$ described by the solution $x(t)$ of equation (1.11) with initial condition $x(\tau_1) = x_1^+$ until a new encounter with the set M_t, and so on.

The relations (a), (b), (c) characterizing the evolutionary process will be called an impulsive differential equation. The curve described by the point P_t in the extended phase space will be called an integral curve, and the function $x(t)$ defining this curve will be called a solution of the impulsive differential equation. The instants τ_k when the point P_t meets the set M_t will be called instants (moments) of the impulse effect. We shall assume that the solution of the impulsive differential equation is a left-continuous function at the instants of the impulse effect, that is, $x(\tau_k^-) = x(\tau_k)$.

The freedom of choice of the sets M_t and N_t and the operator A_t leads to a great variety of impulsive equations. In this book, equations for which the instants of the impulse effect are determined by probability laws (Mil'man and Myshkis [58]) will not be considered. However, the equations which will be considered are those for which the instants of the impulse effect occur when some space–time relation $\phi(t, x) = 0$ is satisfied, that is, when the mapping point (t, x) meets the surface σ of the equation $\phi(t, x) = 0$. We shall write such equations in the form

$$\begin{aligned} \frac{dx}{dt} &= f(t, x), \quad \text{if } \phi(t, x) \neq 0, \\ \Delta x &= I(t, x), \quad \text{if } \phi(t, x) = 0. \end{aligned} \tag{1.12}$$

The sets M_t and N_t and the operator A_t are defined by the relations

$$M_t = \{(t, x) \in \mathbb{R} \times \Omega : \phi(t, x) = 0\}, \quad N_t = \mathbb{R} \times \Omega$$

$$A_t : M_t \to N_t, \quad (t, x) \to (t, x + I(t, x))$$

where $I : \mathbb{R} \times \Omega \to \Omega$ and $t = \tau_k$ is an instant of the impulse effect for the solution $x(t)$ if $\phi(\tau_k, x(\tau_k)) = 0$. Then

$$\Delta x(\tau_k) = I(\tau_k, x(\tau_k)).$$

Denote by $x(t; t_0, x_0)$ the solution of (1.12) for which $x(t_0^+; t_0, x_0) = x_0$. We shall consider mainly the following three classes of impulsive differential equations which are particular cases of equation (1.12).

Class I: *Equations with fixed moments of the impulse effect.* The equations of this class are written as follows:

$$\frac{dx}{dt} = f(t, x), \quad t \neq \tau_k,$$
$$\Delta x = I_k(x), \quad t = \tau_k. \tag{1.13}$$

The moments of the impulse effect are fixed beforehand by defining the sequence $\tau_k : \tau_k < \tau_{k+1}$, $k \in K \subset \mathbb{Z}$. For $t \in (\tau_k, \tau_{k+1}]$ the solution $x(t)$ of (1.13) satisfies the equation $dx/dt = f(t, x)$, and for $t = \tau_k$, $x(t)$ satisfies the relation $x(\tau_k^+) = \psi_k(x(\tau_k)) \equiv x(\tau_k) + I_k(x(\tau_k))$.

Sometimes equation (1.13) is written in the form

$$\frac{dx}{dt} = f(t, x), \quad t \neq \tau_k,$$
$$x(\tau_k^+) = \psi_k(x(\tau_k)) \tag{1.14}$$

where $\psi_k : \Omega \to \Omega$ $(k \in K)$.

Class II: *Equations with unfixed moments of the impulse effect.* These equations have the form

$$\frac{dx}{dt} = f(t, x), \quad t \neq \tau_k(x),$$
$$\Delta x = I_k(x), \quad t = \tau_k(x) \tag{1.15}$$

where $\tau_k : \Omega \to \mathbb{R}$ and $\tau_k(x) < \tau_{k+1}(x)$ $(k \in K \subset \mathbb{Z}, x \in \Omega)$. The moments of the impulse effect for these equations occur when the mapping point (t, x) meets some of the hypersurfaces σ_k of the equation $t = \tau_k(x)$, that is, when $t = \tau_k(x(t))$ for some $k \in K$.

Class III: *Autonomous impulsive equations.* Autonomous impulsive equations have the form

$$\frac{dx}{dt} = f(x), \quad x \notin \sigma,$$
$$\Delta x = I(x), \quad x \in \sigma \tag{1.16}$$

where σ is an $(n-1)$-dimensional manifold contained in the phase space $\Omega \subset \mathbb{R}^n$.

The moments of the impulse effect for equation (1.16) occur when the point $x(t)$ of the phase space meets the manifold σ. If σ is given by the equation $\phi(x) = 0$ then (1.16) can be written in the form

$$\frac{dx}{dt} = f(x), \quad \phi(x) \neq 0,$$
$$\Delta x = I(x), \quad \phi(x) = 0. \tag{1.17}$$

We shall note some peculiarities of these classes of impulsive differential equations.

The general feature of these equations is that their solutions are piecewise continuous functions with points of discontinuity at the moments of the impulse effect. That is why all the solutions of equation (1.13) have the same points of discontinuity while the distinct solutions of equation (1.15) (or (1.16)) have different points of discontinuity. This peculiarity of the solutions of equations (1.15) and (1.16) makes their investigation difficult.

The second peculiarity is that equation (1.16) has the property of autonomy $(x(t; t_0, x_0) = x(t - t_0; 0, x_0)$ for all $t_0 \in \mathbb{R}$, $x_0 \in \Omega$ and $t > t_0)$, while equations (1.13) and (1.15) do not enjoy this property even in the cases when $f(t, x) = f(x)$ and $I_k(x) = I(x) \not\equiv 0$ for $t \in \mathbb{R}$, $k \in \mathbb{Z}$ and $x \in \Omega$.

Example 1.3   Consider the impulsive equation

$$\frac{dx}{dt} = 0, \quad \text{for } t \neq 2,$$
$$\Delta x = 1, \quad \text{for } t = 2. \tag{1.18}$$

Since

$$x(t; 1, 0) = \begin{cases} 0, & 1 < t \leq 2, \\ 1, & t > 2, \end{cases}$$

and

$$x(t - 1; 0, 0) = \begin{cases} 0, & 1 < t \leq 3, \\ 1, & t > 3, \end{cases}$$

then $x(t; 1, 0) \neq x(t - 1; 0, 0)$ for $2 < t \leq 3$ and equation (1.18) does not enjoy the property of autonomy.

One of the phenomena which occur in equations of the type (1.15) is the so-called 'beating' of the solutions. This happens when the mapping point $P_t = (t, x(t))$ meets one and the same hypersurface σ_k several or infinitely many times.

Example 1.4   Consider the impulsive equation

$$\frac{dx}{dt} = 0, \quad\quad\quad t \neq \tau_k(x),$$
$$\Delta x = x^2 \operatorname{sgn} x - x, \quad t = \tau_k(x) \tag{1.19}$$

where $t \geq 0$, $x \in \mathbb{R}$, $\tau_k(x) = x + 6k$ for $|x| < 3$ and $k = 0, 1, 2, \ldots$.

Each motion of the mapping point P_t which starts from an initial point $(0, x_0)$ with $|x_0| \geq 3$ is not subject to an impulse effect since the integral curve does not intersect the hypersurface σ_k (Fig. 1.7).

The motion which starts from a point $(0, x_0)$ with $1 < |x_0| < 3$ is subject to an impulse effect finitely many times. For instance, the motion which starts from the point $(0, \sqrt[4]{2})$ is subject to an impulse effect three times, and after the moment $\tau_3 = 2$ the integral curve no longer intersects the hypersurfaces σ_k (Fig. 1.8).

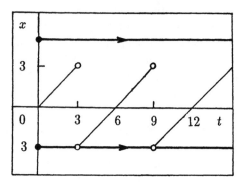

Fig. 1.7

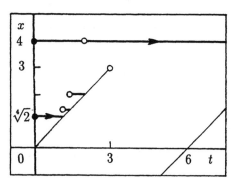

Fig. 1.8

A motion which starts from a point $(0, x_0)$ with $0 < x_0 < 1$ is subject to an impulse effect infinitely many times at the moments τ_k for which we have $\lim_{k \to \infty} \tau_k = \infty$, $\lim_{k \to \infty} x(\tau_k) = 0$ (Fig. 1.9).

The motion which starts from a point $(0, x_0)$ with $-1 < x_0 < 0$ is also subject to an impulse effect infinitely many times but $\lim_{k \to \infty} \tau_k = 6$, $\lim_{k \to \infty} x(\tau_k) = 0$ (Fig. 1.10). In this case there is a beating of the solution against the hypersurface $\sigma_1 \equiv t = x + 6$ $(|x| < 3)$.

The integral curves which start from the points $(0, 0)$, $(0, 1)$ and $(0, -1)$ also intersect the hypersurfaces σ_k infinitely many times, but at fixed points of the operator $A_t x = x^2 \text{sgn}\, x$. That is why the respective motion of the point P_t is not subject to an impulse effect (Fig. 1.10).

Besides the phenomenon of 'beating' in this example, the *merging of the solutions* from a given moment on is observed. Thus, for instance, the solutions with initial points $(0, \sqrt[4]{2})$ and $(0, 4)$ coincide for $t > 2$.

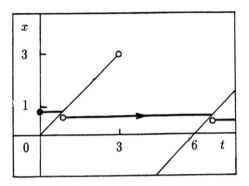

Fig. 1.9

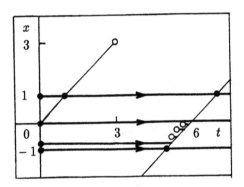

Fig. 1.10

2. Auxiliary assertions

2.1 The phenomenon of 'beating'

The presence of 'beating' for impulsive differential equations makes their investigation considerably more difficult. That is why the results of the subsequent chapters are obtained under the assumption that, for the equations considered, the phenomenon of 'beating' is absent.

Many sufficient conditions for the absence of beating are known for impulsive equations of the form

$$\frac{\mathrm{d}x}{\mathrm{d}t} = f(t, x), \quad t \neq \tau_k(x),$$

$$\Delta x = I_k(x), \quad t = \tau_k(x). \tag{2.1}$$

where $f : \mathbb{R}_+ \times \Omega \to \mathbb{R}^n$, $\tau_k : \Omega \to \mathbb{R}_+$, $I_k : \Omega \to \mathbb{R}^n$, $\Omega \subset \mathbb{R}^n$, and $0 < \tau_1(x) < \tau_2(x) < \cdots$, $\lim_{k \to \infty} \tau_k(x) = \infty$ for $x \in \Omega$. We shall give one such result.

Let $\mu > 0$, $(t_0, x_0) \in \mathbb{R}_+ \times \mathbb{R}^n$ and $K(t_0, x_0, \mu) = \{(t, x) \in \mathbb{R}^{n+1} : \mu \mid x - x_0 \mid < t - t_0\}$.

Theorem 2.1 *Let the following conditions hold.*

1. *The function $f(t, x)$ is continuous in $\mathbb{R}_+ \times \Omega$ and is locally Lipschitz continuous with respect to x in $\mathbb{R}_+ \times \Omega$.*
2. *The integral curve $(t, x(t))$, $t \geq 0$, of equation (2.1) is contained in $\mathbb{R}_+ \times \Omega$ and*

$$(\tau_{k+1}, x_{k+1}) \in K(\tau_k, x_k^+, \mu),$$

$$\tau_k < \tau_{k+1}, \tag{2.2}$$

   where (τ_k, x_k) are the points at which the integral curve $(t, x(t))$, $t \geq 0$, meets the hypersurfaces $\sigma_j \equiv t = \tau_j(x)$ and $x_i^+ = x_i + I_j(x_i)$ if $(\tau_i, x_i) \in \sigma_j$.
3. *The functions $\tau_k(x)$ are Lipschitz continuous with respect to $x \in \Omega$ with constants $L_k \leq \mu$ and*

$$\tau_k(x + I_k(x)) \leq \tau_k(x) \quad \text{for } x \in \Omega, k \in \mathbb{N}.$$

Then the integral curve meets each hypersurface σ_k at most once.

Remark 2.1 Condition 1 of Theorem 2.1 is imposed in order to guarantee the existence and uniqueness of the solution.

Corollary 2.1 *Theorem 2.1 still holds if condition (2.2) is replaced by the condition*

$$\mid f(t, x) \mid \leq M \quad \text{for } (t, x) \in \mathbb{R}_+ \times \Omega \tag{2.3}$$

where $ML_k < 1$ $(k \in \mathbb{N})$.

Corollary 2.2 *Theorem 2.1 still holds if condition (2.2) is replaced by condition (2.3), and condition 3 is replaced by the following condition.*

4. *The functions $\tau_k(x)$ are differentiable in Ω and*

$$\left| \frac{\partial \tau_k}{\partial x}(x) \right| \leq L_k < \frac{1}{M} \quad \text{for } x \in \Omega, k \in \mathbb{N},$$

$$\sup_{\substack{0 \leq s \leq 1 \\ x \in \Omega}} \left(\frac{\partial \tau_k}{\partial x}(x + sI_k(x)) \mid I_k(x) \right) \leq 0 \quad \text{for } k \in \mathbb{N}.$$

Finally we shall note the following. In the investigation of the autonomous equation (1.6), those solutions for which the point $x(t)$ meets the hypersurface σ infinitely many times are of interest. For instance, this is observed for periodic solutions subject to an impulse effect. We shall therefore say that there is beating of a given solution of equation (1.6) if this solution has moments of an impulse effect which form an increasing sequence with a finite accumulation point.

2.2 Existence, uniqueness and continuability of the solutions

Let $\Omega \subset \mathbb{R}^n$ be an open set. Suppose that for each $k \in \mathbb{Z}$ the functions $\tau_k : \Omega \to \mathbb{R}$ are continuous in Ω and are such that

$$\tau_k(x) < \tau_{k+1}(x), \qquad \lim_{k \to \pm\infty} \tau_k(x) = \pm\infty \qquad (x \in \Omega).$$

Let $f : \mathbb{R} \times \Omega \to \mathbb{R}^n$, $I_k : \Omega \to \mathbb{R}^n$, $(t_0, x_0) \in \mathbb{R} \times \Omega$ and $\alpha < \beta$.

Consider the impulsive differential equation

$$\frac{dx}{dt} = f(t, x), \quad t \neq \tau_k(x),$$
$$\Delta x = I_k(x), \qquad t = \tau_k(x). \tag{2.4}$$

with initial condition

$$x(t_0^+) = x_0. \tag{2.5}$$

Definition 2.1 The function $\varphi : \langle \alpha, \beta \rangle \to \mathbb{R}^n$ is said to be a *solution* of (2.4) if:

1. $(t, \varphi(t)) \in \mathbb{R} \times \Omega$ for $t \in \langle \alpha, \beta \rangle$;
2. for $t \in (\alpha, \beta)$, $t \neq \tau_k(\varphi(t))$, $k \in \mathbb{Z}$ the function $\varphi(t)$ is differentiable and $\frac{d\varphi}{dt}(t) = f(t, \varphi(t))$;
3. the function $\varphi(t)$ is continuous from the left in $\langle \alpha, \beta \rangle$ and if $t \in \langle \alpha, \beta \rangle$, $t = \tau_k(\varphi(t))$ and $t \neq \beta$, then $\varphi(t^+) = \varphi(t) + I_k(\varphi(t))$ and, for each $j \in \mathbb{Z}$ and some $\delta > 0$, $s \neq \tau_j(\varphi(s))$ for $t < s < t + \delta$.

Definition 2.2 Each solution $\varphi(t)$ of (2.4) which is defined in an interval of the form (t_0, β) and satisfies the condition $\varphi(t_0^+) = x_0$ is said to be a *solution of the initial value problem* (2.4), (2.5) (or a *solution of equation* (2.4) *with initial values* (t_0, x_0)).

We note that, instead of the initial condition $x(t_0) = x_0$, we have imposed the limiting condition $x(t_0^+) = x_0$ which, in general, is natural for equation (2.4) since (t_0, x_0) may be such that $t_0 = \tau_k(x_0)$ for some k. Whenever $t_0 \neq \tau_k(x_0)$, for all k, we shall understand the initial condition $x(t_0^+) = x_0$ in the usual sense, that is, $x(t_0) = x_0$.

It is clear that if $t_0 \neq \tau_k(x_0)$, for $k \in \mathbb{Z}$, then the existence and uniqueness of the solution of the initial value problem (2.4), (2.5) depends only on the properties of the function f. Thus, for instance, if the function f is continuous in a neighbourhood of the point (t_0, x_0), then there exists a solution of the initial value problem (2.4), (2.5) and this solution is unique if f is Lipschitz continuous in this neighbourhood.

If, however, $t_0 = \tau_k(x_0)$, for some $k \in \mathbb{Z}$, that is, (t_0, x_0) belongs to the hypersurface $\sigma_k \equiv t = \tau_k(x)$, then it is possible that the solution $x(t)$ of the initial value problem

$$x' = f(t, x), \qquad x(t_0) = x_0 \tag{2.6}$$

lies entirely in σ_k. Consequently, we need some additional conditions on f and τ_k to guarantee the existence of a solution $x(t)$ of the initial value problem (2.6) in some interval $[t_0, \beta)$, and the validity of the condition

$$t \neq \tau_k(x(t)) \quad \text{for } t \in (t_0, \beta) \text{ and } k \in \mathbb{Z}.$$

Conditions of this type are imposed in the following theorem.

Theorem 2.2   *Let the following conditions hold.*

1. *The function $f : \mathbb{R} \times \Omega \to \mathbb{R}^n$ is continuous in $t \neq \tau_k(x)$ $(k \in \mathbb{Z})$.*
2. *For any $(t, x) \in \mathbb{R} \times \Omega$ there exists a locally integrable function l such that in a small neighbourhood of (t, x)*

$$|f(s, y)| \leq l(s).$$

3. *For each $k \in \mathbb{Z}$ the condition $t_1 = \tau_k(x_1)$ implies the existence of $\delta > 0$ such that*

$$t \neq \tau_k(x)$$

   for all $0 < t - t_1 < \delta$ and $|x - x_1| < \delta$.

Then for each $(t_0, x_0) \in \mathbb{R} \times \Omega$ there exists a solution $x : (t_0, \beta) \to \mathbb{R}^n$ of the initial value problem (2.4), (2.5) for some $\beta > t_0$.

Remark 2.2   Condition 2 of Theorem 2.2 can be replaced by the condition:

2'. *For any $k \in \mathbb{Z}$ and $(t, x) \in \sigma_k$ there exists the finite limit of $f(s, y)$ as $(s, y) \to (t, x)$, $s > \tau_k(y)$.*

Remark 2.3   The solution $x(t)$ of the initial value problem (2.4), (2.5) is unique if the function f is such that the solution of the initial value problem (2.6) is unique. This requirement is met if, for instance, f is (locally) Lipschitz continuous with respect to x in a neighbourhood of (t_0, x_0).

If the initial value problem (2.4), (2.5) has a unique solution, then we shall denote this solution by $x(t; t_0, x_0)$.

Consider in more detail the impulsive equation with fixed moments of an impulse effect:

$$\frac{dx}{dt} = f(t, x), \quad t \neq \tau_k,$$
$$\Delta x = I_k(x), \quad t = \tau_k \tag{2.7}$$

where $\tau_k < \tau_{k+1}$ $(k \in \mathbb{Z})$ and $\lim_{k \to \pm\infty} = \pm\infty$.

The following theorem is valid.

Theorem 2.3   *Let the function $f : \mathbb{R} \times \Omega \to \mathbb{R}^n$ be continuous in the sets $(\tau_k, \tau_{k+1}] \times \Omega$ ($k \in \mathbb{Z}$) and for each $k \in \mathbb{Z}$ and $x \in \Omega$ suppose there exists the finite limit of $f(t, y)$ as $(t, y) \to (\tau_k, x)$, $t > \tau_k$.*

Then for each $(t_0, x_0) \in \mathbb{R} \times \Omega$ there exists $\beta > t_0$ and a solution $x : (t_0, \beta) \to \mathbb{R}^n$ of the initial value problem (2.7), (2.5).

If, moreover, the function f is locally Lipschitz continuous with respect to x in $\mathbb{R} \times \Omega$ then this solution is unique.

Let us consider the problem of the continuability to the right of a given solution $\varphi(t)$ of equation (2.7).

Theorem 2.4   *Let the following conditions hold.*

1. *The function $f : \mathbb{R} \times \Omega \to \mathbb{R}^n$ is continuous in the sets $(\tau_k, \tau_{k+1}] \times \Omega$ ($k \in \mathbb{Z}$) and for each $k \in \mathbb{Z}$ and $x \in \Omega$ there exists the finite limit of $f(t, y)$ as $(t, y) \to (\tau_k, x)$, $t > \tau_k$.*
2. *The function $\varphi : (\alpha, \beta) \to \mathbb{R}^n$ is a solution of (2.7).*

Then the solution $\varphi(t)$ is continuable to the right of β if and only if there exists the limit

$$\lim_{t \to \beta-} \varphi(t) = \eta$$

and one of the following conditions is fulfilled:

(a) $\beta \neq \tau_k$ for each $k \in \mathbb{Z}$ and $\eta \in \Omega$;
(b) $\beta = \tau_k$ for some $k \in \mathbb{Z}$ and $\eta + I_k(\eta) \in \Omega$.

Theorem 2.5   *Let the following conditions hold.*

1. *Condition 1 of Theorem 2.4 is satisfied.*
2. *The function f is locally Lipschitz continuous with respect to x in $\mathbb{R} \times \Omega$.*
3. *$\eta + I_k(\eta) \in \Omega$ for each $k \in \mathbb{Z}$ and $\eta \in \Omega$.*

Then for any $(t_0, x_0) \in \mathbb{R} \times \Omega$ there exists a unique solution of the initial value problem (2.7), (2.5) which is defined in an interval of the form (t_0, ω) and is not continuable to the right of ω.

Let the conditions of Theorem 2.5 be satisfied and let $(t_0, x_0) \in \mathbb{R} \times \Omega$. Denote by $J^+ = J^+(t_0, x_0)$ the maximal interval of the form (t_0, ω) in which the solution $x(t; t_0, x_0)$ is defined.

Theorem 2.6   *Let the following conditions hold.*

1. *Conditions 1, 2 and 3 of Theorem 2.5 are met.*
2. $\varphi(t)$ *is a solution of the initial value problem (2.7), (2.5).*
3. *There exists a compact $Q \subset \Omega$ such that $\varphi(t) \in Q$ for $t \in J^+(t_0, x_0)$.*

Then $J^+(t_0, x_0) = (t_0, +\infty)$.

Let $\varphi : (\alpha, \omega) \to \mathbb{R}^n$ be a solution of equation (2.7) and consider the question of the continuability of this solution to the left of α

If $\alpha \neq \tau_k$ $(k \in \mathbb{Z})$ then the problem of continuability to the left of α is solved as for ordinary differential equations [21]. In this case such an extension is possible if and only if there exists the limit

$$\lim_{t \to \alpha+} \varphi(t) = \eta \qquad (2.8)$$

and $\eta \in \Omega$.

If $\alpha = \tau_k$, for some $k \in \mathbb{Z}$, then the solution $\varphi(t)$ will be continuable to the left of τ_k when there exists the limit (2.8), $\eta \in \Omega$, and the equation $x + I_k(x) = \eta$ has a unique solution $x_k \in \Omega$. In this case the extension $\psi(t)$ of $\varphi(t)$ for $t \in (\tau_{k-1}, \tau_k]$ coincides with the solution of the initial value problem

$$\frac{d\psi}{dt} = f(t, \psi), \quad \tau_{k-1} < t \leq \tau_k,$$
$$\psi(\tau_k) = x_k.$$

If the solution $\varphi(t)$ can be continued up to τ_{k-1}, then the above procedure is repeated, and so on. Under the conditions of Theorem 2.5 for each $(t_0, x_0) \in \mathbb{R} \times \Omega$ there exists a unique solution $x(t; t_0, x_0)$ of the initial value problem (2.7), (2.5) which is defined in an interval of the form (α, ω) and is not continuable to the right of ω and to the left of α. Denote by $J(t_0, x_0)$ this maximal interval of existence of the solution $x(t; t_0, x_0)$ and set $J^- = J^-(t_0, x_0) = (\alpha, t_0]$. A straightforward verification shows that the solution $x(t) = x(t; t_0, x_0)$ of the initial value problem (2.7), (2.5) satisfies the following integro-summary equation

$$x(t) = \begin{cases} x_0 + \int_{t_0}^{t} f(s, x(s))\, ds + \sum_{t_0 < \tau_k < t} I_k(x(\tau_k)) & \text{for } t \in J^+, \\[2mm] x_0 + \int_{t_0}^{t} f(s, x(s))\, ds - \sum_{t \leq \tau_k \leq t_0} I_k(x(\tau_k)) & \text{for } t \in J^-. \end{cases} \qquad (2.9)$$

2.3 Linear impulsive equations

Let the sequence τ_k $(k \in \mathbb{Z})$ be fixed and satisfy the condition:

H2.1 $\tau_k < \tau_{k+1}$ $(k \in \mathbb{Z})$ and $\lim_{k \to \pm\infty} \tau_k = \pm\infty$.

Let $m, n, r \in \mathbb{N}$, $D \subset \mathbb{R}$ and $F \subset \mathbb{R}^{n \times m}$ (or $F \subset \mathbb{C}^{n \times m}$). Denote by $PC(D, F)$ the set of functions $\psi : D \to F$ which are continuous for $t \in D$, $t \neq \tau_k$, are continuous from the left for $t \in D$, and have discontinuities of the first kind at the points $\tau_k \in D$.

Denote by $PC^r(D, F)$ the set of functions $\psi : D \to F$ with a derivative $\frac{d^r \psi}{dt^r} \in PC(D, F)$.

Consider the linear homogeneous impulsive equation

$$\frac{dx}{dt} = A(t)x, \quad t \neq \tau_k,$$
$$\Delta x = B_k x, \quad t = \tau_k \tag{2.10}$$

under the assumption that the following condition holds:

H2.2 $A(\,\cdot\,) \in PC(\mathbb{R}, \mathbb{C}^{n \times n})$, $B_k \in \mathbb{C}^{n \times n}$ $(k \in \mathbb{Z})$.

Theorem 2.7 *Let conditions H2.1 and H2.2 hold. Then for any $(t_0, x_0) \in \mathbb{R} \times \mathbb{C}^n$ there exists a unique solution $x(t)$ of equation (2.10) with $x(t_0^+) = x_0$ and this solution is defined for $t > t_0$.*

If, moreover, $\det (E + B_k) \neq 0$ $(k \in \mathbb{Z})$, then this solution is defined for all $t \in \mathbb{R}$.

Let $\det (E + B_k) \neq 0$ $(k \in \mathbb{Z})$ and let the functions

$$x_1(t), \ldots, x_n(t) \tag{2.11}$$

be solutions of equation (2.10) defined in $\mathbb{R}$. Let $X(t) = [x_1(t), \ldots, x_n(t)]$ be a matrix-valued function whose columns are the solutions (2.11). We shall note that the solutions $x_1(t), \ldots, x_n(t)$ are linearly independent in $\mathbb{R}$ if and only if $\det X(t_0^+) \neq 0$ for some $t_0 \in \mathbb{R}$. In this case we say that $X(t)$ is a fundamental matrix of equation (2.10).

The following theorem is valid.

Theorem 2.8 *Let conditions H2.1 and H2.2 hold, let $\det (E + B_k) \neq 0$ $(k \in \mathbb{Z})$, and let $X(t)$ be a fundamental matrix of equation (2.10). Then:*

1. *For any constant matrix $B \in \mathbb{C}^{n \times n}$ the function $X(t)B$ is a solution of equation (2.10).*
2. *If $Y : \mathbb{R} \to \mathbb{C}^{n \times n}$ is a solution of equation (2.10) then there exists a unique matrix $B \in \mathbb{C}^{n \times n}$ such that $Y(t) = X(t)B$. Moreover, if $Y(t)$ is a fundamental matrix of equation (2.10) then $\det B \neq 0$.*

Corollary 2.3 *The space L of solutions of equation (2.10) defined in $\mathbb{R}$ is n-dimensional.*

Indeed, if $x_1(t), \ldots, x_n(t)$ are n linearly independent solutions of equation (2.10) in the interval $\mathbb{R} = (-\infty, \infty)$ then for any solution $y : \mathbb{R} \to \mathbb{C}^n$ of equation (2.10) there exists a unique $b = \text{col}(b_1, \ldots, b_n) \in \mathbb{C}^n$ such that $y(t) = X(t)b$, that is,

$$y(t) = b_1 x_1(t) + \cdots + b_n x_n(t) \quad (t \in \mathbb{R}).$$

If we denote by $U_k(t, s)$ ($k \in \mathbb{Z}$, $t, s \in (\tau_{k-1}, \tau_k]$) the Cauchy matrix for the linear equation

$$\frac{dx}{dt} = A(t)x \quad (\tau_{k-1} < t \le \tau_k)$$

and conditions H2.1 and H2.2 are valid and $\det(E + B_k) \ne 0$ ($k \in \mathbb{Z}$), then the solutions of the linear equation (2.10) can be written in the form

$$x(t; t_0, x_0) = W(t, t_0^+)x_0 \tag{2.12}$$

where

$$W(t, s) = \begin{cases} U_k(t, s) \\ \qquad\qquad \text{for } t, s \in (\tau_{k-1}, \tau_k], \\[4pt] U_{k+1}(t, \tau_k^+)(E + B_k)U_k(\tau_k, s) \\ \qquad\qquad \text{for } \tau_{k-1} < s \le \tau_k < t \le \tau_{k+1}, \\[4pt] U_k(t, \tau_k)(E + B_k)^{-1}U_{k+1}(\tau_k^+, s) \\ \qquad\qquad \text{for } \tau_{k-1} < t \le \tau_k < s \le \tau_{k+1}, \\[4pt] U_{k+1}(t, \tau_k^+)\prod_{j=k}^{i+1}(E + B_j)U_j(\tau_j, \tau_{j-1}^+)(E + B_i)U_i(\tau_i, s) \\ \qquad\qquad \text{for } \tau_{i-1} < s \le \tau_i < \tau_k < t \le \tau_{k+1}, \\[4pt] U_i(t, \tau_i)\prod_{j=i}^{k-1}(E + B_j)^{-1}U_{j+1}(\tau_j^+, \tau_{j+1})(E + B_k)^{-1}U_{k+1}(\tau_k^+, s) \\ \qquad\qquad \text{for } \tau_{i-1} < t \le \tau_i < \tau_k < s \le \tau_{k+1}. \end{cases} \tag{2.13}$$

The matrix $W(t, s)$ is called the *Cauchy matrix* for equation (2.10). A straightforward verification shows that

$$\begin{aligned} W(t, t) &= E, \\ W(\tau_k^-, \tau_k) &= W(\tau_k, \tau_k^-) = E, \\ W(\tau_k^+, s) &= (E + B_k)W(\tau_k, s), \\ W(s, \tau_k^+) &= W(s, \tau_k)(E + B_k)^{-1}, \\ \frac{\partial W}{\partial t}(t, s) &= A(t)W(t, s) \quad (t \ne \tau_k), \\ \frac{\partial W}{\partial s}(t, s) &= -W(t, s)A(s) \quad (s \ne \tau_k). \end{aligned} \tag{2.14}$$

Moreover, from (2.12) and (2.13) it follows that for each $\lambda \in \mathbb{C}$, $x_0, y_0 \in \mathbb{C}^n$ and $t_0 \in \mathbb{R}$ the following equalities are valid:

$$\begin{aligned} x(t; t_0, x_0 + y_0) &= x(t; t_0, x_0) + x(t; t_0, y_0) \quad (t > t_0), \\ x(t; t_0, \lambda x_0) &= \lambda x(t; t_0, x_0) \qquad\qquad\quad (t > t_0), \\ W(t, t_0^+) &= X(t)X^{-1}(t_0^+) \qquad\qquad\quad (t > t_0) \end{aligned} \tag{2.15}$$

where $X(t)$ is the matrix solution of (2.10) for which $\det X(t_0^+) \ne 0$. If $\det(E+B_k) \ne 0$ ($k \in \mathbb{Z}$) then equalities (2.15) are valid for all $t \in \mathbb{R}$.

An immediate consequence of (2.13) is the following analogue of Liouville's formula for linear impulsive equations:

$$\det W(t, t_0) = \begin{cases} \displaystyle\prod_{t_0 \le \tau_k < t} \det(E + B_k)\exp\left(\int_{t_0}^{t} \operatorname{Tr} A(s)\,ds\right) & t > t_0 \\[2ex] \displaystyle\prod_{t \le \tau_k < t_0} \det(E + B_k)^{-1}\exp\left(\int_{t_0}^{t} \operatorname{Tr} A(s)\,ds\right) & t \le t_0. \end{cases} \tag{2.16}$$

Consider the linear non-homogeneous impulsive equation

$$\frac{dx}{dt} = A(t)x + g(t), \quad t \ne \tau_k, \tag{2.17}$$
$$\Delta x = B_k x + h_k, \qquad t = \tau_k$$

where $g(\,\cdot\,) \in \operatorname{PC}(\mathbb{R}, \mathbb{C}^n)$, $h_k \in \mathbb{R}^n$ $(k \in \mathbb{Z})$.

For the solution $x(t)$ of (2.17) the following formula for the variation of parameters is valid:

$$x(t) = \begin{cases} \displaystyle W(t, t_0^+)x(t_0^+) + \int_{t_0}^{t} W(t, s)g(s)\,ds + \sum_{t_0 < \tau_k < t} W(t, \tau_k^+)h_k & t > t_0 \\[2ex] \displaystyle W(t, t_0^+)x(t_0^+) + \int_{t_0}^{t} W(t, s)g(s)\,ds - \sum_{t \le \tau_k \le t_0} W(t, \tau_k^+)h_k & t \le t_0. \end{cases} \tag{2.18}$$

Note that equalities (2.12), (2.13), (2.15), (2.16) and (2.18) are valid for $t > t_0$ even if $\det(E + B_k) = 0$ for $\tau_k > t_0$, while for $t \le t_0$ these formulae are valid if $\det(E + B_k) \ne 0$ for $\tau_k \le t_0$.

Let $\det(E + B_k) \ne 0$ $(k \in \mathbb{Z})$ and $X(t) = [x_1(t), \dots, x_n(t)]$ be a fundamental matrix of equation (2.10) in $\mathbb{R}$. Then the function

$$x = c_1 x_1(t) + \cdots + c_n x_n(t)$$

is a general solution of (2.10), and the general solution of (2.17) has the form

$$x = c_1 x_1(t) + \cdots + c_n x_n(t) + \eta(t)$$

where $\eta(t)$ is a particular solution of (2.17).

Let $x(t) = x(t; t_0, x_0)$ be a solution of the equation

$$\frac{dx}{dt} = A(t)x + g(t, x), \quad t \ne \tau_k, \tag{2.19}$$
$$\Delta x = B_k x + h_k(x), \qquad t = \tau_k$$

where $A(\,\cdot\,) \in \operatorname{PC}(\mathbb{R}, \mathbb{R}^{n \times n})$, $B_k \in \mathbb{R}^{n \times n}$ $(k \in \mathbb{Z})$, and the functions $g : \mathbb{R} \times \Omega \to \mathbb{R}^n$ and $h_k : \Omega \to \mathbb{R}^n$ are such that the conditions of Theorem 2.5 are met.

Then the following representation is valid for $x(t)$:

$$x(t) = \begin{cases} \displaystyle W(t, t_0^+)x_0 + \int_{t_0}^{t} W(t, s)g(s, x(s))\,ds + \sum_{t_0 < \tau_k < t} W(t, \tau_k^+)h(x(\tau_k)) & (t \in J^+), \\[2ex] \displaystyle W(t, t_0^+)x_0 + \int_{t_0}^{t} W(t, s)g(s, x(s))\,ds - \sum_{t \le \tau_k \le t_0} W(t, \tau_k^+)h(x(\tau_k)) & (t \in J^-). \end{cases}$$

$$\tag{2.20}$$

2.4 Continuity, differentiability and equations in variations

Consider the impulsive equation

$$\frac{dx}{dt} = f(t, x, \lambda), \quad \phi(t, x, \lambda) \neq 0, \tag{2.21}$$

$$\Delta x = I(t, x, \lambda), \quad \phi(t, x, \lambda) = 0 \tag{2.22}$$

where $t \in \mathbb{R}$, $x \in \mathbb{R}^n$ and $\lambda \in \Lambda \subset \mathbb{R}^m$ is a parameter.

Denote by $x(t; \tau, y, \lambda)$ the solution of equation (2.21), (2.22) satisfying the initial condition

$$x(\tau^+; \tau, y, \lambda) = y. \tag{2.23}$$

Let $\phi(t_0, x_0, \lambda_0) \neq 0$ and $\varphi(t) = x(t; t_0, x_0, \lambda_0)$ be a solution of (2.21), (2.22) for $\lambda = \lambda_0$ which is defined in the interval $[t_0, t_1]$ and satisfies the relations

$$\varphi(t_0) = x_0, \qquad \phi(t_1, \varphi(t_1), \lambda_0) \neq 0.$$

We shall discuss the conditions under which the solution $x(t; \tau, y, \lambda)$ is a continuous function at the point $(t_1, t_0, x_0, \lambda_0)$. Denote by $u(t; \tau, y, \lambda)$ the solution of the initial value problem

$$\frac{du}{dt} = f_1(t, u, \lambda), \quad u(\tau) = y$$

where $f_1 : \mathbb{R} \times \mathbb{R}^n \times \Lambda \to \mathbb{R}^n$, and consider the equation

$$\phi(t, u(t; \tau, y, \lambda), \lambda) = 0$$

with respect to t.

An immediate consequence of the implicit function theorem is the following lemma.

Lemma 2.1   *Let the following conditions hold.*

1. *The function $f_1 : \mathbb{R} \times \mathbb{R}^n \times \Lambda \to \mathbb{R}^n$ is continuous in $\mathbb{R} \times \mathbb{R}^n \times \Lambda$.*
2. *The function $\phi : \mathbb{R} \times \mathbb{R}^n \times \Lambda \to \mathbb{R}$ is continuously differentiable in a neighbourhood of the point (τ_1, x_1, λ_0), where $\tau_1 > t_0$ and $x_1 = u(\tau_1; t_0, x_0, \lambda_0)$.*
3. $\phi(\tau_1, x_1, \lambda_0) = 0.$
4. $\frac{\partial \phi}{\partial t}(\tau_1, x_1, \lambda_0) + \frac{\partial \phi}{\partial x}(\tau_1, x_1, \lambda_0) f_1(\tau_1, x_1, \lambda_0) \neq 0.$

Then there exists a neighbourhood U of the point (t_0, x_0, λ_0) and a unique function $T = T(\tau, y, \lambda)$ which is continuous in U and such that

$$T(t_0, x_0, \lambda_0) = \tau_1,$$

$$\phi(T(\tau, y, \lambda), u(T(\tau, y, \lambda); \tau, y, \lambda), \lambda) \equiv 0 \quad (\tau, y, \lambda) \in U.$$

Moreover, if the function f_1 is continuously differentiable in $\mathbb{R} \times \mathbb{R}^n \times \Lambda$ then the function T is also continuously differentiable in U and the relations

$$\frac{\partial \phi}{\partial t} \frac{\partial T}{\partial \tau} + \frac{\partial \phi}{\partial x}\left[f_1 \frac{\partial T}{\partial \tau} + \frac{\partial u}{\partial \tau}\right] = 0,$$

$$\frac{\partial \phi}{\partial t} \frac{\partial T}{\partial y} + \frac{\partial \phi}{\partial x}\left[f_1 \frac{\partial T}{\partial y} + \frac{\partial u}{\partial y}\right] = 0, \qquad (2.24)$$

$$\frac{\partial \phi}{\partial t} \frac{\partial T}{\partial \lambda} + \frac{\partial \phi}{\partial x}\left[f_1 \frac{\partial T}{\partial \lambda} + \frac{\partial u}{\partial \lambda}\right] + \frac{\partial \phi}{\partial \lambda} = 0$$

are valid, where $\frac{\partial \phi}{\partial t}, \frac{\partial \phi}{\partial x}, \frac{\partial \phi}{\partial \lambda}$ and f_1 are computed at (τ_1, x_1, λ_0), $\frac{\partial T}{\partial \tau}, \frac{\partial T}{\partial y}, \frac{\partial T}{\partial \lambda}$ are computed at (t_0, x_0, λ_0), and $\frac{\partial u}{\partial \tau}, \frac{\partial u}{\partial y}, \frac{\partial u}{\partial \lambda}$ are computed at $(\tau_1, t_0, x_0, \lambda_0)$.

Corollary 2.4 *The function $T(\tau, y, \lambda)$ is Lipschitz continuous with respect to $(\tau, y, \lambda) \in U$ if f_1 is continuously differentiable in $\mathbb{R} \times \mathbb{R}^n \times \Lambda$.*

Introduce the following conditions.

H2.3 The function $\phi : \mathbb{R} \times \mathbb{R}^n \times \Lambda \to \mathbb{R}$ is continuously differentiable in $\mathbb{R} \times \mathbb{R}^n \times \Lambda$ and there exists a $\delta > 0$ such that for $|\lambda - \lambda_0| < \delta$ the equation $\phi(t, x, \lambda) = 0$ defines a smooth hypersurface $S(\lambda)$ which partitions the space $\mathbb{R} \times \mathbb{R}^n$ into a finite number of disjoint domains $D_1(\lambda), \ldots, D_{N+1}(\lambda)$:

$$D_1(\lambda) \cup \cdots \cup D_{N+1}(\lambda) \cup S(\lambda) = \mathbb{R} \times \mathbb{R}^n.$$

H2.4 The solution $\varphi(t) = x(t; t_0, x_0, \lambda_0)$ is defined for $t \in [t_0, t_1]$, has moments of the impulse effect $\tau_k = \tau_k(t_0, x_0, \lambda_0)$ $(k = 1, \ldots, N)$ and the relations

$$(t, \varphi(t)) \in D_k(\lambda_0), \qquad t \in \Delta_k \ (k = 1, \ldots, N+1)$$

hold, where $\Delta_1 = [t_0, \tau_1)$, $\Delta_{N+1} = (\tau_N, t_1]$, $\Delta_k = (\tau_{k-1}, \tau_k)$ $(k = 2, \ldots, N)$.

H2.5 $\frac{\partial \phi}{\partial t}(\tau_k, \varphi(\tau_k), \lambda_0) + \frac{\partial \phi}{\partial x}(\tau_k, \varphi(\tau_k), \lambda_0)f(\tau_k, \varphi(\tau_k), \lambda_0) \neq 0$ $(k = 1, \ldots, N)$.

Under these conditions there exists a $\delta > 0$ such that for $|\lambda - \lambda_0| < \delta$ the sets

$$S_k(\lambda) = \{(t, x) \in \mathbb{R} \times \mathbb{R}^n : \phi(t, x, \lambda) = 0, |t - \tau_k| < \delta, |x - \varphi(\tau_k)| < \delta\} \quad (k = 1, \ldots, N)$$

are open smooth n-dimensional manifolds.

H2.6 There exists a $\delta > 0$ such that for $|\lambda - \lambda_0| < \delta$ and $|y - x_0| < \delta$ the solution $x(t) = x(t; t_0, y, \lambda)$ is defined for $t \in [t_0, t_1]$ and the integral curve $(t, x(t))$, $t \in [t_0, t_1]$ meets successively each $S_k(\lambda)$ $(k = 1, \ldots, N)$ just once.

H2.7 In the domain $D_k(\lambda)$ $(k = 1, \ldots, N+1)$ the function $f(t, x, \lambda)$ coincides with the function $f_k : \mathbb{R} \times \mathbb{R}^n \times \Lambda \to \mathbb{R}^n$ which is continuous in $\mathbb{R} \times \mathbb{R}^n \times \Lambda$.

H2.8 The function $I : \mathbb{R} \times \mathbb{R}^n \times \Lambda \to \mathbb{R}^n$ is continuous in $\mathbb{R} \times \mathbb{R}^n \times \Lambda$.

The following theorem is based on Lemma 2.1.

Theorem 2.9 *Let conditions H2.3–H2.8 hold. Then the solution $x(t; \tau, y, \lambda)$ of the initial value problem (2.21)–(2.23) is a continuous function in some neighbourhood of $(t_1, t_0, x_0, \lambda_0)$.*

Moreover, the moments of the impulse effect $\tau_k(\tau, y, \lambda)$ $(k = 1, \ldots, N)$ of this solution are continuous functions in some neighbourhood of (t_0, x_0, λ_0).

Now let us consider the question of the differentiability of the solution $x(t; \tau, y, \lambda)$ at the point $(t_1, t_0, x_0, \lambda_0)$.

Introduce the following condition.

H2.9 In the domain $D_k(\lambda)$ $(k = 1, \ldots, N+1)$ the function $f(t, x, \lambda)$ coincides with the function $f_k : \mathbb{R} \times \mathbb{R}^n \times \Lambda \to \mathbb{R}^n$ which is continuously differentiable in $\mathbb{R} \times \mathbb{R}^n \times \Lambda$.

H2.10 The function $I : \mathbb{R} \times \mathbb{R}^n \times \Lambda \to \mathbb{R}^n$ is continuously differentiable in $\mathbb{R} \times \mathbb{R}^n \times \Lambda$.

The following theorem is based on Lemma 2.1 as well.

Theorem 2.10 *Let conditions H2.3–H2.6, H2.9 and H2.10 hold. Then:*

1. *The solution $x(t; \tau, y, \lambda)$ of the initial value problem (2.21)–(2.23) is a continuously differentiable function in some neighbourhood of $(t_1, t_0, x_0, \lambda_0)$ and the moments of the impulse effect $\tau_k(\tau, y, \lambda)$ of this solution are continuously differentiable functions in some neighbourhood of (t_0, x_0, λ_0).*

2. *The derivative $u = \frac{\partial x}{\partial x_0}(t; t_0, x_0, \lambda_0)$ is a solution of the initial value problem*

$$\frac{du}{dt} = \frac{\partial f}{\partial x}(t, \varphi(t), \lambda_0)u, \qquad\qquad t \neq \tau_k,$$

$$\Delta u = \frac{\partial I}{\partial x}u + \left[f^+ - f - \frac{\partial I}{\partial t} - \frac{\partial I}{\partial x}f\right]\frac{\frac{\partial \phi}{\partial x}u}{\frac{\partial \phi}{\partial x}f + \frac{\partial \phi}{\partial t}}, \qquad t = \tau_k, \qquad (2.25)$$

$$u(t_0^+) = E_n.$$

3. *The derivative $v = \frac{\partial x}{\partial \lambda}(t; t_0, x_0, \lambda_0)$ is a solution of the initial value problem*

$$\frac{dv}{dt} = \frac{\partial f}{\partial x}(t, \varphi(t), \lambda_0)v + \frac{\partial f}{\partial \lambda}(t, \varphi(t), \lambda_0), \qquad t \neq \tau_k,$$

$$\Delta v = \frac{\partial I}{\partial x}v + \frac{\partial I}{\partial \lambda} + \left[f^+ - f - \frac{\partial I}{\partial t} - \frac{\partial I}{\partial x}f\right]\frac{\frac{\partial \phi}{\partial x}v + \frac{\partial \phi}{\partial \lambda}}{\frac{\partial \phi}{\partial x}f + \frac{\partial \phi}{\partial t}}, \qquad t = \tau_k, \qquad (2.26)$$

$$v(t_0^+) = O_{nm}.$$

4. *The derivative $\frac{\partial x}{\partial t_0}$ satisfies the relation*

$$\frac{\partial x}{\partial t_0}(t; t_0, x_0, \lambda_0) = -\frac{\partial x}{\partial x_0}(t; t_0, x_0, \lambda_0)f(t_0, x_0, \lambda_0). \qquad (2.27)$$

Here $\frac{\partial I}{\partial x}$, $\frac{\partial I}{\partial \lambda}$, $\frac{\partial I}{\partial t}$, $\frac{\partial \phi}{\partial x}$, $\frac{\partial \phi}{\partial \lambda}$, $\frac{\partial \phi}{\partial t}$ are computed at the point $(\tau_k, \varphi(\tau_k), \lambda_0)$ and
$f = f_k(\tau_k, \varphi(\tau_k), \lambda_0)$, $f^+ = f_{k+1}(\tau_k, \varphi(\tau_k^+), \lambda_0)$.

Equations (2.25) and (2.26) are called *equations in variations*.

Note that the jump conditions in these equations have a different form when equation (2.21), (2.22) coincides with some of the equations (1.13), (1.15) or (1.16). We shall give the formulae for computation of the jump Δu of the derivative $u = \frac{\partial x}{\partial x_0}(t; t_0, x_0, \lambda_0)$ at the point τ_k in the following particular cases.

I. If $\phi = t - \tau_k$, $I = I_k(x)$ (equation (1.13)), then

$$\Delta u = \frac{\partial I_k}{\partial x} u. \tag{2.28}$$

II. If $\phi = t - \tau_k(x)$, $I = I_k(x)$ (equation (1.15)), then

$$\Delta u = L_k u = \frac{\partial I_k}{\partial x} u + \left[\frac{\partial I_k}{\partial x} f + f - f^+ \right] \frac{\frac{\partial \tau_k}{\partial x} u}{1 - \frac{\partial \tau_k}{\partial x} f}. \tag{2.29}$$

III. If $\phi = \phi(x)$, $I = I(x)$ (equation (1.16)), then

$$\Delta u = N_k u = \frac{\partial I}{\partial x} u + \left[f^+ - f - \frac{\partial I}{\partial x} f \right] \frac{\frac{\partial \phi}{\partial x} u}{\frac{\partial \phi}{\partial x} f}. \tag{2.30}$$

2.5 Impulsive inequalities
We shall need the following two lemmas about impulsive differential and integral inequalities.

Lemma 2.2   Let the function $u \in PC^1(\mathbb{R}_+, \mathbb{R})$ satisfy the inequalities

$$\frac{du}{dt}(t) \le p(t)u(t) + f(t), \quad t \ne \tau_k, \ t > 0,$$

$$u(\tau_k^+) \le d_k u(\tau_k) + h_k, \qquad \tau_k > 0,$$

$$u(0+) \le u_0$$

where $p, f \in PC(\mathbb{R}_+, \mathbb{R})$ and $d_k \ge 0$, h_k and u_0 are constants. Then for $t > 0$

$$u(t) \le u_0 \prod_{0 < \tau_k < t} d_k \exp\left(\int_0^t p(s) \, ds \right)$$
$$+ \int_0^t \prod_{s \le \tau_k < t} d_k \exp\left(\int_s^t p(\tau) \, d\tau \right) f(s) \, ds$$
$$+ \sum_{0 < \tau_k < t} \prod_{\tau_k < \tau_j < t} d_j \exp\left(\int_{\tau_k}^t p(\tau) \, d\tau \right) h_k.$$

Lemma 2.3 *For $t > 0$ let the function $u \in PC(\mathbb{R}_+, \mathbb{R}_+)$ satisfy the inequality*

$$u(t) \leq u_0 + \int_0^t p(s)u(s) \, ds + \sum_{0 < \tau_k < t} \beta_k u(\tau_k)$$

where $p \in PC(\mathbb{R}_+, \mathbb{R}_+)$ and $\beta_k \geq 0$ and u_0 are constants. Then for $t > 0$

$$u(t) \leq u_0 \prod_{0 < \tau_k < t} (1 + \beta_k) \exp \left(\int_0^t p(s) \, ds \right).$$

2.6 Compactness criterion. B-convergence

Let a finite number of points of the sequence $\{\tau_k\}$, $k \in \mathbb{Z}$, lie in the interval $[0, T]$. Let $PC([0, T], \mathbb{R}^n)$ be the set of functions $x : [0, T] \to \mathbb{R}^n$ which are piecewise continuous in $[0, T]$ and have points of discontinuity $\tau_k \in [0, T]$, where they are continuous from the left. In the set $PC([0, T], \mathbb{R}^n)$ introduce the norm

$$\|x\| = \sup\{|x(t)| : t \in [0, T]\}$$

with which $PC([0, T], \mathbb{R}^n)$ becomes a Banach space with the uniform convergence topology.

Applying the Schauder–Tychonoff fixed point theorem we shall need a compactness criterion for a set $\mathcal{F} \subset PC([0, T], \mathbb{R}^n)$.

Definition 2.3 The set $\mathcal{F}$ is said to be *quasiequicontinuous* in $[0, T]$ if for any $\epsilon > 0$ there exists a $\delta > 0$ such that if $x \in \mathcal{F}$; $k \in \mathbb{Z}$; $t_1, t_2 \in (\tau_{k-1}, \tau_k] \cap [0, T]$, and $|t_1 - t_2| < \delta$, then

$$|x(t_1) - x(t_2)| < \epsilon.$$

The following lemma gives a necessary and sufficient condition for relative compactness in $PC([0, T], \mathbb{R}^n)$.

Lemma 2.4 (Compactness criterion) *The set $\mathcal{F} \subset PC([0, T], \mathbb{R}^n)$ is relatively compact if and only if:*

1. *$\mathcal{F}$ is bounded, that is, $\|x\| \leq c$ for each $x \in \mathcal{F}$ and some $c > 0$;*
2. *$\mathcal{F}$ is quasiequicontinuous in $[0, T]$.*

Now suppose that each one of the functions $x_j : D_j \to \mathbb{R}^n$ $(j = 0, 1, 2, \ldots)$ is piecewise continuous in $D_j \subset \mathbb{R}$ and has points of discontinuity τ_k^j $(k = 1, \ldots, q)$, where it is continuous from the left. The situation when $\tau_k^i \neq \tau_k^j$ $(i \neq j)$ is usual if the functions $x_j(t)$ are distinct solutions of an impulsive differential equation with unfixed moments of the impulse effect. For functional sequences of this type we shall use the following type of convergence.

Definition 2.4   We say that the sequence $x_j(t)$ is B-convergent to $x_0(t)$ as $j \to \infty$ and we shall denote

$$x_j(t) \overset{B}{\to} x_0(t) \quad \text{as } j \to \infty$$

if for any $\epsilon > 0$ there exists a $\nu \in \mathbb{N}$ such that for $j \geq \nu$

1. $\operatorname{mes}\left[(D_j \backslash D_0) \cup (D_0 \backslash D_j)\right] < \epsilon$;
2. $|\tau_k^j - \tau_k^0| < \epsilon \ (k = 1, \dots, q)$;
3. $|x_j(t) - x_0(t)| < \epsilon$ for $t \in D_j \cap D_0$, $|t - \tau_k^0| > \epsilon$.

Notes and comments for Chapter I

Example 1.2 was taken from A. A. Andronov et al. [6].

     Conditions for the absence of the phenomenon of beating were first obtained by A. M. Samoilenko and N. A. Perestyuk [70]. Theorem 2.1 is due to A. B. Dishliev and D. D. Bainov [24] and Corollary 2.2 reflects the contents of [70], Lemma 3. A number of sufficient conditions for the absence of beating were found by A. B. Dishliev and D. D. Bainov [24, 25], V. Lakshmikantham, D. D. Bainov and P. S. Simeonov [55], Yu. V. Rogovchenko and S. I. Trofimchuk [66] and by A. M. Samoilenko, N. A. Perestyuk and S. I. Trofimchuk [74].

     The question of the existence and uniqueness of the solutions of the initial value problem for impulsive differential equations was discussed by A. D. Myshkis and A. M. Samoilenko [60], A. M. Samoilenko and N. A. Perestyuk [70] and by P. S. Simeonov [75, 76]. Theorem 2.2 was taken from [55] while Theorems 2.3–2.8 were taken from [15].

     The results on continuity and differentiability of the solutions of impulsive differential equations with respect to initial values and a parameter were first obtained by P. S. Simeonov and D. D. Bainov [77]. In [77] the three basic variational equations were obtained. For similar results see [55] and M. U. Ahmetov and N. A. Perestyuk [2,4].

     Lemma 2.2 was taken from P. S. Simeonov and D. D. Bainov [80] while Lemma 2.3 was taken from A. M. Samoilenko and N. A. Perestyuk [70]. For other results on differential and integral inequalities see also P. S. Simeonov and D. D. Bainov [78–80] and D. D. Bainov and P. S. Simeonov [16]. The spaces PC and PC^1 were introduced in [81], where Lemma 2.4 was proved. The definition of B-convergence was taken from [2].

Chapter II

Linear Impulsive Periodic Equations

This chapter is devoted to the main problem traditionally considered in the linear theory of periodic systems of differential equations. The results obtained are of independent interest and, moreover, are repeatedly applied in the later chapters of the book.

3. Linear homogeneous periodic equations. Floquet theory

We shall present the Floquet theory for the linear T-periodic impulsive equation

$$
\begin{aligned}
\frac{dx}{dt} &= A(t)x, \quad t \neq \tau_k, t \in \mathbb{R}, \\
\Delta x &= B_k x, \quad t = \tau_k, k \in \mathbb{Z}.
\end{aligned}
\tag{3.1}
$$

Introduce the following conditions (H3).

H3.1 $A(\,\cdot\,) \in PC(\mathbb{R}, \mathbb{C}^{n \times n})$ and $A(t+T) = A(t)$ $(t \in \mathbb{R})$.
H3.2 $B_k \in \mathbb{C}^{n \times n}$, $\det(E + B_k) \neq 0$, $\tau_k < \tau_{k+1}$ $(k \in \mathbb{Z})$.
H3.3 There exists a $q \in \mathbb{N}$ such that

$$
B_{k+q} = B_k, \qquad \tau_{k+q} = \tau_k + T \qquad (k \in \mathbb{Z}).
$$

Without loss of generality we assume that $\tau_0 \leq 0 < \tau_1$. The following theorem is a generalization of Floquet's theorem.

Theorem 3.1 *Let conditions (H3) hold. Then each fundamental matrix of (3.1) can be represented in the form*

$$
X(t) = \phi(t)e^{\Lambda t} \quad (t \in \mathbb{R})
\tag{3.2}
$$

where the matrix $\Lambda \in \mathbb{C}^{n \times n}$ *is constant and the matrix* $\phi(\,\cdot\,) \in PC^1(\mathbb{R}, \mathbb{C}^{n \times n})$ *is non-singular and* T-*periodic.*

Proof   The proof is exactly the same as that in standard Floquet theory but for the sake of completeness we give the details.

Let $X(t)$ be a fundamental matrix of (3.1). Then the matrix $Y(t) = X(t+T)$ is also fundamental. Indeed,

$$\frac{dY}{dt}(t) = \frac{dX}{dt}(t+T) = A(t+T)X(t+T) = A(t)Y(t), \quad t \neq \tau_k,$$

$$\Delta Y(\tau_k) = \Delta X(\tau_k + T) = \Delta X(\tau_{k+q}) = B_{k+q}X(\tau_{k+q}) = B_k X(\tau_k + T) = B_k Y(\tau_k).$$

By Theorem 2.8 there exists a unique non-singular matrix $M \in \mathbb{C}^{n \times n}$ such that

$$X(t+T) = X(t)M \quad (t \in \mathbb{R}). \tag{3.3}$$

Set

$$\Lambda = \frac{1}{T}\ln M, \tag{3.4}$$

$$\phi(t) = X(t)e^{-\Lambda t}. \tag{3.5}$$

From (3.5) it follows that equality (3.2) holds, and the matrix $\phi(t)$ is non-singular and belongs to the class $PC^1(\mathbb{R}, \mathbb{C}^{n \times n})$. In view of $Me^{-\Lambda T} = E$ we obtain

$$\phi(t+T) = X(t+T)e^{-\Lambda(t+T)} = X(t)Me^{-\Lambda T}e^{-\Lambda t} = \phi(t),$$

that is, $\phi(t)$ is T-periodic. $\Box$

Remark 3.1   From Theorem 3.1 and conditions (H3) it follows that the matrix $\phi(t)$ in the Floquet representation (3.2) is a Lyapunov matrix, that is, it satisfies the conditions:

(i) $\phi(\,\cdot\,) \in PC^1(\mathbb{R}, \mathbb{C}^{n \times n})$;

(ii) $\phi(t)$ and $\frac{d\phi}{dt}(t)$ are bounded in $\mathbb{R}$;

(iii) $\inf_{t \in \mathbb{R}} |\det \phi(t)| > 0$.

Moreover, the T-periodic matrix $\phi^{-1}(t)$ is defined for each $t \in \mathbb{R}$ and is also a Lyapunov matrix.

Remark 3.2   The matrices Λ and $\phi(t)$ can be complex even if the matrices $A(t)$ and B_k $(t \in \mathbb{R}, k \in \mathbb{Z})$ are real. In the case when equation (3.1) is real, we can set

$$\Lambda_1 = \frac{1}{2T}\ln M^2, \qquad \phi_1(t) = X(t)e^{-\Lambda_1 t}$$

and obtain the representation

$$X(t) = \phi_1(t)e^{\Lambda_1 t}$$

where the matrices Λ_1 and $\phi_1(t)$ are real and $\phi_1(t)$ is $2T$-periodic.

By equality (3.3) there corresponds to the fundamental matrix $X(t)$ the constant matrix M which we call the *monodromy matrix* of (3.1) (corresponding to the fundamental matrix $X(t)$). Let $X_1(t)$ be another fundamental matrix of (3.1) and let M_1 be the corresponding monodromy matrix:

$$X_1(t+T) = X_1(t)M_1.$$

By Theorem 2.8

$$X_1(t) = X(t)S, \qquad \det S \neq 0.$$

Hence

$$X(t+T)S = X(t)SM_1. \tag{3.6}$$

We compare (3.3) and (3.6) and conclude that

$$M = SM_1S^{-1},$$

that is, all monodromy matrices of (3.1) are similar and have the same eigenvalues.

The eigenvalues $\mu_1, \ldots, \mu_n$ of the monodromy matrices are called (*Floquet*) *multipliers* of equation (3.1) and the eigenvalues $\lambda_1, \ldots, \lambda_n$ of the matrix Λ in Theorem 3.1 are called *characteristic exponents* (or *Floquet exponents*) of (3.1). From (3.4) we obtain

$$\lambda_j = \frac{1}{T}\ln \mu_j \quad (j = 1, \ldots, n).$$

Moreover, to the simple multipliers there correspond simple characteristic exponents and to the multiple multipliers there correspond exponents having elementary divisors of the same multiplicity. The multipliers of (3.1) are determined uniquely but its characteristic exponents are determined only modulo $2k\pi i$ $(k \in \mathbb{Z})$.

Remark 3.3 In order to calculate the multipliers $\mu_1, \ldots, \mu_n$ of (3.1) we have to choose an arbitrary fundamental matrix $X(t)$ of (3.1) and calculate the eigenvalues of the matrix

$$M = W(t_0 + T, t_0) = X(t_0 + T)X^{-1}(t_0) \tag{3.7}$$

where $t_0 \in \mathbb{R}$ is fixed.

If $X(0) = E$ (or $X(0_+) = E$) then we can choose $M = X(T)$ (or $M = X(T+)$) as the monodromy matrix of (3.1).

From (3.7) and Liouville's formula (2.16) it follows that for any monodromy matrix M of (3.1) we have

$$\det M = \mu_1 \cdot \mu_2 \cdot \; \cdots \; \cdot \mu_n = \prod_{k=1}^{q} \det (E + B_k)\exp\left(\int_0^T \mathrm{Tr}\; A(s)\; ds\right). \tag{3.8}$$

The following two theorems are simple corollaries of Theorem 3.1.

Theorem 3.2 *Let conditions (H3) hold. Then the number $\mu \in \mathbb{C}$ is a multiplier of (3.1) if and only if there exists a non-trivial solution $\varphi(t)$ of (3.1) such that $\varphi(t+T) = \mu\varphi(t)$ $(t \in \mathbb{R})$.*

Theorem 3.3 *Let conditions (H3) hold. Then (3.1) has a non-trivial kT-periodic solution if and only if the kth power of some of its multipliers equals 1.*

We shall prove the following theorem of reducibility.

Theorem 3.4 *Let conditions (H3) hold. Then the linear T-periodic impulsive equation (3.1) is reducible to an equation with constant coefficients*

$$\frac{dy}{dt} = \Lambda y$$

by a T-periodic Lyapunov transformation $x = \phi(t)y$.

Proof By Theorem 3.1 and Remark 3.1 the fundamental matrix $X(t)$ of equation (3.1) has the form

$$X(t) = \phi(t)e^{\Lambda t} \tag{3.9}$$

where $\phi(t)$ is a T-periodic Lyapunov matrix.

We transform (3.1) by Lyapunov's transformation

$$x = \phi(t)y. \tag{3.10}$$

Then for $t \neq \tau_k$, $t \in \mathbb{R}$, we have

$$\frac{d\phi}{dt}(t)y + \phi(t)\frac{dy}{dt} = A(t)\phi(t)y$$

and in view of (3.9) we obtain

$$\frac{dX}{dt}(t)e^{-\Lambda t}y - X(t)\Lambda e^{-\Lambda t}y + X(t)e^{-\Lambda t}\frac{dy}{dt} = A(t)X(t)e^{-\Lambda t}y$$

or

$$\frac{dy}{dt} = \Lambda y. \tag{3.11}$$

For $t = \tau_k$ we obtain successively

$$\phi(\tau_k^+)y(\tau_k^+) - \phi(\tau_k)y(\tau_k) = B_k\phi(\tau_k)y(\tau_k),$$

$$X(\tau_k^+)e^{-\Lambda\tau_k}y(\tau_k^+) - X(\tau_k)e^{-\Lambda\tau_k}y(\tau_k) = B_kX(\tau_k)e^{-\Lambda\tau_k}y(\tau_k),$$

$$X(\tau_k^+)e^{-\Lambda\tau_k}y(\tau_k^+) = X(\tau_k^+)e^{-\Lambda\tau_k}y(\tau_k),$$

$$y(\tau_k^+) = y(\tau_k),$$

that is, by transformation (3.10), equation (3.1) is reduced to the linear equation with constant coefficients (3.11). □

The multipliers of equation (3.1) completely characterize its stability. This is seen from the following theorem which is a consequence of Theorem 3.4 and the relation

$$\frac{1}{T}\ln|\mu_j| = \Re\lambda_j \quad (j = 1,\dots,n) \tag{3.12}$$

between the multipliers μ_j of (3.1) and the real parts of the eigenvalues λ_j of the matrix Λ.

Theorem 3.5   *Let conditions (H3) hold. Then the linear T-periodic impulsive equation (3.1) is:*

1. *stable if and only if all multipliers μ_j ($j = 1,\dots,n$) of equation (3.1) satisfy the inequality $|\mu_j| \leq 1$ and, moreover, to those μ_j for which $|\mu_j| = 1$ there correspond simple elementary divisors;*
2. *asymptotically stable if and only if all multipliers μ_j ($j = 1,\dots,n$) of equation (3.1) satisfy the inequality $|\mu_j| < 1$;*
3. *unstable if $|\mu_j| > 1$ for some $j = 1,\dots,n$.*

Remark 3.4   When the impulsive differential equation is with fixed moments of the impulse effect, the definitions of various types of stability of the solutions coincide with the known notions of stability of the solutions of ordinary differential equations without impulses.

In the case when the moments of the impulse effect are not fixed, the definitions of stability need a modification.

Remark 3.5   If the T-periodic equation (3.1) is asymptotically stable ($|\mu_j| < 1$, $j = 1,\dots,n$) then equation (3.1) is exponentially stable in $\mathbb{R}$, that is, there exist constants $K \geq 1$ and $\alpha > 0$ such that for any fundamental matrix $X(t)$ of (3.1) we have

$$|X(t)X^{-1}(s)| \leq Ke^{-\alpha(t-s)} \quad \text{for} \ -\infty < s \leq t < \infty. \tag{3.13}$$

The property of exponential stability is a particular case of the property of exponential dichotomy.

Let $X(t)$, $t \in \mathcal{J} \subset \mathbb{R}$ be a fundamental matrix of (3.1).

Definition 3.1   Equation (3.1) is said to have *exponential dichotomy* in $\mathcal{J}$ if there exists a projection operator P, that is, a matrix P for which $P^2 = P$, and positive constants K, L, α, and β, such that

$$\begin{aligned}
|X(t)PX^{-1}(s)| &\leq Ke^{-\alpha(t-s)} &\quad (t,s \in \mathcal{J}, s \leq t), \\
|X(t)(E-P)X^{-1}(s)| &\leq Le^{-\beta(s-t)} &\quad (t,s \in \mathcal{J}, s \geq t).
\end{aligned} \tag{3.14}$$

Theorem 3.6   *Let conditions (H3) hold. Then the linear T-periodic equation (3.1) has an exponential dichotomy in $\mathbb{R}$ if and only if the multipliers of equation (3.1) do not lie on the unit circle, that is, if*

$$|\mu_j| \neq 1 \quad (j = 1, \ldots, n). \tag{3.15}$$

Proof   By Theorem 3.4, equation (3.1) is reduced, by Lyapunov's transformation (3.10), to equation (3.11) with constant coefficients. Equation (3.11) has an exponential dichotomy [22] if and only if the eigenvalues λ_j of the matrix Λ satisfy the condition

$$\Re\lambda_j \neq 0 \quad (j = 1, \ldots, n).$$

But from (3.12) it follows that this condition is met if and only if the multipliers μ_j of equation (3.1) satisfy condition (3.15), and Theorem 3.6 follows from the fact that Lyapunov's transformation preserves the property of exponential dichotomy. $\square$

Example 3.1   At the moments $\tau_k = \tau_0 + kT$ $(k \in \mathbb{Z}, T > 0)$ let the linear oscillator with equation $\ddot{x} + \omega^2 x = 0$ be subject to an impulse effect as a result of which the velocity $\dot{x}$ obtains an increment $\Delta\dot{x} = ax + b\dot{x}$. The motion of such an oscillator can be described by the T-periodic linear impulsive system

$$\begin{aligned} \dot{x} = \omega y, \quad \dot{y} = -\omega x, \qquad & t \neq \tau_k, t \in \mathbb{R}, \\ \Delta x = 0, \quad \Delta y = \frac{a}{\omega}x + by, \quad & t = \tau_k, k \in \mathbb{Z}. \end{aligned} \tag{3.16}$$

The monodromy matrix of (3.16) equals

$$\begin{aligned} M &= W(\tau_0 + T+, \tau_0^+) = (E + B)e^{AT} \\ &= \begin{bmatrix} 1 & 0 \\ \frac{a}{\omega} & b+1 \end{bmatrix} \begin{bmatrix} \cos\omega T & \sin\omega T \\ -\sin\omega T & \cos\omega T \end{bmatrix} \\ &= \begin{bmatrix} \cos\omega T & \sin\omega T \\ \frac{a}{\omega}\cos\omega T - (b+1)\sin\omega T & \frac{a}{\omega}\sin\omega T + (b+1)\cos\omega T \end{bmatrix} \end{aligned}$$

and its multipliers μ_j $(j = 1, 2)$ satisfy the equation

$$\mu^2 - \left[\frac{a}{\omega}\sin\omega T + (b+2)\cos\omega T\right]\mu + b + 1 = 0. \tag{3.17}$$

By means of Theorems 3.3 and 3.5 we shall investigate the existence of T- and $2T$-periodic motions of the oscillator and their stability.

(i) *Existence of T-periodic solutions.* System (3.16) has non-trivial T-periodic solutions if and only if $\mu = 1$ is a root of (3.17), that is, if

$$\frac{a}{\omega}\sin\omega T + (b+2)\cos\omega T = b + 2$$

or

$$\sin \frac{1}{2}\omega T \left[\frac{a}{\omega} \cos \frac{1}{2}\omega T - (b+2) \sin \frac{1}{2}\omega T \right] = 0. \tag{3.18}$$

Let $x = x(t)$, $y = y(t)$ be a T-periodic solution of (3.16) and $x(\tau_0^+) = x_0$, $y(\tau_0^+) = y_0$. Then $z_0 = \text{col}\,(x_0, y_0)$ must satisfy the system $M z_0 = z_0$ or, in detail, the system

$$(\cos \omega T - 1)x_0 + \sin \omega T \, y_0 = 0,$$

$$\left[\frac{a}{\omega} \cos \omega T - (b+1) \sin \omega T \right] x_0 + \left[\frac{a}{\omega} \sin \omega T + (b+1) \cos \omega T - 1 \right] y_0 = 0. \tag{3.19}$$

The following cases are possible.

(i.1) Let $\sin \frac{1}{2}\omega T = 0$, that is, $T = 2k\pi/\omega$ $(k \in \mathbb{N})$. The condition (3.18) is satisfied and system (3.19) is reduced to the equation

$$\frac{a}{\omega} x_0 + b y_0 = 0.$$

Since for $\tau_j < t \leq \tau_{j+1}$

$$x(t) = x_0 \cos \omega(t - \tau_j) + y_0 \sin \omega(t - \tau_j),$$

$$y(t) = -x_0 \sin \omega(t - \tau_j) + y_0 \cos \omega(t - \tau_j),$$

then $x(\tau_{j+1}) = x_0$, $y(\tau_{j+1}) = y_0$. Thus for $t = \tau_{j+1}$

$$\Delta x = 0, \qquad \Delta y = \frac{a}{\omega} x_0 + b y_0 = 0,$$

that is, the T-periodic solutions starting at $t = \tau_0^+$ from the point (x_0, y_0) of the straight line $(a/\omega)x + by = 0$ are continuous.

(i.2) Let $\sin \frac{1}{2}\omega T \neq 0$, $a = 0$ and $b = -2$. The condition (3.18) is satisfied for each $T > 0$, $T \neq 2k\pi/\omega$, and in this case the T-periodic motions of the oscillator start at $t = \tau_0^+$ from the point (x_0, y_0) of the straight line

$$-x_0 + \cot \frac{1}{2}\omega T \, y_0 = 0.$$

(i.3) Let $\sin \frac{1}{2}\omega T \neq 0$, $a \neq 0$ and let T satisfy the condition

$$\cot \frac{1}{2}\omega T = \frac{(b+2)\omega}{a}.$$

The the T-periodic motions of (3.16) start from the points (x_0, y_0) for which

$$-a x_0 + (b+2)\omega y_0 = 0.$$

(ii) *$2T$-periodic solutions of system (3.16).* The eigenvalues of the matrix M^2 are $m_j = \mu_j^2$ $(j = 1, 2)$, where μ_j $(j = 1, 2)$ are the eigenvalues of the monodromy matrix M and satisfy equation (3.17). Hence

$$m_1 + m_2 = \mu_1^2 + \mu_2^2 = (\mu_1 + \mu_2)^2 - 2\mu_1\mu_2$$

$$= \left[\frac{a}{\omega} \sin \omega T + (b+2) \cos \omega T \right]^2 - 2(b+1),$$

$$m_1 m_2 = \mu_1^2 \mu_2^2 = (b+1)^2,$$

that is, m_j $(j = 1, 2)$ satisfy the equation

$$m^2 - \left\{ \left[\frac{a}{\omega} \sin \omega T + (b+2) \cos \omega T \right]^2 - 2(b+1) \right\} m + (b+1)^2 = 0. \qquad (3.20)$$

System (3.16) has non-trivial $2T$-periodic solutions if and only if $m = 1$ is a solution of (3.20), that is, if

$$\left[\frac{a}{\omega} \sin \omega T + (b+2) \cos \omega T \right]^2 = (b+2)^2. \qquad (3.21)$$

Then $z_0 = \mathrm{col}\,(x_0, y_0)$ must satisfy the system $M^2 z_0 = z_0$ or $(M - M^{-1})z_0 = 0$. In view of

$$M^{-1} = \frac{1}{b+1} \begin{bmatrix} \frac{a}{\omega} \sin \omega T + (b+1) \cos \omega T & -\sin \omega T \\ -\frac{a}{\omega} \cos \omega T + (b+1) \sin \omega T & \cos \omega T \end{bmatrix}$$

the system $(M - M^{-1})z_0 = 0$ takes the form

$$\sin \omega T \left[\frac{a}{\omega} x_0 - (b+2)y_0 \right] = 0,$$

$$(b+2) \left[\frac{a}{\omega} \cos \omega T - (b+1) \sin \omega T \right] x_0 + \left[(b+1) \frac{a}{\omega} \sin \omega T + b(b+2) \cos \omega T \right] y_0 = 0. \qquad (3.22)$$

The following cases are possible.

(ii.1) Let $\sin \omega T = 0$, that is, $T = k\pi/\omega$ $(k \in \mathbb{N})$. In this case condition (3.21) is satisfied and system (3.22) is reduced to the equation

$$(b+2)\left(\frac{a}{\omega} x_0 + by_0 \right) = 0.$$

Then:

(ii.1.1) If $b = -2$, then all solutions of (3.16) are periodic with period $T_1 = 2T = 2k\pi/\omega$.

(ii.1.2) If $b \neq -2$, then $T_1 = 2k\pi/\omega$ is the period of the motions of the oscillator starting at $t = \tau_0^+$ from the points (x_0, y_0) of the straight line $(a/\omega)x + by = 0$. These motions are continuous (case (i.1)).

(ii.2) Let $\sin \omega T \neq 0$, $a = 0$ and $b = -2$. Then conditions (3.21) and (3.22) are satisfied and all solutions of (3.16) have period $T_1 = 2T \neq 2k\pi/\omega$.

(ii.3) Let $\sin \omega T \neq 0$, $a \neq 0$, $b \neq -2$ and let T satisfy condition (3.21) which can be written in the form

$$\cot \omega T = \frac{\omega^2(b+2)^2 - a^2}{2\omega a(b+2)}.$$

The (3.16) has $2T$-periodic solutions starting from the points (x_0, y_0) of the straight line

$$\frac{a}{\omega} x - (b+2)y = 0.$$

When the oscillator has periodic motions with period T (or $2T$), then the multipliers of (3.16) are $\mu_1 = 1, \mu_2 = b+1$ (or $m_1 = 1, m_2 = (b+1)^2$). These periodic motions are stable if $|b+1| < 1$ or $-2 < b < 0$.

Example 3.2   Consider the mathematical model of the mechanism represented in Fig. 3.1.

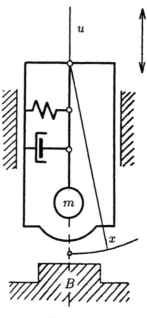

Fig. 3.1

A body of mass m is hinged by an absolutely rigid rod of length l to a moveable frame whose mass is much greater than m. The small oscillations of the body about the equilibrium point in the case when the frame is motionless are described by an equation of the form

$$\ddot{x} + b\dot{x} + \frac{g}{l}x = 0 \quad (b \geq 0).$$

Let the frame perform periodic motions, and at the moments $\tau_n = nT$ $(n \in \mathbb{Z}, T > 0)$ let it collide with the motionless stopping device B and the law of motion is (Fig. 3.2):

$$u(t) = \begin{cases} -\frac{1}{2}\mu(t^2 - Tt) = u_0(t), & 0 < t \leq T, \\ u_0(t - nT), & nT < t \leq nT + T. \end{cases}$$

In this case the oscillations of the body about the equilibrium point are described by the equation with impulse effect

$$\ddot{x} + b\dot{x} + \frac{g - \mu}{l}x = 0, \qquad t \neq \tau_n,$$

$$\Delta\dot{x} = -\frac{\mu T}{l}x, \quad t = \tau_n. \tag{3.23}$$

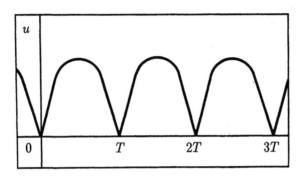

Fig. 3.2

Assume that $b = 0$, $g > \mu$, and consider the question of the existence of non-trivial periodic solutions of equation (3.23). We write (3.23) as a system

$$\dot{x} = \omega y, \qquad \dot{y} = -\omega x, \qquad t \neq nT,$$
$$\Delta x = 0, \qquad \Delta y = -kx, \qquad t = nT, \qquad (3.24)$$

where $\omega = \left(\dfrac{g - \mu}{l}\right)^{1/2}$, $k = \dfrac{\mu T}{\omega l}$.

The monodromy matrix of (3.24) is

$$
M = \begin{bmatrix} 1 & 0 \\ -k & 1 \end{bmatrix} \begin{bmatrix} \cos \omega T & \sin \omega T \\ -\sin \omega T & \cos \omega T \end{bmatrix}
$$
$$
= \begin{bmatrix} \cos \omega T & \sin \omega T \\ -k \cos \omega T - \sin \omega T & -k \sin \omega T + \cos \omega T \end{bmatrix}
$$

and its multipliers μ_j $(j = 1, 2)$ satisfy the equation

$$\mu^2 - (2 \cos \omega T - k \sin \omega T)\mu + 1 = 0. \qquad (3.25)$$

We apply Theorems 3.3 and 3.5 and obtain the following conclusions.

(i) System (3.24) has a non-trivial T-periodic solution if and only if $\mu = 1$ satisfies (3.25), that is, if

$$2 \cos \omega T - k \sin \omega T = 2.$$

In this case, there corresponds to the eigenvalue $\mu = 1$ of the monodromy matrix M no simple elementary divisors, and the non-trivial T-periodic solutions of (3.24) are unstable.

(ii) System (3.24) has a non-trivial $2T$-periodic solution if and only if $\mu = -1$ satisfies (3.25), that is, if

$$2 \cos \omega T - k \sin \omega T = -2.$$

In this case, there correspond to the eigenvalue $\mu = 1$ of the matrix M^2 no simple elementary divisors, and the non-trivial $2T$-periodic solutions of (3.24) are unstable.

(iii) System (3.24) has a non-trivial $4T$-periodic solution if and only if $\mu_1 = i$ and $\mu_2 = -i$ are roots of (3.25), that is, if

$$2 \cos \omega T - k \sin \omega T = 0.$$

In this case, there correspond to the eigenvalue $\mu = 1$ of the matrix M^4 simple elementary divisors, and the non-trivial $4T$-periodic solutions of (3.24) are stable.

(iv) The multipliers μ_j $(j = 1, 2)$ of (3.24) are complex conjugate (and distinct) if

$$|2 \cos \omega T - k \sin \omega T| < 2.$$

Since $\mu_1 \mu_2 = 1$, in this case $|\mu_j| = 1$ $(j = 1, 2)$ and the trivial solution of (3.24) is stable.

(v) The multipliers μ_j $(j = 1, 2)$ of (3.24) are real and distinct if

$$|2 \cos \omega T - k \sin \omega T| > 2. \tag{3.26}$$

Since $\mu_1 \mu_2 = 1$, then in this case the modulus of one of the multipliers is greater than 1 and the trivial solution of (3.24) is unstable, that is, a parametric resonance arises in the system. The condition (3.26) of parametric resonance can be written in the form

$$-\frac{1}{2}k < \tan \frac{1}{2}\omega T < 0, \qquad \text{or} \qquad 0 < \cot \frac{1}{2}\omega T < \frac{1}{2}k. \tag{3.27}$$

Example 3.3   Consider the oscillations of a physical pendulum which is a rigid body rotating freely in a definite vertical plane about its point of suspension (Fig. 3.3). Assume that the point of suspension is subject to a vertical periodic motion, and its vertical movement $F(t)$ is given by the T-periodic extension of the function

$$f(t) = \begin{cases} at & \text{for } 0 \leq t \leq \frac{1}{4}T \\ \frac{1}{2}aT - at & \text{for } \frac{1}{4}T \leq t \leq \frac{3}{4}T \\ at - aT & \text{for } \frac{3}{4}T \leq t \leq T. \end{cases}$$

For small oscillations near the upper equilibrium point the equation of motion of the pendulum has the form

$$\frac{\mathrm{d}^2 x}{\mathrm{d}t^2} - \frac{gl}{r_0^2}x = 0, \qquad\qquad t \neq \tau_k,$$
$$\Delta \frac{\mathrm{d}x}{\mathrm{d}t} = (-1)^k \frac{2al}{r_0^2}x, \quad t = \tau_k, \tag{3.28}$$

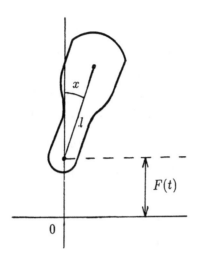

Fig. 3.3

where x is the angular shift, t is the time, l is the distance from the mass centre of the pendulum to the point of suspension, r_0 is the inertial radius of the pendulum with respect to the point of suspension, g is the acceleration of the gravitational force, and $t_k = \frac{1}{4}T + \frac{1}{2}(k-1)T \ (k \in \mathbb{Z})$.

After the change of variables

$$t = \frac{T}{2\pi}\tau, \quad \omega = \frac{T\sqrt{gl}}{2\pi r_0}, \quad b = \frac{alT}{\pi r_0^2},$$

equation (3.28) becomes

$$\frac{d^2x}{d\tau^2} - \omega^2 x = 0, \qquad t \neq \tau_k = \tfrac{1}{2}\pi + (k-1)\pi,$$

$$\Delta\frac{dx}{d\tau} = (-1)^k bx, \quad t = \tau_k, k \in \mathbb{Z}. \tag{3.29}$$

We write (3.29) as a system

$$\frac{dx}{d\tau} = \omega y, \quad \frac{dy}{d\tau} = \omega x, \qquad t \neq \tau_k, \tag{3.30}$$

$$\Delta x = 0, \quad \Delta y = (-1)^k \frac{b}{\omega} x, \quad t = \tau_k. \tag{3.31}$$

System (3.30), (3.31) is 2π-periodic since $\tau_{k+2} = \tau_k + 2\pi$, $\Delta y(\tau_{k+2}) = \Delta y(\tau_k)$.
Since

$$W(\tau, s) = \begin{bmatrix} \cosh\omega(\tau - s) & \sinh\omega(\tau - s) \\ \sinh\omega(\tau - s) & \cosh\omega(\tau - s) \end{bmatrix}$$

is the Cauchy matrix for system (3.30), we obtain the monodromy matrix of (3.30), (3.31):

$$M = \begin{bmatrix} 1 & 0 \\ \frac{b}{\omega} & 1 \end{bmatrix} \begin{bmatrix} \cosh\omega\pi & \sinh\omega\pi \\ \sinh\omega\pi & \cosh\omega\pi \end{bmatrix} \begin{bmatrix} 1 & 0 \\ -\frac{b}{\omega} & 1 \end{bmatrix} \begin{bmatrix} \cosh\omega\pi & \sinh\omega\pi \\ \sinh\omega\pi & \cosh\omega\pi \end{bmatrix}$$

$$= \begin{bmatrix} \cosh 2\omega\pi - \frac{b}{2\omega}\sinh 2\omega\pi & \sinh 2\omega\pi - \frac{b}{\omega}\sinh^2\omega\pi \\ \frac{b}{\omega}\sinh^2\omega\pi + \sinh 2\omega\pi - \frac{b^2}{2\omega^2}\sinh 2\omega\pi & \cosh 2\omega\pi + \frac{b}{2\omega}\sinh 2\omega\pi - \frac{b^2}{\omega^2}\sinh^2\omega\pi \end{bmatrix}$$

The multipliers μ_j $(j = 1, 2)$ of (3.30), (3.31) satisfy the equation

$$\mu^2 - 2\left(\cosh 2\omega\pi - \frac{b^2}{2\omega^2}\sinh^2\omega\pi\right)\mu + 1 = 0. \tag{3.32}$$

System (3.30), (3.31) has a non-trivial T-periodic solution if and only if $\mu = 1$ satisfies (3.32), that is, if

$$\cosh 2\omega\pi - \frac{b^2}{2\omega^2}\sinh^2\omega\pi = 1. \tag{3.33}$$

Condition (3.33) is satisfied only if $b = 2\omega$. In this case, there correspond to the eigenvalue $\mu = 1$ of the matrix M no simple elementary divisors. Hence the non-trivial T-periodic solutions of (3.30), (3.31) are unstable.

We analogously conclude that (3.30), (3.31) has a non-trivial $2T$-periodic solution if and only if

$$\cosh 2\omega\pi - \frac{b^2}{2\omega^2}\sinh^2\omega\pi = -1,$$

that is, if

$$\coth^2\omega\pi = \frac{b^2}{4\omega^2}.$$

This $2T$-periodic solution is also unstable.

The roots of (3.32) are real and distinct if the following inequality holds:

$$\left|\cosh 2\omega\pi - \frac{b^2}{2\omega^2}\sinh^2\omega\pi\right| > 1,$$

which is equivalent to the conditions

$$\frac{b^2}{4\omega^2} < 1 \qquad \text{or} \qquad \coth^2\omega\pi < \frac{b^2}{4\omega^2}. \tag{3.34}$$

Since $\mu_1\mu_2 = 1$, then in this case the modulus of one of the multipliers is greater than 1. Consequently, under condition (3.34) the upper equilibrium point of the pendulum is unstable.

The roots of (3.32) are complex conjugate and distinct if

$$\left|\cosh 2\omega\pi - \frac{b^2}{2\omega^2}\sinh^2\omega\pi\right| < 1,$$

that is, if

$$1 < \frac{b^2}{4\omega^2} < \coth^2\omega\pi. \tag{3.35}$$

Under this condition $|\mu_1| = |\mu_2| = 1$ and the upper equilibrium point of the pendulum is stable.

4. Linear non-homogeneous periodic equations

Consider the linear non-homogeneous T-periodic impulsive equation

$$\frac{\mathrm{d}x}{\mathrm{d}t} = A(t)x + g(t), \quad t \neq \tau_k,$$
$$\Delta x = B_k x + h_k, \qquad t = \tau_k, k \in \mathbb{Z}. \tag{4.1}$$

Suppose that conditions (H3) hold as well as the following condition

H4 $g(\,\cdot\,) \in PC(\mathbb{R}, \mathbb{C}^n)$, $h_k \in \mathbb{C}^n (k \in \mathbb{Z})$ and
$g(t+T) = g(t)$, $h_{k+q} = h_k$ $(t \in \mathbb{R}, k \in \mathbb{Z})$.

We shall investigate the question of the existence of a T-periodic solution of equation (4.1).

From the variation of parameters formula (2.18) it follows that the solution $x(t)$ of (4.1) has the form

$$x(t) = X(t)x(0) + \int_0^t X(t)X^{-1}(s)g(s)\,\mathrm{d}s + \sum_{0 \leq \tau_k < t} X(t)X^{-1}(\tau_k^+)h_k$$

where $X(t) = W(t,0)$ is the normalized (at $t = 0$) fundamental matrix of equation (3.1).

The solution $x(t)$ will be T-periodic if $x(T) = x(0)$, or if

$$(E - X(T))x(0) = \int_0^T X(T)X^{-1}(s)g(s)\,\mathrm{d}s + \sum_{0 \leq \tau_k < T} X(T)X^{-1}(\tau_k^+)h_k. \tag{4.2}$$

4.1 Non-critical case   First consider the so-called *non-critical case* when the following condition is valid

$$\det\,(E - X(T)) \neq 0.$$

Since $M = X(T)$ is a monodromy matrix of equation (3.1) this condition means that all multipliers of (3.1) are distinct from 1, that is, equation (3.1) has no T-periodic solutions other than $x \equiv 0$. In this case equation (4.2) has a unique solution

$$x(0) = [E - X(T)]^{-1}\left[\int_0^T X(T)X^{-1}(s)g(s)\,\mathrm{d}s + \sum_{0 \leq \tau_k < T} X(T)X^{-1}(\tau_k^+)h_k\right]. \tag{4.3}$$

Hence equation (4.1) has a unique T-periodic solution

$$\tilde{x}(t) = X(T)[E - X(T)]^{-1}\left[\int_0^T X(T)X^{-1}(s)g(s)\,\mathrm{d}s + \sum_{0 \leq \tau_k < T} X(T)X^{-1}(\tau_k^+)h_k\right]$$
$$+ \int_0^t X(t)X^{-1}(s)g(s)\,\mathrm{d}s + \sum_{0 \leq \tau_k < t} X(t)X^{-1}(\tau_k^+)h_k \tag{4.4}$$

or

$$\tilde{x}(t) = \int_0^T G(t,s)g(s)\,\mathrm{d}s + \sum_{0 \leq \tau_k < T} G(t,\tau_k^+)h_k, \tag{4.5}$$

where

$$G(t,s) = \begin{cases} X(t)[E - X(T)]^{-1}X^{-1}(s) & (0 < s < t \le T), \\ X(t+T)[E - X(T)]^{-1}X^{-1}(s) & (0 < t \le s \le T), \\ G(t - kT, s - jT) & (kT < t \le kT + T, \\ & jT < s \le jT + T, \\ & k \in \mathbb{Z}, j \in \mathbb{Z}) \end{cases} \qquad (4.6)$$

is Green's function for the periodic solutions of equation (4.1).

The function $G(t,s)$ enjoys the following properties:

1. $G(t,t^-) - G(t,t^+) = E$ $(t \ne \tau_k, t \in \mathbb{R})$,
2. $G(t+T,s) = G(t,s)$ $(t \in \mathbb{R}, s \in \mathbb{R})$,
3. $\frac{\partial G}{\partial t}(t,s) = A(t)G(t,s)$ $(t \ne \tau_k, s \in \mathbb{R})$,
4. $G(\tau_k^+, s) = (E + B_k)G(\tau_k, s)$ $(s \ne \tau_k^+)$,
5. $G(\tau_k^+, \tau_k^+) = \lim_{t \to \tau_k^+} G(t, \tau_k^+) = (E + B_k)G(\tau_k, \tau_k^+) + E$.

These properties of the function $G(t,s)$ follow immediately from properties (2.14) of the fundamental matrix $X(t) = W(t,0)$.

Let

$$K = \sup_{t,s \in [0,T]} |G(t,s)|. \qquad (4.7)$$

Then (4.5) and (4.7) imply the estimate

$$|\tilde{x}(t)| \le K\left(\int_0^T |g(s)|\, ds + \sum_{k=1}^q |h_k|\right). \qquad (4.8)$$

Thus the following assertion is proved.

Theorem 4.1   *Let conditions (H3) and H4 hold and let the homogeneous equation (3.1) have no non-trivial T-periodic solutions.*

Then the non-homogeneous equation (4.1) has a unique T-periodic solution $\tilde{x}(t)$ for which the representation (4.5) and estimate (4.8) are valid.

Remark 4.1   By simple transformations of (4.4) we can obtain the following representation of the T-periodic solution $\tilde{x}(t)$ of (4.1):

$$\tilde{x}(t) = \int_t^{t+T} D(t,s)g(s)\, ds + \sum_{t \le \tau_k < t+T} D(t, \tau_k^+)h_k \qquad (4.9)$$

where

$$D(t,s) = X(t+T)[E - X(T)]^{-1}X^{-1}(s) \qquad (4.10)$$

and $X(t)$ is the fundamental matrix for equation (3.1) with $X(0) = E$.

Note that the function $D(t, s)$ enjoys the following properties:

1. $\dfrac{\partial D}{\partial t}(t, s) = A(t)D(t, s)$ $(t \neq \tau_k)$,

2. $\dfrac{\partial D}{\partial s}(t, s) = -D(t, s)A(s)$ $(s \neq \tau_k)$,

3. $D(t + T, s + T) = D(t, s)$,   (4.11)

4. $D(t, t + T) - D(t, t) = E$,

5. $D(\tau_k^+, s) = (E + B_k)D(\tau_k, s)$,

6. $D(t, \tau_k^-) = D(t, \tau_k^+)(E + B_k)$.

Remark 4.2   Representation (4.9) and properties (4.11) are still valid if the function $D(t, s)$ of (4.10) is replaced by the function $D(t, s) = X(t + T)[E - X(T+)]^{-1}X^{-1}(s)$ where $X(t)$ is the fundamental matrix for equation (3.1) with $X(0+) = E$.

Remark 4.3   If all multipliers μ_j of equation (3.1) are by modulus smaller than 1, that is, $|\mu_j| < 1$, $j = 1, \ldots, n$, then the T-periodic solution $\tilde{x}(t)$ of (4.1) is exponentially stable.

If equation (3.1) has an exponential dichotomy then there exist constants $K \geq 1$ and $\alpha > 0$ and linear manifolds S_m^+ and S_{n-m}^- for which

$$\dim S_m^+ = m, \qquad \dim S_{n-m}^- = n - m, \qquad S_m^+ \cap S_{n-m}^- = \tilde{x}(0)$$

such that for the solution $x(t)$ of equation (4.1) we have

$$|x(t) - \tilde{x}(t)| \leq K\, e^{-\alpha t} \quad (t \in \mathbb{R}_+)$$ (4.12)

whenever $x(0) \in S_m^+$ and

$$|x(t) - \tilde{x}(t)| \leq K\, e^{\alpha t} \quad (t \in \mathbb{R}_-)$$

whenever $x(0) \in S_{n-m}^-$.

Estimate (4.12) shows that the solution $\tilde{x}(t)$ is conditionally exponentially stable.

4.2 Critical case   Consider the so-called critical case when equation (3.1) has a non-trivial T-periodic solution, that is, when at least one of the multipliers of (3.1) equals 1. The $\det (E - X(T)) = 0$ and equation (4.2) may have no solution. In this case equation (4.1) will not have a T-periodic solution either. The existence of such a solution is determined by relations between the free terms $g(t)$ and h_k of equation (4.1) and the T-periodic solutions of the adjoint equation with respect to (3.1):

$$
\begin{aligned}
\frac{dy}{dt} &= -A^*(t)y, & t &\neq \tau_k, \\
\Delta y &= -(E + B_k^*)^{-1}B_k^* y, & t &= \tau_k.
\end{aligned}
$$ (4.13)

Since $(E - B(E + B)^{-1})^{-1}B(E + B)^{-1} = B$ for $B \in \mathbb{C}^{n \times n}$, it is easily verified that equation (3.1) is adjoint to equation (4.13). We shall therefore call equations (3.1) and (4.13) adjoint to each other, or mutually adjoint.

We shall need the following auxiliary assertions.

Lemma 4.1   Let $A(\,\cdot\,) \in PC(\mathbb{R}, \mathbb{C}^{n \times n})$, $B_k \in \mathbb{C}^{n \times n}$ and $\det(E + B_k) \neq 0$ $(k \in \mathbb{Z})$. Then:

1. For any two solutions $x(t)$ and $y(t)$ of the mutually adjoint equations (3.1) and (4.13) the following identity is valid

$$(x(t)|y(t)) = \text{const.} = (x(0)|y(0)) \quad (t \in \mathbb{R}). \tag{4.14}$$

2. The fundamental matrices $X(t)$ and $Y(t)$ of the mutually adjoint equations (3.1) and (4.13) satisfy the identity

$$Y^*(t)X(t) \equiv C \quad (t \in \mathbb{R}) \tag{4.15}$$

where $C \in \mathbb{C}^{n \times n}$ is a constant matrix.
3. If identity (4.15) is valid, where $X(t)$ is a fundamental matrix of equation (3.1), and $C \in \mathbb{C}^{n \times n}$ is a non-singular matrix, then $Y(t)$ is a fundamental matrix of equation (4.13).

Proof   For each $t \in (\tau_k, \tau_{k+1}]$ $(k \in \mathbb{Z})$ we have

$$\begin{aligned}(x(t)|y(t))' &= (x'(t)|y(t)) + (x(t)|y'(t))\\ &= (A(t)x(t)|y(t)) - (x(t)|A^*(t)y(t))\\ &= (A(t)x(t)|y(t)) - (A(t)x(t)|y(t)) = 0.\end{aligned}$$

Consequently, $(x(t)|y(t)) \equiv c_k = \text{const.}$ in each interval $(\tau_k, \tau_{k+1}]$ $(k \in \mathbb{Z})$. But since

$$\begin{aligned}(x(\tau_k^+)|y(\tau_k^+)) &= ((E + B_k)x(\tau_k)|(E - (E + B_k^*)^{-1}B_k^*)y(\tau_k))\\ &= ((E + B_k)x(\tau_k)|(E + B_k^*)^{-1}y(\tau_k)) = (x(\tau_k)|y(\tau_k))\end{aligned}$$

then all c_k are equal, that is, (4.14) is valid.

In the same way (4.15) is proved.

A straightforward verification shows that for $t \neq \tau_k$ the matrix $Y(t)$ defined by (4.15) satisfies the equation

$$\frac{dY}{dt} = -A^*(t)Y.$$

Moreover, for $t \neq \tau_k$ we have

$$Y(\tau_k^+) = [X^*(\tau_k^+)]^{-1}C^* = (E + B_k^*)^{-1}[X^*(\tau_k)]^{-1}C^* = (E + B_k^*)^{-1}Y(\tau_k)$$

or

$$\Delta Y(\tau_k) = [(E + B_k^*)^{-1} - E]Y(\tau_k) = -(E + B_k^*)^{-1}B_k^*Y(\tau_k).$$

Hence $Y(t)$ is a matrix solution of (4.13) and since $\det Y(t) \neq 0$ $(t \in \mathbb{R})$, then $Y(t)$ is a fundamental matrix of (4.13). $\square$

The second auxiliary assertion concerns the following linear algebraic equations:

$$Ax = b, \tag{4.16}$$
$$Ay = 0, \tag{4.17}$$
$$A^*z = 0 \tag{4.18}$$

where $A \in \mathbb{C}^{n \times n}$ and $b, x, y, z \in \mathbb{C}^{n \times 1}$.

Lemma 4.2   *The following assertions are valid.*

1. *The mutually adjoint homogeneous equations (4.17) and (4.18) have the same number, m, of linearly independent solutions and*

$$m = n - \operatorname{rank} A = n - \operatorname{rank} A^*.$$

2. *If $z_1, \ldots, z_m$ are the linearly independent solutions of equation (4.18), then the non-homogeneous equation (4.16) has a solution if and only if*

$$(z_j | b) = 0 \quad (j = 1, \ldots, m).$$

3. *If $(z_j | b) = 0$ $(j = 1, \ldots, m)$ then there exists a unique solution $x = \tilde{x}$ of equation (4.16) which satisfies the conditions*

$$(y_j | \tilde{x}) = 0 \quad (j = 1, \ldots, m)$$

where $y_1, \ldots, y_m$ are the linearly independent solutions of equation (4.17). This solution is defined by the formula

$$\tilde{x} = Q^{-1}b,$$

where $Q = A - ZY^*$, $Z = [z_1, \ldots, z_m] \in \mathbb{C}^{n \times m}$, and $Y = [y_1, \ldots, y_m] \in \mathbb{C}^{n \times m}$. Moreover, $\det Q \neq 0$ and there exists a constant $K > 0$ independent of b such that

$$|\tilde{x}| \leq K|b|.$$

We omit the proofs of the assertions of Lemma 4.2 which are well known and proceed to the following theorem.

Theorem 4.2   *Let conditions (H3) and H4 hold and let the homogeneous equation (3.1) have m linearly independent T-periodic solutions $\varphi_1(t), \ldots, \varphi_m(t)$ ($1 \le m \le n$). Then:*

1. *The adjoint equation (4.13) also has m linearly independent T-periodic solutions $\psi_1(t), \ldots, \psi_m(t)$.*
2. *Equation (4.1) has a T-periodic solution if and only if the following conditions are met*

$$\int_0^T \psi_j^*(t)g(t)\,\mathrm{d}t + \sum_{0 \le \tau_k < T} \psi_j^*(\tau_k^+)h_k = 0 \quad (j = 1, \ldots, m). \tag{4.19}$$

3. *If condition (4.19) is met then each T-periodic solution of equation (4.1) has the form*

$$x(t) = c_1\varphi_1(t) + \cdots + c_m\varphi_m(t) + x_0(t)$$

   where $x_0(t)$ is a particular T-periodic solution of (4.1).
4. *If condition (4.19) is met then equation (4.1) has a unique T-periodic solution $\tilde{x}(t)$ which satisfies the conditions*

$$(\varphi_i(0)|\tilde{x}(0)) = 0 \quad (i = 1, \ldots, m). \tag{4.20}$$

This solution is defined by the formula

$$\tilde{x}(t) = \int_0^T \tilde{G}(t,s)g(s)\,\mathrm{d}s + \sum_{k=1}^q \tilde{G}(t,\tau_k^+)h_k \tag{4.21}$$

where

$$\tilde{G}(t,s) = \begin{cases} W(t,0)B^{-1}W(T,s) + W(t,s) & (0 < s < t \le T), \\ W(t,0)B^{-1}W(T,s) & (0 < t \le s \le T), \\ \tilde{G}(t - kT, s - jT) & (kT < t \le kT + T, \\ & \quad jT < s \le jT + T, \\ & \quad k \in \mathbb{Z}, j \in \mathbb{Z}) \end{cases} \tag{4.22}$$

is the generalized Green's function, $W(t,s)$ is the Cauchy matrix for equation (3.1), the matrix

$$B = [E - M - \Psi_1(0)\Phi_1^*(0)] \tag{4.23}$$

is non-singular, and

$$M = W(T,0), \quad \Phi_1(t) = [\varphi_1(t), \ldots, \varphi_m(t)], \quad \Psi_1(t) = [\psi_1(t), \ldots, \psi_m(t)]. \tag{4.24}$$

Moreover, there exists a constant $K > 0$ independent of g and h_k such that

$$\sup_{t \in \mathbb{R}} |\tilde{x}(t)| \le K\big[\sup_{t \in [0,T]} |g(t)| + \max_{k=1,\ldots,q} |h_k|\big]. \tag{4.25}$$

Proof 1. From the condition of Theorem 4.2 it follows that the equation

$$(E - M)x = 0$$

has m linearly independent solutions x_i $(i = 1, \ldots, m)$ to which there correspond the linearly independent T-periodic solutions $\varphi_1(t), \ldots, \varphi_m(t)$ of equation (3.1): $\varphi_i(0) = x_i$ $(i = 1, \ldots, m)$. By Lemma 4.2 the adjoint equation

$$(E - M^*)y = 0$$

also has m linearly independent solutions y_j $(j = 1, \ldots, m)$ to which there correspond m linearly independent T-periodic solutions $\psi_1(t), \ldots, \psi_m(t)$ of equation (4.13): $\psi_j(0) = y_j$ $(j = 1, \ldots, m)$. Thus assertion 1 is proved.

In the notation introduced by (4.24) conditions (4.19) and (4.20) take the form

$$\int_0^T \Psi_1^*(t)g(t)\,\mathrm{d}t + \sum_{0 \le \tau_k < T} \Psi_1^*(\tau_k^+)h_k = 0, \tag{4.26}$$

$$\Phi_1^*(0)\tilde{x}(0) = 0. \tag{4.27}$$

Let $\Phi(t) = [\Phi_1(t), \Phi_2(t)]$ be a fundamental matrix of equation (3.1).

2. Equation (4.1) has a T-periodic solution $x(t)$ if and only if the equation

$$[E - M]x(0) = \int_0^T W(T, s)g(s)\,\mathrm{d}s + \sum_{k=1}^q W(T, \tau_k^+)h_k \equiv b$$

has a solution $x(0)$. By Lemma 4.2 this is valid if and only if the compatibility condition

$$y_j^* b = 0 \quad (j = 1, \ldots, m)$$

is met, or in matrix form,

$$\Psi_1^*(0)\left[\int_0^T W(T, s)g(s)\,\mathrm{d}s + \sum_{0 \le \tau_k < T} W(T, \tau_k^+)h_k\right] = 0.$$

But from the relations

$$W(T, s) = \Phi(T)\Phi^{-1}(s),$$
$$\Psi_1^*(0) = \Psi_1^*(T),$$
$$\Psi_1^*(T)\Phi(T) = \Psi_1^*(s)\Phi(s) \quad \text{(by Lemma 4.1)}$$

it follows that this condition is equivalent to (4.26).

3. Assertion 3 is obvious.

4. Let condition (4.19) hold and consider the system

$$(E - M)\tilde{x}(0) = \int_0^T W(T, s)g(s)\,\mathrm{d}s + \sum_{0 \le \tau_k < T} W(T, \tau_k^+)h_k, \tag{4.28}$$

$$\Phi_1^*(0)\tilde{x}(0) = 0.$$

By Lemma 4.2 the matrix $B = E - M - \Psi_1(0)\Phi_1^*(0)$ is non-singular and system (4.28) has a unique solution

$$\tilde{x}(0) = B^{-1}b = B^{-1}\left[\int_0^T W(T,s)g(s)\,\mathrm{d}s + \sum_{0\le\tau_k<T} W(T,\tau_k^+)h_k\right]$$

to which there corresponds the unique T-periodic solution $\tilde{x}(t)$ of (4.1) satisfying (4.27):

$$\tilde{x}(t) = W(t,0)B^{-1}\left[\int_0^T W(T,s)g(s)\,\mathrm{d}s + \sum_{0\le\tau_k<T} W(T,\tau_k^+)h_k\right]$$
$$+ \int_0^t W(t,s)g(s)\,\mathrm{d}s + \sum_{0\le\tau_k<t} W(t,\tau_k^+)h_k. \tag{4.29}$$

Taking into account (4.22) we obtain that for $\tilde{x}(t)$ the representation (4.21) is valid. Since by Lemma 4.2 there exists a constant $K_1 > 0$ such that

$$|\tilde{x}(0)| \le K_1\left|\int_0^T W(T,s)g(s)\,\mathrm{d}s + \sum_{0\le\tau_k<T} W(T,\tau_k^+)h_k\right|,$$

we easily conclude from (4.29) that estimate (4.25) is valid with constant $K > 0$ independent of g and h_k. $\square$

The existence of T-periodic solutions of the linear non-homogeneous equation (4.1) depends on the existence of bounded solutions of this equation. This dependence is given in the following theorem which is a generalization of the theorem of Massera [23].

Theorem 4.3   Let conditions (H3) and H4 hold. If the linear non-homogeneous T-periodic equation (4.1) has a bounded solution for $t \ge 0$ then this equation has a T-periodic solution.

Proof   Let $\hat{y}(t)$ be a bounded solution of (4.1) for $t \ge 0$. Then for $t \in \mathbb{R}_+$ we have

$$\hat{y}(t) = W(t,0)\hat{y}(0) + \int_0^t W(t,s)g(s)\,\mathrm{d}s + \sum_{0\le\tau_k<t} W(t,\tau_k^+)h_k$$

where $W(t,s)$ is the Cauchy matrix for equation (3.1). Hence

$$\hat{y}(T) = M\hat{y}(0) + b \tag{4.30}$$

where $M = W(T,0)$ is a monodromy matrix for (3.1) and

$$b = \int_0^T W(T,s)g(s)\,\mathrm{d}s + \sum_{0\le\tau_k<T} W(T,\tau_k^+)h_k.$$

From the T-periodicity of equation (4.1) it follows that the functions $z_\nu(t) = \hat{y}(t+\nu T)$ are also bounded solutions of (4.1) and in view of (4.30) we conclude that

$$\hat{y}(\nu T) = M^\nu \hat{y}(0) + \sum_{k=0}^{\nu-1} M^k b \quad (\nu \in \mathbb{N}). \tag{4.31}$$

Suppose that the T-periodic equation (4.1) has no T-periodic solution. Then the linear algebraic equation

$$(E - M)y = b$$

has no solution and by Lemma 4.2 there exists a $z \in \mathbb{C}^n$ such that

$$(E - M^*)z = 0$$

and

$$(z|b) \neq 0. \tag{4.32}$$

Hence $z = M^* z$, whence

$$z = (M^k)^* z \quad (k \in \mathbb{N}). \tag{4.33}$$

We take the scalar product of both sides of (4.31) with z and find

$$(z|\hat{y}(\nu T)) = (z|M^\nu \hat{y}(0)) + \sum_{k=0}^{\nu-1}(z|M^k b).$$

Now in view of (4.33) and (4.32) we conclude that

$$(z|\hat{y}(\nu T)) = (z|\hat{y}(0)) + \nu(z|b) \to \infty \quad \text{as } \nu \to \infty$$

and arrive at a contradiction with the boundedness of $\hat{y}(t)$. Consequently, the assumption is not true and equation (4.1) has at least one T-periodic solution. $\square$

Corollary 4.1   *If the linear T-periodic equation (4.1) has no T-periodic solutions then all solutions of this equation are unbounded both for $t \geq 0$ and for $t \leq 0$.*

Corollary 4.2   *If the linear T-periodic equation (4.1) has a unique bounded solution for $t \geq 0$ then this solution is T-periodic.*

Consider the case when the T-periodic equation (4.1) is scalar, that is, it has the form

$$\begin{aligned} \frac{dx}{dt} &= a(t)x + g(t), \quad t \neq \tau_k, \\ \Delta x &= b_k x + h_k, \qquad t = \tau_k \end{aligned} \tag{4.34}$$

where $a(\,\cdot\,), g(\,\cdot\,) \in \mathrm{PC}(\mathbb{R}, \mathbb{C})$; $b_k, h_k \in \mathbb{C}$, $1 + b_k \neq 0$ $(k \in \mathbb{Z})$.
The Cauchy matrix for the corresponding homogeneous equation

$$\begin{aligned} \frac{dx}{dt} &= a(t)x, \quad t \neq \tau_k, \\ \Delta x &= b_k x, \qquad t = \tau_k \end{aligned} \tag{4.35}$$

is given by the function

$$X(t,s) = \prod_{s \le \tau_k < t} (1 + b_k) \exp\left(\int_s^t a(\tau)\, d\tau\right) \quad (-\infty < s \le t < \infty).$$

The unique multiplier of equation (4.35) is equal to

$$\mu = X(\tau_q^+, \tau_0^+) = \prod_{k=1}^q (1 + b_k) \exp\left(\int_{\tau_0}^{\tau_q} a(\tau)\, d\tau\right). \tag{4.36}$$

If $\mu \ne 1$ then equation (4.34) has a unique T-periodic solution

$$\tilde{x}(t) = X(t, \tau_0^+) \frac{1}{1 - \mu}\left[\int_{\tau_0}^{\tau_q} X(\tau_q^+, s)g(s)\, ds + \sum_{k=1}^q X(\tau_q^+, \tau_k^+)h_k\right]$$

$$+ \int_{\tau_0}^t X(t, s)g(s)\, ds + \sum_{\tau_0 < \tau_k < t} X(t, \tau_k^+)h_k.$$

Moreover, the solution $\tilde{x}(t)$ is exponentially stable if $|\mu| < 1$ and unstable if $|\mu| > 1$.

If $\mu = 1$ then all solutions of (4.35) are T-periodic. Then all solutions of the adjoint equation

$$\frac{dy}{dt} = -a^*(t)y, \quad t \ne \tau_k,$$

$$\Delta y = -\frac{b_k^*}{1 + b_k^*} y, \quad t = \tau_k \tag{4.37}$$

are also T-periodic. In particular, the function

$$y = \Psi(t) = \prod_{\tau_0 < \tau_k < t} \frac{1}{1 + b_k^*} \exp\left(-\int_{\tau_0}^t a^*(\tau)\, d\tau\right)$$

is a T-periodic solution of (4.37) for which $\Psi(\tau_q^+) = \Psi(\tau_0^+) = 1$.

If in this case the condition

$$\int_{\tau_0}^{\tau_q} \Psi^*(t)g(t)\, dt + \sum_{k=1}^q \Psi^*(\tau_k^+)h_k = 0 \tag{4.38}$$

is met, then all solutions of equation (4.34)

$$x(t) = X(t, \tau_0^+)x(0) + \int_{\tau_0}^t X(t, s)g(s)\, ds + \sum_{\tau_0 < \tau_k < t} X(t, \tau_k^+)h_k$$

are T-periodic and stable.

In particular, if $q = 1$, $b_k \equiv b$, $h_k \equiv h$ $(k \in \mathbb{Z})$, then formula (4.36) for the computation of the multiplier μ takes the form

$$\mu = (1 + b)\exp\left(\int_{\tau_0}^{\tau_1} a(\tau)\, d\tau\right),$$

and condition (4.38) for the existence of a T-periodic solution of equation (4.34) in the critical case $\mu = 1$ takes the form

$$\int_{\tau_0}^{\tau_1} \exp\left(-\int_{\tau_0}^{t} a(\tau)\,d\tau\right) g(t)\,dt + h = 0.$$

Consider the second-order impulsive differential equation

$$\ddot{x} + 2\alpha\dot{x} + \beta^2 x = f(t), \quad t \neq \tau_k,$$

$$\Delta x = g_k, \quad \Delta\dot{x} = h_k, \quad t = \tau_k, \tag{4.39}$$

where $x, \alpha, \beta, g_k, h_k \in \mathbb{R}$, $f(\,\cdot\,) \in PC(\mathbb{R}, \mathbb{R})$.

In matrix form equation (4.39) has the form

$$\frac{dz}{dt} = Az + F(t), \quad t \neq \tau_k,$$

$$\Delta z = I_k, \quad t = \tau_k, \tag{4.40}$$

where

$$z = \begin{bmatrix} x \\ \dot{x} \end{bmatrix}, \quad A = \begin{bmatrix} 0 & 1 \\ -\beta^2 & -2\alpha \end{bmatrix}, \quad F(t) = \begin{bmatrix} 0 \\ f(t) \end{bmatrix}, \quad I_k = \begin{bmatrix} g_k \\ h_k \end{bmatrix}.$$

The respective eigenvalues $\lambda_{1,2}$ of the matrix A are determined from the characteristic equation

$$\lambda^2 + 2\alpha\lambda + \beta^2 = 0, \tag{4.41}$$

and the normalized (at $t = 0$) fundamental matrix $\Phi(t)$ of the homogeneous equation

$$\frac{d\Phi}{dt} = A\Phi \tag{4.42}$$

is determined in the various cases by the following formulae.

(i) If $\alpha^2 - \beta^2 > 0$ then the roots λ_1, λ_2 of (4.41) are real and distinct and

$$\Phi(t) = \begin{bmatrix} \dfrac{\lambda_1 e^{\lambda_2 t} - \lambda_2 e^{\lambda_1 t}}{\lambda_1 - \lambda_2} & \dfrac{e^{\lambda_1 t} - e^{\lambda_2 t}}{\lambda_1 - \lambda_2} \\ -\lambda_1\lambda_2 \dfrac{e^{\lambda_1 t} - e^{\lambda_2 t}}{\lambda_1 - \lambda_2} & \dfrac{\lambda_1 e^{\lambda_1 t} - \lambda_2 e^{\lambda_2 t}}{\lambda_1 - \lambda_2} \end{bmatrix}. \tag{4.43}$$

(ii) If $\beta^2 = \alpha^2$ then $\lambda_1 = \lambda_2 = -\alpha$ and

$$\Phi(t) = \begin{bmatrix} (1 + \alpha t)e^{-\alpha t} & te^{-\alpha t} \\ -\alpha^2 te^{-\alpha t} & (1 - \alpha t)e^{-\alpha t} \end{bmatrix}. \tag{4.44}$$

(iii) If $\omega^2 = \beta^2 - \alpha^2 > 0$ then $\lambda_1 = \lambda_2^* = -\alpha + i\omega$ and

$$\Phi(t) = \begin{bmatrix} \dfrac{e^{-\alpha t}}{\omega}[\omega\cos\omega t + \alpha\sin\omega t] & \dfrac{e^{-\alpha t}}{\omega}\sin\omega t \\ -\dfrac{\omega^2 + \alpha^2}{\omega}e^{-\alpha t}\sin\omega t & \dfrac{e^{-\alpha t}}{\omega}[\omega\cos\omega t - \alpha\sin\omega t] \end{bmatrix}. \tag{4.45}$$

The solution $z(t)$ of (4.40) with $z(\tau_0^+) = \begin{bmatrix} x_0 \\ \dot{x}_0 \end{bmatrix}$ has the form

$$z(t) = \Phi(t - \tau_0)z(\tau_0^+) + \int_{\tau_0}^t \Phi(t - s)F(s) \, ds + \sum_{\tau_0 < \tau_k < t} \Phi(t - \tau_k)I_k \quad (t > \tau_0)$$

from which we find that the solution $x(t)$ of (4.39) with $x(\tau_0^+) = x_0$, $\dot{x}(\tau_0^+) = \dot{x}_0$ is

$$\begin{aligned}
x(t) =& \varphi_{11}(t - \tau_0)x_0 + \varphi_{12}(t - \tau_0)\dot{x}_0 + \int_{\tau_0}^t \varphi_{12}(t - s)f(s) \, ds \\
& + \sum_{\tau_0 < \tau_k < t} [\varphi_{11}(t - \tau_k)g_k + \varphi_{12}(t - \tau_k)h_k] \quad (t > \tau_0)
\end{aligned} \tag{4.46}$$

where $\varphi_{ij}(t)$, $i, j = 1, 2$, are the entries of the matrix $\Phi(t)$ defined by one of the formulae (4.43), (4.44) or (4.45).

Let us discuss the question of the existence of T-periodic solutions of equation (4.39) under the assumption that it is T-periodic, that is, that there exists a $q \in N$ such that

$$\tau_{k+q} = \tau_k + T, \quad g_{k+q} = g_k, \quad h_{k+q} = h_k, \quad f(t + T) = f(t) \quad (t \in \mathbb{R}, k \in \mathbb{Z}).$$

The equation for the initial value $z(\tau_0^+)$ of the T-periodic solution of (4.40) is

$$(E - \Phi(T))z(\tau_0^+) = \int_{\tau_0}^{\tau_q} \Phi(\tau_q - s)F(s) \, ds + \sum_{k=1}^{q} \Phi(\tau_q - \tau_k)I_k$$

or in coordinate form

$$\begin{aligned}
(1 - \varphi_{11}(T))x_0 \quad\quad - \varphi_{12}(T)\dot{x}_0 &= Q_1, \\
-\varphi_{21}(T)x_0 + (1 - \varphi_{22}(T))\dot{x}_0 &= Q_2,
\end{aligned} \tag{4.47}$$

where

$$Q_1 = \int_{\tau_0}^{\tau_q} \varphi_{12}(\tau_q - s)f(s) \, ds + \sum_{k=1}^{q}[\varphi_{11}(\tau_q - \tau_k)g_k + \varphi_{12}(\tau_q - \tau_k)h_k],$$

$$Q_2 = \int_{\tau_0}^{\tau_q} \varphi_{22}(\tau_q - s)f(s) \, ds + \sum_{k=1}^{q}[\varphi_{21}(\tau_q - \tau_k)g_k + \varphi_{22}(\tau_q - \tau_k)h_k].$$

The following cases are possible.

(i) $\lambda_1 \neq 2m\pi i/T$, $\lambda_2 \neq 2m\pi i/T$ for each $m \in \mathbb{Z}$. Then the multipliers $\mu_i = e^{\lambda_i T}$ of equation (4.42) are distinct from 1, and equation (4.39) has a unique T-periodic solution. The initial values x_0, $\dot{x}_0$ of this equation are determined from system (4.47).

(ii) λ_1 and λ_2 are real and $\lambda_1 \lambda_2 = 0$. The following subcases are possible.

(ii.a) $\lambda_1 = \lambda \neq 0$, $\lambda_2 = 0$. Then equation (4.42) has one linearly independent T-periodic solution

$$\Phi_1(t) = \begin{bmatrix} 1 \\ 0 \end{bmatrix}.$$

The adjoint equation to (4.42), $\Psi' = -A^*\Psi$, also has one linearly independent T-periodic solution

$$\Psi_1(t) = \begin{bmatrix} 1 \\ -\frac{1}{\lambda} \end{bmatrix}$$

and the compatibility condition (4.19) takes the form

$$-\int_{\tau_0}^{\tau_q} \frac{f(s)}{\lambda}\, ds + \sum_{k=1}^q g_k - \sum_{k=1}^q \frac{h_k}{\lambda} = 0. \tag{4.48}$$

If condition (4.48) is met then the system (4.47) for the determination of x_0, $\dot{x}_0$ is reduced to the equation

$$(1 - e^{\lambda T})\dot{x}_0 = \int_{\tau_0}^{\tau_q} e^{\lambda(\tau_q - s)} f(s)\, ds + \sum_{k=1}^q e^{\lambda(\tau_q - \tau_k)} h_k, \tag{4.49}$$

and x_0 can be arbitrary. Thus in this case equation (4.39) has a one-parameter family of T-periodic solutions.

(ii.b) $\lambda_1 = \lambda_2 = 0$. In this case, equation (4.42) and the adjoint equation $\Psi' = -A^*\Psi$ each have one linearly independent T-periodic solution

$$\Phi_1(t) = \begin{bmatrix} 1 \\ 0 \end{bmatrix} \qquad \text{and} \qquad \Psi_1(t) = \begin{bmatrix} 0 \\ 1 \end{bmatrix},$$

respectively. The compatibility condition (4.19) takes the form

$$\int_{\tau_0}^{\tau_q} f(s)\, ds + \sum_{k=1}^q h_k = 0. \tag{4.50}$$

If condition (4.50) is met then x_0 can be arbitrary and for $\dot{x}_0$ we have

$$T\dot{x}_0 = \int_{\tau_0}^{\tau_q} s f(s)\, ds + \sum_{k=1}^q \tau_k h_k - \sum_{k=1}^q g_k. \tag{4.51}$$

Consequently, also in this case equation (4.39) has a one-parameter family of T-periodic solutions.

(iii) $\lambda_1 = \lambda_2^* = i\omega = 2m\pi i/T$ for some $m \in \mathbb{Z}$. The equation (4.42) and the adjoint equation $\Psi' = -A^*\Psi$ each have two linearly independent T-periodic solutions, or more concretely,

$$\Phi(t) = \begin{bmatrix} \cos \omega t & \frac{1}{\omega}\sin \omega t \\ -\omega \sin \omega t & \cos \omega t \end{bmatrix}, \qquad \Psi(t) = \begin{bmatrix} \cos \omega t & \omega \sin \omega t \\ -\frac{1}{\omega} \sin \omega t & \cos \omega t \end{bmatrix}.$$

The compatibility condition (4.19) takes the form

$$\int_{\tau_0}^{\tau_q} \cos\omega\tau \cdot f(\tau)\,d\tau + \sum_{k=1}^{q}\omega\sin\omega\tau_k \cdot g_k + \sum_{k=1}^{q}\cos\omega\tau_k \cdot h_k = 0,$$

$$\int_{\tau_0}^{\tau_q} \sin\omega\tau \cdot f(\tau)\,d\tau - \sum_{k=1}^{q}\omega\cos\omega\tau_k \cdot g_k + \sum_{k=1}^{q}\sin\omega\tau_k \cdot h_k = 0.$$

(4.52)

If condition (4.52) is met then system (4.39) is satisfied for all x_0, $\dot{x}_0$. Consequently, in this case all solutions of equation (4.47) are T-periodic.

As an illustration of the results obtained in this section we shall consider the following examples.

Example 4.1   Consider the impulsive equation

$$\frac{dx}{dt} = r(t)\left[1 - \frac{x}{K(t)}\right]x, \quad t \neq \tau_k,$$

(4.53)

$$\Delta x = c_k x, \qquad\qquad t = \tau_k.$$

(4.54)

Suppose that equation (4.53), (4.54) is periodic, that is, there exist $T > 0$ and $q \in \mathbb{N}$ such that

$$r(t+T) = r(t), \qquad K(t+T) = K(t), \qquad (t \in \mathbb{R}),$$
$$\tau_{k+q} = \tau_k + T, \qquad c_{k+q} = c_k, \qquad (k \in \mathbb{Z}).$$

(4.55)

The logistic equation (4.53) describes the variation of the population number $x(t)$ of an isolated species in a periodically changing environment. The intrinsic rate of change $r(t)$ is related to the periodically changing possibility of regeneration of the species, and the carrying capacity of the system $K(t)$ is related to the periodic change of the resources maintaining the evolution of the population. The jump condition (4.54) reflects the possibility of an impulse effect on the population.

The following additional restrictions on system (4.53), (4.54) are natural:

$$K(\,\cdot\,),\ r(\,\cdot\,) \in PC(\mathbb{R},\mathbb{R}),$$

(4.56)

$$\inf_{t\in[0,T]} K(t) > 0, \quad r(t) > 0 \quad (t \in \mathbb{R}),$$

(4.57)

$$1 + c_k > 0 \quad (k \in \mathbb{Z}).$$

(4.58)

Theorem 4.4   *Let conditions (4.55)–(4.58) hold and let*

$$\mu = \prod_{k=1}^{q} \frac{1}{1+c_k}\exp\left(-\int_0^T r(\tau)\,d\tau\right) < 1.$$

(4.59)

Then system (4.53), (4.54) has a unique T-periodic solution $x(t,x_0)$ for which

$$x(0,x_0) = x_0 > 0 \quad \text{and} \quad x(t,x_0) > 0 \quad \text{for } t \in \mathbb{R}.$$

Proof   In (4.53), (4.54) we carry out the change of variable $x = z^{-1}$ and obtain the linear non-homogeneous impulsive equation

$$\frac{dz}{dt} = -r(t)z + \frac{r(t)}{K(t)}, \quad t \neq \tau_k,$$
$$z(\tau_k^+) = \frac{1}{1+c_k} z(\tau_k).$$
(4.60)

Let

$$W(t,s) = \prod_{s \leq \tau_k < t} \frac{1}{1+c_k} \exp\left(-\int_s^t r(\tau)\, d\tau\right)$$

be the Cauchy matrix for the respective homogeneous equation. Then

$$z(t) = W(t,0)z(0) + \int_0^t W(t,s)\frac{r(s)}{K(s)}\, ds$$
(4.61)

is a solution of (4.60). This solution is T-periodic if $z(T) = z(0)$, or if

$$(1 - W(T,0))z(0) = \int_0^T W(T,s)\frac{r(s)}{K(s)}\, ds.$$
(4.62)

Since the multiplier $\mu = W(T,0)$ of the homogeneous equation

$$\frac{dz}{dt} = -r(t)z, \quad t \neq \tau_k,$$
$$\Delta z = \frac{1}{1+c_k}z, \quad t = \tau_k$$
(4.63)

is less than 1, and

$$\int_0^T W(T,s)\frac{r(s)}{K(s)}\, ds > 0,$$

equation (4.62) has a unique solution $z(0) = p_0 > 0$. To the initial value $z(0) = p_0$ so obtained there corresponds the unique T-periodic solution of (4.60) which in view of (4.61) is positive for each $t \in \mathbb{R}$. Denote this solution by $p(t)$. Then the function $x = \pi(t) = 1/p(t)$ is the unique T-periodic solution of (4.53), (4.54) which is also positive. $\square$

Theorem 4.5   *If conditions (4.55)–(4.59) hold then all positive solutions of system (4.53), (4.54) are asymptotically stable.*

Proof   Since μ is the multiplier of the linear homogeneous periodic equation (4.63) and $\mu \in (0,1)$, then, by Theorem 3.5, equation (4.63) is exponentially stable, that is, there exist constants $N \geq 1$ and $\alpha > 0$ such that for any $z_1 > 0$ and $z_2 > 0$

$$|z(t,z_1) - z(t,z_2)| \leq Ne^{-\alpha t}|z_1 - z_2| \quad (t \geq 0).$$

Since $\inf_{t \in \mathbb{R}_+} z(t,z_i) = d(z_i) > 0$ for $z_i > 0$ and $i = 1,2$, then

$$\left|\frac{1}{z(t,z_1)} - \frac{1}{z(t,z_2)}\right| = \left|\frac{z(t,z_1) - z(t,z_2)}{z(t,z_1)z(t,z_2)}\right| \leq \frac{Ne^{-\alpha t}|z_1 - z_2|}{d(z_1)d(z_2)} \quad (t \geq 0).$$

Then for the solutions $x(t, x_1)$ and $x(t, x_2)$ of (4.53), (4.54) with $x_1 > 0$ and $x_2 > 0$ we have

$$|x(t, x_1) - x(t, x_2)| \leq \frac{N e^{-\alpha t} \left| \frac{1}{x_1} - \frac{1}{x_2} \right|}{d(\frac{1}{x_1}) d(\frac{1}{x_2})} \quad (t \geq 0).$$

Finally we shall note two particular cases.

(i) If $r(t) \equiv r > 0$, $K(t) \equiv K > 0$, $q = 1$, and $c_k \equiv c > -1$, then condition (4.59) for the existence of a positive T-periodic solution of system (4.53), (4.54) takes the form

$$\beta \equiv 1 + c - e^{-rT} > 0$$

and the T-periodic solution itself is

$$p(t) = \begin{cases} \dfrac{K\beta}{\beta - ce^{-rt}} \equiv P_0(t) & \text{for } 0 < t \leq T, \\ P_0(t - nT) & \text{for } nT < t \leq nT + T. \end{cases}$$

(ii) If the equation is without impulses, that is, $c = 0$, then the unique T-periodic solution of (4.53) has the form

$$p(t) = \frac{1 - e(T)}{b(T)e(t) + b(t) - b(t)e(T)},$$

where

$$e(t) = \exp\left(- \int_0^t r(s)\, ds\right), \qquad b(t) = \int_0^t \exp\left(- \int_s^t r(\tau)\, d\tau\right) \frac{r(s)}{K(s)}\, ds.$$

It can be shown that

$$p(t) = \frac{1 - e(T)}{\displaystyle\int_0^T \frac{r(t - s)}{K(t - s)} \exp\left(- \int_0^s r(t - \sigma)\, d\sigma\right) ds}.$$

Example 4.2   Consider a simple two-compartmental model of drug distribution in the human body. It is assumed that the drug which is administered orally is first dissolved in the gastro-intestinal tract. The drug is then absorbed into the so-called apparent volume of distribution (a lumped compartment which accounts for blood, muscle, tissue, etc.), and finally is eliminated from the system by the kidneys. Let $x(t)$ and $y(t)$ denote respectively the amount of the drug at time t in the gastro-intestinal tract and the apparent volume of distribution and let $\mu > 0$ and $\lambda > 0$ be relevant rate constants. Then the dynamical system of this model is

$$\frac{dx}{dt} = -\mu x, \qquad \frac{dy}{dt} = \mu x - \lambda y.$$

At the moments of time τ_n $(\tau_n < \tau_{n+1}, n \in \mathbb{Z})$ let the sick person take in doses $\delta_n > 0$ of the medicine so that

$$\Delta x = \delta_n, \qquad \Delta y = 0, \qquad t = \tau_n.$$

For some diseases it is necessary that the doses of medicine be taken in periodically $(\tau_n = nT)$ and be fixed: $\delta_n = \delta$ $(n \in \mathbb{Z})$. Some constraints are usually imposed on the value of the dose δ. On the one hand, it must be large enough to produce the desired therapeutic effect, and on the other hand, it cannot exceed a certain bound to avoid harmful effects of overdose of the medicine. Naturally the following problem arises: *Determine the dose $\delta > 0$ so that the T-periodic solution $x(t)$, $y(t)$ of the impulsive system*

$$\begin{array}{lll}
\frac{dx}{dt} = -\mu x, & \frac{dy}{dt} = \mu x - \lambda y, & t \neq nT, \\
\Delta x = \delta, & \Delta y = 0, & t = nT
\end{array} \tag{4.64}$$

satisfies the inequalities

$$x(t) \leq x^*, \qquad y(t) \geq y_* \tag{4.65}$$

where $0 < y_* < x^*$.

Denote

$$z = \begin{bmatrix} x \\ y \end{bmatrix}, \qquad A = \begin{bmatrix} -\mu & 0 \\ \mu & -\lambda \end{bmatrix}, \qquad b = \begin{bmatrix} \delta \\ 0 \end{bmatrix}, \qquad z_0 = z(0+) = \begin{bmatrix} x_0 \\ y_0 \end{bmatrix}.$$

First note that the eigenvalues $-\mu$ and $-\lambda$ of the matrix A are negative. Then the homogeneous system $z' = Az$ is exponentially stable and has no non-trivial T-periodic solutions. Consequently, by Theorem 4.1 the non-homogeneous system (4.64) has a unique T-periodic solution and this solution is exponentially stable. In order to determine its initial value, we use the condition of T-periodicity $z(T+) = z(0+)$ which implies that

$$(E - e^{AT})z_0 = b. \tag{4.66}$$

In view of

$$e^{At} = \begin{bmatrix} e^{-\mu t} & 0 \\ \frac{\mu}{\lambda - \mu}(e^{-\mu t} - e^{-\lambda t}) & e^{-\lambda t} \end{bmatrix},$$

from (4.66) we determine z_0 after which we find that for $t \in (0, T]$ the T-periodic solution of (4.64) has the form

$$x(t) = \frac{\delta}{1 - e^{-\mu T}} e^{-\mu t},$$

$$y(t) = \frac{\delta \mu}{\lambda - \mu} \left[\frac{e^{-\mu t}}{1 - e^{-\mu T}} - \frac{e^{-\lambda t}}{1 - e^{-\lambda T}} \right]. \tag{4.67}$$

Since $x(t) \leq x(0+)$ and $y(t) \geq y(0+)$, conditions (4.65) are met if

$$y_* \frac{(\lambda - \mu)(1 - e^{-\mu T})(1 - e^{-\lambda T})}{\mu(e^{-\mu T} - e^{-\lambda T})} \leq \delta \leq x^*(1 - e^{-\mu T}). \tag{4.68}$$

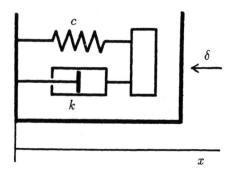

Fig. 4.1

Example 4.3 (A buffer)   Consider a buffer with linear rigidity and linear damping, that is, a buffer with a characteristic $u(x, \dot{x}) = cx + k\dot{x}$ where $c > 0$ is the coefficient of rigidity and $k > 0$ is the coefficient of damping (Fig. 4.1).

Assume that at equal intervals of time the buffer is subject to a shock effect, as a result of which at each shock the velocity $y = \dot{x}$ obtains a constant increment, that is, at the instants $\tau_n = nT$ $(T > 0, n \in \mathbb{Z})$ we shall have $\Delta y = \delta$. Then the system which describes the work of the buffer takes the form

$$\begin{aligned} \frac{dx}{dt} = y, \qquad & \frac{dy}{dt} = -cx - ky, \qquad t \neq nT, \\ \Delta x = 0, \qquad & \Delta y = \delta, \qquad\qquad t = nT, n \in \mathbb{Z}. \end{aligned} \tag{4.69}$$

Set

$$z = \begin{bmatrix} x \\ y \end{bmatrix}, \qquad z_0 = \begin{bmatrix} x_0 \\ y_0 \end{bmatrix}, \qquad A = \begin{bmatrix} 0 & 1 \\ -c & -k \end{bmatrix}, \qquad b = \begin{bmatrix} 0 \\ \delta \end{bmatrix}.$$

Then, by the variation of parameters formula (2.18) we get

$$z(t) = z(t; 0, z_0) = e^{At}\left(z_0 + \sum_{0 < \tau_k < t} e^{-AT k} b\right). \tag{4.70}$$

If $nT < t \leq nT + T$ and $U = e^{AT}$, then from (4.70) it follows that

$$z(t) = e^{A(t-nT)}(U^n z_0 + (E + U + \cdots + U^{n-1})b). \tag{4.71}$$

The eigenvalues λ_1, λ_2 of the matrix A are determined from the equation $\lambda^2 + k\lambda + c = 0$. Since $k > 0$ and $c > 0$, then $\Re\lambda_i < 0$ $(i = 1, 2)$. Consequently, the homogeneous system

$$\begin{aligned} \frac{dx}{dt} = y, \qquad & \frac{dy}{dt} = -cx - ky, \qquad t \neq nT, \\ \Delta x = 0, \qquad & \Delta y = 0, \qquad\qquad t = nT \end{aligned}$$

is exponentially stable and has no non-trivial T-periodic solutions. Thus by Theorems 4.1 and 3.5 system (4.69) has a unique T-periodic solution $p(t) = \mathrm{col}\,(\xi(t), \eta(t))$ which is exponentially stable. In view of (4.71) and the fact that $\lim_{n \to \infty} U^n = 0$ we conclude that there exist constants $K \geq 1$ and $\alpha > 0$ such that

$$|z(t) - p(t)| \leq Ke^{-\alpha t} \quad (t \geq 0)$$

where $p(t)$ is the T-periodic solution of (4.69):

$$p(t) = e^{A(t-nT)}(E - e^{AT})^{-1}b \quad \text{for } nT < t \le nT + T, n \in \mathbb{Z}.$$

In more detail, for $\xi(t)$ and $\eta(t)$ we have

$$\xi(t) = \frac{\delta}{\lambda_1 - \lambda_2}\left[\frac{e^{\lambda_1 t}}{1 - e^{\lambda_1 T}} - \frac{e^{\lambda_2 t}}{1 - e^{\lambda_2 T}}\right],$$

$$\eta(t) = \frac{\delta}{\lambda_1 - \lambda_2}\left[\frac{\lambda_1 e^{\lambda_1 t}}{1 - e^{\lambda_1 T}} - \frac{\lambda_2 e^{\lambda_2 t}}{1 - e^{\lambda_2 T}}\right]$$

(4.72)

if $k^2 \ne 4c$ and $0 < t \le T$. The graphs of $\xi(t)$ and $\eta(t)$ are given in Fig. 4.2.

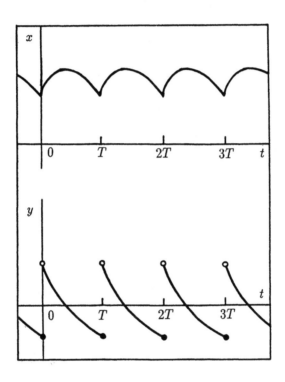

Fig. 4.2

Example 4.4   Consider the buffer with linear rigidity and linear damping (Fig. 4.1). Let the buffer be subject to the action of a T-periodic force. At the moments $\tau_n = nT$ $(n \in \mathbb{Z}, T > 0)$ let the buffer be additionally subject to shock effects, as a result of which its velocity $\dot{x}$ obtains an increment $\Delta\dot{x} = \gamma\dot{x} + b$. The motion of the buffer is described by the system

$$\dot{x} = y, \qquad \dot{y} = -cx - ky + F(t), \qquad t \ne \tau_n,$$

$$\Delta x = 0, \qquad \Delta y = \gamma y + b, \qquad t = \tau_n$$

(4.73)

where $c > 0$, $k > 0$, γ and b are constants, $k^2 \neq 4c$, $\tau_n = nT$ $(n \in \mathbb{Z})$ and $F(\,\cdot\,) \in PC(\mathbb{R}, \mathbb{R})$ is T-periodic.

The homogeneous system corresponding to (4.73) is

$$\frac{dz}{dt} = Az, \qquad t \neq \tau_n,$$
$$\Delta z = Bz, \qquad t = \tau_n$$

(4.74)

where

$$z = \begin{bmatrix} x \\ y \end{bmatrix}, \qquad A = \begin{bmatrix} 0 & 1 \\ -c & -k \end{bmatrix}, \qquad B = \begin{bmatrix} 0 & 0 \\ 0 & \gamma \end{bmatrix}.$$

Since the entries $u_{ij}(t)$ of the matrix e^{At} are

$$u_{11}(t) = \frac{\lambda_1 e^{\lambda_2 t} - \lambda_2 e^{\lambda_1 t}}{\lambda_1 - \lambda_2}, \qquad u_{12}(t) = \frac{e^{\lambda_1 t} - e^{\lambda_2 t}}{\lambda_1 - \lambda_2},$$

$$u_{21}(t) = -\lambda_1 \lambda_2 \frac{e^{\lambda_1 t} - e^{\lambda_2 t}}{\lambda_1 - \lambda_2}, \qquad u_{22}(t) = \frac{\lambda_1 e^{\lambda_1 t} - \lambda_2 e^{\lambda_2 t}}{\lambda_1 - \lambda_2},$$

where $\lambda_{1,2} = \frac{1}{2}(-k \pm \sqrt{k^2 - 4c})$, the monodromy matrix of (4.74) equals

$$M = (E + B)e^{AT} = \begin{bmatrix} u_{11}(T) & u_{12}(T) \\ (1+\gamma)u_{21}(T) & (1+\gamma)u_{22}(T) \end{bmatrix}$$

and the equation for the multipliers is

$$\mu^2 - [u_{11}(T) + (1+\gamma)u_{22}(T)]\mu + (1+\gamma)e^{(\lambda_1 + \lambda_2)T} = 0. \qquad (4.75)$$

System (4.73) has a non-trivial T-periodic solution if and only if $\mu = 1$ satisfies (4.74), that is, if

$$(1+\gamma)e^{(\lambda_1 + \lambda_2)T} + 1 = u_{11}(T) + (1+\gamma)u_{22}(T)$$

or

$$1 + \gamma = \frac{\lambda_1 - \lambda_2 + \lambda_2 e^{\lambda_1 T} - \lambda_1 e^{\lambda_2 T}}{\lambda_1 e^{\lambda_1 T} - \lambda_2 e^{\lambda_2 T} - (\lambda_1 - \lambda_2)e^{(\lambda_1 + \lambda_2)T}}. \qquad (4.76)$$

The following cases are possible.

(i) Condition (4.76) is satisfied. Then rank $(M - E) = 1$ and system (4.74) has one linearly independent T-periodic solution. By Theorem 4.2 the system adjoint to (4.74):

$$\frac{d\psi}{dt} = -A^*\psi, \qquad\qquad t \neq \tau_n,$$
$$\Delta\psi = -(E + B^*)^{-1}B^*\psi, \quad t = \tau_n$$

(4.77)

also has one linearly independent T-periodic solution $\psi(t) = \text{col}\,(\psi_1(t), \psi_2(t))$. The initial value $\psi(0+) = \psi_0 = \text{col}\,(x_0, y_0)$ of this solution is found from the system

$$Q\psi_0 = \psi_0 \qquad (4.78)$$

where

$$Q = (E + B^*)^{-1}e^{-A^*T} = \begin{bmatrix} v_{11}(T) & v_{12}(T) \\ \dfrac{v_{21}(T)}{1+\gamma} & \dfrac{v_{22}(T)}{1+\gamma} \end{bmatrix}$$

$$v_{11}(t) = \frac{\nu_1 e^{\nu_2 t} - \nu_2 e^{\nu_1 t}}{\nu_1 - \nu_2}, \qquad v_{12}(t) = \frac{e^{\nu_1 t} - e^{\nu_2 t}}{\nu_1 - \nu_2},$$

$$v_{21}(t) = -\nu_1\nu_2 \frac{e^{\nu_1 t} - e^{\nu_2 t}}{\nu_1 - \nu_2}, \qquad v_{22}(t) = \frac{\nu_1 e^{\nu_1 t} - \nu_2 e^{\nu_2 t}}{\nu_1 - \nu_2},$$

$$\nu_{1,2} = \frac{1}{2}(k \pm \sqrt{k^2 - 4c}).$$

System (4.78) has a non-zero solution

$$x_0 = 1, \qquad y_0 = \frac{\nu_1(1 - e^{\nu_2 T}) - \nu_2(1 - e^{\nu_1 T})}{e^{\nu_1 T} - e^{\nu_2 T}}.$$

Then as a non-trivial T-periodic solution of (4.77) we can choose the T-periodic extension of the function

$$\psi(t) = \begin{bmatrix} v_{11}(t) + y_0 v_{12}(t) \\ v_{21}(t) + y_0 v_{22}(t) \end{bmatrix} \qquad (0 < t \le T).$$

Consequently, by Theorem 4.2 the non-homogeneous system (4.73) has a T-periodic solution if and only if the free terms $F(t)$ and b satisfy condition (4.19), that is, if

$$\int_0^T \psi_2(t)F(t)\,\mathrm{d}t + y_0 b = 0$$

where

$$\psi_2(t) = v_{21}(t) + y_0 v_{22}(t) = \frac{\nu_1 e^{\nu_1 t}(1 - e^{\nu_2 T}) - \nu_2 e^{\nu_2 t}(1 - e^{\nu_1 T})}{e^{\nu_1 T} - e^{\nu_2 T}}.$$

(ii) Condition (4.76) is not fulfilled. Then, by Theorem 4.1, system (4.73) has a unique T-periodic solution. If, moreover, the condition

$$|u_{11}(T) + (1+\gamma)u_{22}(T)| < 1 + (1+\gamma)e^{-kT} < 2$$

holds, then the modulus of the multipliers μ_j $(j = 1, 2)$ of system (4.74) are less than 1, and system (4.73) is asymptotically stable.

5. Linear Hamiltonian impulsive equations

The linear impulsive equation

$$\begin{aligned} \frac{\mathrm{d}x}{\mathrm{d}t} &= JA(t)x, & t \ne \tau_k, t \in \mathbb{R}, \\ \Delta x &= JB_k x, & t = \tau_k, k \in \mathbb{Z} \end{aligned} \tag{5.1}$$

is called *Hamiltonian* if $x \in \mathbb{R}^{2n}$, $A(\cdot) \in PC(\mathbb{R}, \mathbb{R}^{2n \times 2n})$, $B_k \in \mathbb{R}^{2n \times 2n}$, $A^*(t) = A(t)$, $B_k^* = B_k$ $(t \in \mathbb{R}, k \in \mathbb{Z})$, the matrix J has the form

$$J = J_{2n} = \begin{bmatrix} 0 & E_n \\ -E_n & 0 \end{bmatrix},$$

and the matrices B_k satisfy the condition $(JB_k)^2 = 0$ $(k \in \mathbb{Z})$.

In particular, for $n = 1$ the matrix

$$B_k = \begin{bmatrix} \alpha_k & \beta_k \\ \beta_k & \gamma_k \end{bmatrix}$$

satisfies the condition $(JB_k)^2 = 0$ if and only if $\alpha_k \gamma_k = \beta_k^2$.

We note that $\text{Tr}\,(JA(t)) = 0$ and $\text{Tr}\,(JB_k) = 0$ for $t \in \mathbb{R}$, $k \in \mathbb{Z}$ since $\text{Tr}\,(JC) = 0$ for any symmetric matrix $C \in \mathbb{R}^{2n \times 2n}$.

For the linear Hamiltonian equation (5.1) the following lemma is valid.

Lemma 5.1   For any two solutions $x(t)$ and $y(t)$ of the linear Hamiltonian impulsive system (5.1) the following identity is valid:

$$(x(t)|Jy(t)) \equiv \text{const.} \quad (t \in \mathbb{R}). \tag{5.2}$$

Moreover, if $X(t)$ and $Y(t)$ are matrix solutions of the Hamiltonian equation (5.1) then

$$X^*(t)JY(t) \equiv C \quad (t \in \mathbb{R}) \tag{5.3}$$

where $C \in \mathbb{R}^{2n \times 2n}$ is a constant matrix.

Proof   Lemma 5.1 is a consequence of Lemma 4.1 and the fact that if $y(t)$ is a solution of (5.1), then $Jy(t)$ is a solution of the equation adjoint to (5.1). $\square$

In the investigation of the properties of periodic Hamiltonian equations we shall use some properties of the reciprocal equations, that is, equations of the form

$$f(\lambda) = \lambda^n + a_1\lambda^{n-1} + \cdots + a_{n-1}\lambda + 1 = 0 \tag{5.4}$$

where $a_j = a_{n-j}$ $(j = 1, \ldots, n-1)$.

We note that the polynomial $F(\lambda)$ is reciprocal if and only if the following condition holds

$$F\left(\frac{1}{\lambda}\right) = \frac{1}{\lambda^n}F(\lambda) \quad (\lambda \neq 0). \tag{5.5}$$

The following lemma is valid.

Lemma 5.2 [23]   If $\lambda = \lambda_0$ is a root of multiplicity ν of the reciprocal equation (5.4) then $\lambda = \lambda_0^{-1}$ is also a root of multiplicity ν of this equation. Moreover, if equation (5.4) has a root $\lambda = 1$ then the multiplicity ν of this root is even; if (5.4) has a root $\lambda = -1$ then the multiplicity of this root is even for n even and odd for n odd.

Let $E = E_{2n}$ and let equation (5.1) be T-periodic.

Theorem 5.1   *If the linear Hamiltonian equation (5.1) is T-periodic and M is its monodromy matrix then the characteristic equation*

$$F(\lambda) = \det(\lambda E - M) = 0$$

is reciprocal.

Proof   Since the monodromy matrices of equation (5.1) have the same characteristic equations we shall prove Theorem 5.1 only for the monodromy matrix

$$M = W(\tau_q^+, \tau_0^+) = M_q M_{q-1} \cdots M_1,$$

where $M_k = (E + JB_k)U_k(\tau_k, \tau_{k-1}^+)$ $(k = 1,\ldots,q)$, $\tau_q - \tau_0 = T$ and $U_k(t,s)$ is the Cauchy matrix for the equation

$$\frac{dx}{dt} = JA(t)x \quad (\tau_{k-1} < t \le \tau_k).$$

Since $X(t) = W(t,\tau_0^+)$ is a matrix solution of (5.1), by Lemma 5.1

$$X^*(t)JX(t) \equiv C \quad (t \in \mathbb{R}). \tag{5.6}$$

In particular, $X^*(\tau_0^+)JX(\tau_0^+) = C$, and since $X(\tau_0^+) = E$, we have $C = J$. Passing to the limit in (5.6) as $t \to \tau_q^+$ we obtain

$$M^*JM = J. \tag{5.7}$$

By virtue of the properties $(JB_k)^2 = 0$, $\mathrm{Tr}\,(JB_k) = 0$, $\mathrm{Tr}\,(JA(t)) = 0$ and the formula $\det e^S = e^{\mathrm{Tr}\,S}$ we find successively

$$e^{JB_k} = E + JB_k + \frac{1}{2!}(JB_k)^2 + \cdots + \frac{1}{m!}(JB_k)^m + \cdots = E + JB_k,$$
$$\det(E + JB_k) = \det e^{JB_k} = e^{\mathrm{Tr}\,(JB_k)} = e^0 = 1,$$
$$\det U_k(\tau_k, \tau_{k-1}^+) = \exp\left(\int_{\tau_{k-1}}^{\tau_k} \mathrm{Tr}\,(JA(s))\,ds\right) = e^0 = 1.$$

Therefore

$$\det M = \prod_{k=1}^q \det M_k = \prod_{k=1}^q \det(E + JB_k)\det U_k(\tau_k, \tau_{k-1}^+) = 1.$$

Then by (5.7) and the equalities $\det J^{-1} = \det J = 1$, $\det M^{-1} = \det M = 1$ we obtain

$$F\left(\frac{1}{\lambda}\right) = \det\left(\frac{1}{\lambda}E - M\right) = \frac{1}{\lambda^{2n}}\det(E - \lambda M) = \frac{1}{\lambda^{2n}}\det(E - \lambda M^*)$$

$$= \frac{1}{\lambda^{2n}}\det(E - \lambda JM^{-1}J^{-1}) = \frac{1}{\lambda^{2n}}\det J \det(E - \lambda M^{-1})\det J^{-1}$$

$$= \frac{\det M^{-1}}{\lambda^{2n}}\det(\lambda E - M) = \frac{F(\lambda)}{\lambda^{2n}}. \quad \square$$

Remark 5.1   A periodic Hamiltonian equation cannot be asymptotically stable since by Lemma 5.1 and Theorem 5.1 its multipliers cannot all satisfy the condition $|\mu_j| < 1$.

As a corollary of Theorem 5.1 and Theorem 3.5 we obtain the following theorem.

Theorem 5.2   *The linear Hamiltonian periodic equation (5.1) is stable if and only if all of its multipliers μ_j ($j = 1, \ldots, 2n$) lie on the unit circle and have simple elementary divisors.*

Consider the T-periodic impulsive equation

$$\begin{aligned}
\ddot{x} + a(t)x &= 0, & t \neq \tau_k, t \in \mathbb{R}, \\
\Delta \dot{x} &= b_k x, & t = \tau_k, k \in \mathbb{Z},
\end{aligned} \tag{5.8}$$

where $a(\,\cdot\,) \in PC(\mathbb{R}, \mathbb{R}^{n \times n})$, $b_k \in \mathbb{R}^{n \times n}$, $a^*(t) = a(t)$, $b_k^* = b_k$, $a(t + T) = a(t)$, $b_{k+q} = b_k$, and $\tau_{k+q} = \tau_k + T$ ($t \in \mathbb{R}, k \in \mathbb{Z}$).

We write equation (5.8) in the form

$$\begin{aligned}
\frac{dx}{dt} &= y, & \frac{dy}{dt} &= -a(t)x, & t \neq \tau_k, \\
\Delta x &= 0, & \Delta y &= b_k x, & t = \tau_k
\end{aligned}$$

or

$$\begin{aligned}
\frac{dz}{dt} &= JA(t)z, & t \neq \tau_k, \\
\Delta z &= JB_k z, & t = \tau_k,
\end{aligned} \tag{5.9}$$

where

$$z = \begin{bmatrix} x \\ y \end{bmatrix}, \qquad A(t) = \begin{bmatrix} a(t) & 0 \\ 0 & E_n \end{bmatrix}, \qquad B_k = \begin{bmatrix} -b_k & 0 \\ 0 & 0 \end{bmatrix}.$$

Since the matrices $A(t)$ and B_k are symmetric and

$$(JB_k)^2 = \begin{bmatrix} 0 & 0 \\ b_k & 0 \end{bmatrix}\begin{bmatrix} 0 & 0 \\ b_k & 0 \end{bmatrix} = \begin{bmatrix} 0 & 0 \\ 0 & 0 \end{bmatrix},$$

equation (5.9) is Hamiltonian and the following theorem is valid.

Theorem 5.3   *The characteristic equation of the monodromy matrix of equation (5.8) is reciprocal.*

Example 5.1   Consider the linear Hamiltonian system of two equations with an impulse effect

$$\frac{dx}{dt} = ax + by, \quad \frac{dy}{dt} = cx - ay, \quad t \neq \tau_k,$$
$$\Delta x = \alpha x + \beta y, \quad \Delta y = \gamma x - \alpha y, \quad t = \tau_k,$$

(5.10)

where $a, b, c, \alpha, \beta, \gamma$ are real numbers, $\alpha^2 + \beta\gamma = 0$ and $\tau_{k+q} = \tau_k + T$ $(k \in \mathbb{Z})$.
   System (5.10) can be written in the form

$$\frac{dz}{dt} = JAz, \quad t \neq \tau_k,$$
$$\Delta z = JBz, \quad t = \tau_k,$$

(5.11)

where

$$z = \begin{bmatrix} x \\ y \end{bmatrix}, \quad A = \begin{bmatrix} -c & a \\ a & b \end{bmatrix}, \quad B = \begin{bmatrix} -\gamma & \alpha \\ \alpha & \beta \end{bmatrix}, \quad J = \begin{bmatrix} 0 & 1 \\ -1 & 0 \end{bmatrix}.$$

The following cases are possible

(i) Let $a^2 + bc > 0$ and $\omega = \sqrt{a^2 + bc}$. In this case the monodromy matrix of equation (5.11) has the form

$$M = \begin{bmatrix} 1+\alpha & \beta \\ \gamma & 1-\alpha \end{bmatrix} \begin{bmatrix} \cosh\omega T + \frac{a}{\omega}\sinh\omega T & \frac{b}{\omega}\sinh\omega T \\ \frac{c}{\omega}\sinh\omega T & \cosh\omega T - \frac{a}{\omega}\sinh\omega T \end{bmatrix}.$$

The multipliers of (5.11) are the roots of the characteristic equation

$$\mu^2 - \operatorname{Tr} M\mu + \det M = 0.$$

(5.12)

Straightforward calculations show that

$$\operatorname{Tr} M = 2\left(\cosh\omega T + \frac{1}{2\omega}(2a\alpha + b\gamma + c\beta)\sinh\omega T\right).$$

Since $\det M = 1$, equation (5.12) becomes

$$\mu^2 - 2\left(\cosh\omega T + \frac{1}{2\omega}(2a\alpha + b\gamma + c\beta)\sinh\omega T\right)\mu + 1 = 0.$$

(5.13)

The solutions of (5.11) are stable if the roots of equation (5.13) are distinct and their moduli are equal to 1. This is satisfied if

$$\left|\cosh\omega T + \frac{1}{2\omega}(2a\alpha + b\gamma + c\beta)\sinh\omega T\right| < 1.$$

(5.14)

If

$$\left|\cosh\omega T + \frac{1}{2\omega}(2a\alpha + b\gamma + c\beta)\sinh\omega T\right| \geq 1$$

(5.15)

then the solutions of system (5.11) are unstable.

(ii) Let $a^2 + bc < 0$ and $\omega = \sqrt{-a^2 - bc}$. Then the monodromy matrix of (5.11) is

$$M = \begin{bmatrix} 1 + \alpha & \beta \\ \gamma & 1 - \alpha \end{bmatrix} \begin{bmatrix} \cos \omega T + \frac{a}{\omega} \sin \omega T & \frac{b}{\omega} \sin \omega T \\ \frac{c}{\omega} \sin \omega T & \cos \omega T - \frac{a}{\omega} \sin \omega T \end{bmatrix}$$

and its multipliers are the roots of the equation

$$\mu^2 - 2\left(\cos \omega T + \frac{1}{2\omega}(2a\alpha + b\gamma + c\beta) \sin \omega T\right)\mu + 1 = 0. \tag{5.16}$$

In this case the solutions of system (5.11) are stable if

$$\left|\cos \omega T + \frac{1}{2\omega}(2a\alpha + b\gamma + c\beta) \sin \omega T\right| < 1 \tag{5.17}$$

and unstable if

$$\left|\cos \omega T + \frac{1}{2\omega}(2a\alpha + b\gamma + c\beta) \sin \omega T\right| \geq 1. \tag{5.18}$$

(iii) Let $a^2 + bc = 0$. Then the monodromy matrix of (5.11) is

$$M = \begin{bmatrix} 1 + \alpha & \beta \\ \gamma & 1 - \alpha \end{bmatrix} \begin{bmatrix} 1 + aT & bT \\ cT & 1 - aT \end{bmatrix}$$

and the equation for the multipliers is

$$\mu^2 - 2(1 + [a\alpha + \tfrac{1}{2}(b\gamma + c\beta)]T)\mu + 1 = 0.$$

In this case the solutions of system (5.11) are stable if

$$|1 + [a\alpha + \tfrac{1}{2}(b\gamma + c\beta)]T| < 1 \tag{5.19}$$

and unstable if

$$|1 + [a\alpha + \tfrac{1}{2}(b\gamma + c\beta)]T| \geq 1. \tag{5.20}$$

If $\operatorname{Tr} M = 2$ (or $\operatorname{Tr} M = -2$) then equation (5.11) has a non-trivial T-periodic (or $2T$-periodic) solution. But by (5.15), (5.18) and (5.20) these solutions are unstable.

Notes and comments for Chapter II

The Floquet theory for periodic linear impulsive equations was described partially by A. M. Samoilenko and N. A. Perestyuk in [72]. The proofs of Theorems 3.2–3.5 are given in [73]; Theorem 3.6 is new. Example 3.1 was taken from [73], Example 3.2 from [10], and Example 3.3 from [36]. The contents of Sections 4 and 5 were adapted from [72] and [73]. Lemma 4.2 was taken from [90]; the proof of Theorem 4.2 is new. The logistic equation in Example 4.1 was considered by B. G. Zhang and K. Gopalsamy in [92] but without impulses. The simple two-compartmental model for drug distribution in Example 4.2 was proposed by E. Kruger-Thiemer in [52]; see also R. Bellman [18]. Example 4.3 was taken from [15] and equation (4.39) from [69]. The results in Section 5 were adapted from [73].

Chapter III

Method of the Small Parameter. Non-critical Case

6. Quasilinear equations with fixed moments of an impulse effect

Consider the T-periodic impulsive equation

$$\frac{dx}{dt} = A(t)x + g(t) + f(t, x, \epsilon), \quad t \neq \tau_k,$$

$$\Delta x = B_k x + h_k + I_k(x, \epsilon), \qquad t = \tau_k,$$

(6.1)

where $t \in \mathbb{R}$, $k \in \mathbb{Z}$, $x \in B_H = B_H(0) \subset \mathbb{R}^n$, and $\epsilon \in J = (-\bar{\epsilon}, \bar{\epsilon})$ is a small parameter, and $A(t)$, B_k, $g(t)$ and h_k satisfy conditions (H3) and H4 in which $\mathbb{C}$ is replaced by $\mathbb{R}$.

Introduce the following conditions.

H6.1 The function $f : \mathbb{R} \times B_H \times J \to \mathbb{R}^n$ is continuous in the sets $(\tau_k, \tau_{k+1}] \times B_H \times J$ $(k \in \mathbb{Z})$ and for each $k \in \mathbb{Z}$, $x \in B_H$ and $\epsilon \in J$ there exists a finite limit of $f(t, y, \mu)$ as $(t, y, \mu) \to (\tau_k, x, \epsilon)$, $t > \tau_k$. Moreover,

$$f(t + T, x, \epsilon) = f(t, x, \epsilon) \quad (t \in \mathbb{R}, x \in B_H, \epsilon \in J).$$

H6.2 The functions $I_k : B_H \times J \to \mathbb{R}^n$ are continuous in $B_H \times J$ and

$$I_{k+q}(x, \epsilon) = I_k(x, \epsilon) \quad (k \in \mathbb{Z}, x \in B_H, \epsilon \in J).$$

H6.3 There exists a non-negative function $\mu(\epsilon)$ such that $\lim_{\epsilon \to 0} \mu(\epsilon) = \mu(0) = 0$ and

$$|f(t, x, \epsilon)| \le \mu(\epsilon), \qquad |I_k(x, \epsilon)| \le \mu(\epsilon)$$

for $t \in \mathbb{R}$, $k \in \mathbb{Z}$, $x \in B_H$ and $\epsilon \in J$.

H6.4 There exists a non-negative function $\lambda(\epsilon)$ such that $\lim_{\epsilon \to 0} \lambda(\epsilon) = \lambda(0) = 0$ and

$$|f(t, x, \epsilon) - f(t, y, \epsilon)| \le \lambda(\epsilon)|x - y|,$$

$$|I_k(x, \epsilon) - I_k(y, \epsilon)| \le \lambda(\epsilon)|x - y|$$

for $t \in \mathbb{R}$, $k \in \mathbb{Z}$, $\epsilon \in J$ and $x, y \in B_H$.

Under the assumption that the linear homogeneous equation (3.1) has no non-trivial T-periodic solutions we shall find sufficient conditions under which equation (6.1) has a T-periodic solution. To this end we shall use the following two theorems on the existence of a fixed point of an operator acting in a Banach space [46].

Let B be a Banach space with a norm $\| \cdot \|$, $x_0 \in B$ and $B(x_0, r) = \{x \in B : \|x - x_0\| \le r\}$.

Theorem 6.1 (Banach's fixed point theorem)   *Let the operator $U : \mathcal{B}(x_0, r) \to \mathcal{B}$ be a contraction, that is, there exists a constant $\rho \in (0, 1)$ such that for $x, y \in \mathcal{B}(x_0, r)$ we have*

$$\|U(x) - U(y)\| \le \rho\|x - y\|, \tag{6.2}$$

$$\|U(x) - x_0\| \le r(1 - \rho). \tag{6.3}$$

Then the operator U has a unique fixed point $\tilde{x} \in \mathcal{B}(x_0, r)$ such that $U(\tilde{x}) = \tilde{x}$. Moreover, the fixed point $\tilde{x}$ can be found as the limit of the successive approximations $x_m = U(x_{m-1})$ $(m \in \mathbb{N})$ and the following estimate is valid

$$\|x_m - \tilde{x}\| \le \frac{\rho^m}{1 - \rho}\|U(x_0) - x_0\|. \tag{6.4}$$

Remark 6.1   Condition (6.3) can be omitted if U maps $\mathcal{B}(x_0, r)$ into itself, that is, $U\mathcal{B}(x_0, r) \subseteq \mathcal{B}(x_0, r)$.

Theorem 6.2 (Schauder–Tychonoff fixed point theorem)   *Let S be a non-empty closed bounded convex subset of the Banach space $\mathcal{B}$ and let the operator $U : S \to S$ be compact. Then U has a fixed point $\tilde{x} \in S$.*

Set $\epsilon = 0$ in equation (6.1). Since $f(t, x, 0) \equiv 0$ and $I_k(x, 0) \equiv 0$, then for $\epsilon = 0$ equation (6.1) coincides with the non-homogeneous linear equation (4.1). By Theorem 4.1 this equation has a unique T-periodic solution

$$x_0 = x_0(t) = \int_0^T G(t, s)g(s) \, ds + \sum_{k=1}^q G(t, \tau_k^+)h_k.$$

We shall show that for small ϵ equation (6.1) has a T-periodic solution which is 'close' to $x_0(t)$. More precisely, the following theorem is valid.

Theorem 6.3   *Let the following conditions hold.*

1. *Conditions (H3), H4 and H6.1–H6.4 are met.*
2. *The homogeneous equation (3.1) has no non-trivial T-periodic solutions.*
3. *The following inequality is valid:*

$$m = \sup_{t \in [0,T]} \left| \int_0^T G(t, s)g(s) \, ds + \sum_{k=1}^q G(t, \tau_k^+)h_k \right| < H \tag{6.5}$$

where $G(t, s)$ is the Green's function defined by (4.6).

Then there exists $\epsilon_0 \in (0, \bar{\epsilon})$ such that for $|\epsilon| \le \epsilon_0$ equation (6.1) has a unique T-periodic solution $x_\epsilon(t)$ satisfying the inequality

$$|x_\epsilon(t) - x_0(t)| < H - m. \tag{6.6}$$

Moreover,

$$\lim_{\epsilon \to 0} x_\epsilon(t) = x_0(t) \quad (\text{uniformly on } t \in \mathbb{R}). \tag{6.7}$$

Proof   Introduce the Banach space $\mathcal{B}$ of T-periodic functions $x(\,\cdot\,) \in PC(\mathbb{R}, \mathbb{R}^n)$ with norm $\|x\| = \sup_{t \in [0,T]} |x(t)|$. Let

$$r = H - m, \qquad \mathcal{B}(x_0, r) = \{x \in \mathcal{B} : \|x - x_0\| \le r\} \qquad \text{and} \qquad K = \sup_{t,s \in [0,T]} |G(t,s)|.$$

Define the operator $U_\epsilon : \mathcal{B}(x_0, r) \to \mathcal{B}$ by the formula

$$U_\epsilon[x](t) = \int_0^T G(t,s)[g(s) + f(s, x(s), \epsilon)] \, ds$$
$$+ \sum_{k=1}^q G(t, \tau_k^+)[h_k + I_k(x(\tau_k), \epsilon)]. \tag{6.8}$$

From (6.5) it follows that if $x \in \mathcal{B}(x_0, r)$ then

$$|x(t)| \le |x(t) - x_0(t)| + |x_0(t)| < r + m = H.$$

Then for $x, y \in \mathcal{B}(x_0, r)$ we have

$$\|U_\epsilon[x] - U_\epsilon[y]\| = \sup_{t \in [0,T]} \left| \int_0^T G(t,s)[f(s, x(s), \epsilon) - f(s, y(s), \epsilon)] \, ds \right.$$
$$\left. + \sum_{k=1}^q G(t, \tau_k^+)[I_k(x(\tau_k), \epsilon) - I_k(y(\tau_k), \epsilon)] \right| \tag{6.9}$$
$$\le K(T + q)\lambda(\epsilon)\|x - y\|.$$

Moreover,

$$\|U_\epsilon[x_0] - x_0\| = \sup_{t \in [0,T]} \left| \int_0^T G(t,s) f(s, x_0(s), \epsilon) \, ds + \sum_{k=1}^q G(t, \tau_k^+) I_k(x_0(\tau_k), \epsilon) \right| \tag{6.10}$$
$$\le K(T + q)\mu(\epsilon).$$

Choose $\epsilon_0 \in (0, \bar{\epsilon})$ so that

$$\rho = K(T + q) \sup_{|\epsilon| \le \epsilon_0} \lambda(\epsilon) < 1, \qquad K(t + q) \sup_{|\epsilon| \le \epsilon_0} \mu(\epsilon) \le r(1 - \rho). \tag{6.11}$$

Let $|\epsilon| \le \epsilon_0$. Then from (6.9), (6.10) and (6.11) it follows that conditions (6.2) and (6.3) of Theorem 6.1 are met and the operator U_ϵ has a unique fixed point $\tilde{x}_\epsilon \in \mathcal{B}(x_0, r)$, that is,

$$x_\epsilon(t) = \int_0^T G(t,s)[g(s) + f(s, x_\epsilon(s), \epsilon)] \, ds$$
$$+ \sum_{k=1}^q G(t, \tau_k^+)[h_k + I_k(x_\epsilon(\tau_k), \epsilon)]. \tag{6.12}$$

From (6.12) we conclude that the periodic function $x_\epsilon(t)$ is a T-periodic solution of (6.1) satisfying the estimate (6.6). This solution can be found as a limit of the uniformly convergent sequence of T-periodic functions

$$x_m(t) = \int_0^T G(t,s)[g(s) + f(s, x_{m-1}(s), \epsilon)] \, ds$$
$$+ \sum_{k=1}^q G(t, \tau_k^+)[h_k + I_k(x_{m-1}(\tau_k), \epsilon)] \quad (m \in \mathbb{N})$$

and the following estimate is valid

$$\sup_{t\in\mathbb{R}} |x_m(t) - x_\epsilon(t)| \leq \frac{\rho^m K(T+q)}{1-\rho} \sup_{|\epsilon|\leq\epsilon_0} \mu(\epsilon) \quad (m \in \mathbb{N})$$

where $\rho = K(T+q)\lambda(\epsilon)$. Finally, from the estimate

$$|x_\epsilon(t) - x_0(t)| \leq \sup_{t\in\mathbb{R}} \left| \int_0^T G(t,s)f(s,x_\epsilon(s),\epsilon)\,ds + \sum_{k=1}^q G(t,\tau_k^+)I_k(x_\epsilon(\tau_k),\epsilon) \right|$$

$$\leq K(T+q)\mu(\epsilon)$$

there follows the validity of (6.7). $\square$

Theorem 6.4   *Let the following conditions hold.*

1. *Conditions (H3), H4 and H6.1–H6.3 are met.*
2. *The homogeneous equation (3.1) has no non-trivial T-periodic solutions.*
3. *Inequality (6.5) is valid.*

Then there exists $\epsilon_0 \in (0,\bar{\epsilon})$ such that for $|\epsilon| \leq \epsilon_0$ equation (6.1) has a T-periodic solution satisfying inequality (6.6).

Proof   As in the proof of Theorem 6.3 we determine successively the number $r = H - m$, the Banach space $\mathcal{B}$, the set $S = \mathcal{B}(x_0, r)$ and the operator $U_\epsilon : S \to \mathcal{B}$ acting by formula (6.8). Obviously, S is a non-empty bounded closed and convex set and if $x \in S$, then $|x(t)| < H$ for $t \in \mathbb{R}$. Choose the number $\epsilon_0 \in (0,\bar{\epsilon})$ so that

$$K(T+q) \sup_{|\epsilon|\leq\epsilon_0} \mu(\epsilon) \leq r.$$

Let $|\epsilon| \leq \epsilon_0$. Then for any $x \in S$ we have

$$\|U_\epsilon[x] - x_0\| = \sup_{t\in[0,T]} \left| \int_0^T G(t,s)f(s,x(s),\epsilon)\,ds \right.$$

$$\left. + \sum_{k=1}^q G(t,\tau_k^+)I_k(x(\tau_k),\epsilon) \right|$$

$$\leq K(T+q)\mu(\epsilon) \leq r.$$

This means that $U_\epsilon[x] \in S$, that is, $U_\epsilon : S \to S$. Let $\mathcal{F} = \{y \in \mathcal{B} : y = U_\epsilon[x], x \in S\}$. From (6.5) it follows that $\|U_\epsilon[x]\| \leq \|U_\epsilon[x] - x_0\| + \|x_0\| \leq r + m = H$, that is, the set S is uniformly bounded.

Let $x \in S$; $t_1, t_2 \in (\tau_{j-1}, \tau_j] \cap [0,T]$; $j = 1,\ldots,q+1$, and $t_1 < t_2$. Then from the estimate

$$|U_\epsilon[x](t_1) - U_\epsilon[x](t_2)| \leq \left| \int_0^{t_1} [G(t_1,s) - G(t_2,s)]f(s,x(s),\epsilon)\,ds \right|$$

$$+ \left| \int_{t_2}^T [G(t_1,s) - G(t_2,s)]f(s,x(s),\epsilon)\,ds \right|$$

$$+ \int_{t_1}^{t_2} (|G(t_1,s)| + |G(t_2,s)|)|f(s,x(s),\epsilon)|\,ds$$

$$+ \left| \sum_{k=1}^q [G(t_1,\tau_k^+) - G(t_2,\tau_k^+)]I_k(x(\tau_k),\epsilon) \right|$$

and from formula (4.6) for the function $G(t, s)$ it follows that the set S is quasiequicontinuous and by Lemma 2.4 $\mathcal{F}$ is relatively compact in $\mathcal{B}$.

Consequently, the conditions of Theorem 6.2 are met and so the operator U_ϵ has a fixed point $x_\epsilon = x_\epsilon(t) \in S$. The function $x_\epsilon(t)$ is T-periodic and satisfies (6.12), whence it follows that $x_\epsilon(t)$ is a T-periodic solution of (6.1). $\square$

Consider the T-periodic equation

$$
\begin{aligned}
\frac{dx}{dt} &= \epsilon A(t)x + \epsilon g(t) + \epsilon f(t, x, \epsilon), \quad t \neq \tau_k, \\
\Delta x &= \epsilon B_k x + \epsilon h_k + \epsilon I_k(x, \epsilon), \quad\quad t = \tau_k
\end{aligned}
\tag{6.13}
$$

where $\epsilon \in J = (-\bar{\epsilon}, \bar{\epsilon})$ is a small parameter.

Before we formulate sufficient conditions for the existence of T-periodic solutions of equation (6.13) we shall prove some auxiliary assertions about the matrix

$$
D_\epsilon(t, s) = X_\epsilon(t + T)[E - X_\epsilon(T)]^{-1}X_\epsilon^{-1}(s)
$$

where $X_\epsilon(t)$ is the fundamental matrix with $X_\epsilon(0) = E$ for the equation

$$
\begin{aligned}
\frac{dx}{dt} &= \epsilon A(t)x, \quad t \neq \tau_k, \\
\Delta x &= \epsilon B_k x, \quad t = \tau_k.
\end{aligned}
\tag{6.14}
$$

For the Cauchy matrix $W_\epsilon(t, s)$ of equation (6.14) we have

$$
W_\epsilon(t, s) = E + \epsilon \int_s^t A(\tau)W_\epsilon(\tau, s) \, d\tau + \epsilon \sum_{s \leq \tau_k < t} B_k W_\epsilon(\tau_k, s) \quad (s \leq t).
\tag{6.15}
$$

Since $\sup_{t \in \mathbb{R}} |A(t)| \leq c < \infty$ and $\sup_{k \in \mathbb{Z}} |B_k| \leq c < \infty$, the function $u(t) = |W_\epsilon(t, s)|$ satisfies the inequality

$$
u(t) \leq 1 + |\epsilon|c \int_s^t u(\tau) \, d\tau + |\epsilon|c \sum_{s \leq \tau_k < t} u(\tau_k) \quad (s \leq t).
\tag{6.16}
$$

We apply Lemma 2.3 to inequality (6.16) and obtain the estimate

$$
|W_\epsilon(t, s)| \leq (1 + |\epsilon|c)^{i[s,t)}e^{|\epsilon|c(t-s)} \quad (-\infty < s \leq t < +\infty)
$$

which implies the uniform boundedness of the matrix $W_\epsilon(t, s)$ with respect to $\epsilon \in J$, $t \in \mathbb{R}$ and $s \in [t - T, t)$.

Lemma 6.1   *Let conditions (H3) hold. Then*

$$
\lim_{\epsilon \to 0} W_\epsilon(t, s) = E,
\tag{6.17}
$$

and

$$
\lim_{\epsilon \to 0} \frac{1}{\epsilon}[W_\epsilon(t, s) - E] = \int_s^t A(\tau) \, d\tau + \sum_{s \leq \tau_k < t} B_k
\tag{6.18}
$$

uniformly with respect to $t \in \mathbb{R}$, $s \in [t - T, t)$.

Proof   Relation (6.17) follows from (6.15) and the boundedness of $W_\epsilon(t, s)$, and (6.18) follows from (6.15) and (6.17). $\square$

Corollary 6.1

$$\lim_{\epsilon \to 0} \frac{1}{\epsilon}[W_\epsilon(T, 0) - E] = \int_0^T A(\tau)\, d\tau + \sum_{0 \leq \tau_k < T} B_k.$$

Lemma 6.2   Let conditions (H3) hold and let the matrix

$$\int_0^T A(\tau)\, d\tau + \sum_{0 \leq \tau_k < T} B_k$$

be non-singular. Then the matrix $D_\epsilon(t, s)$ is defined for small $\epsilon \neq 0$ and

$$\lim_{\epsilon \to 0} \epsilon D_\epsilon(t, s) = -\left[\int_0^T A(\tau)\, d\tau + \sum_{0 \leq \tau_k < T} B_k\right]^{-1}$$

uniformly with respect to $t \in \mathbb{R}$, $s \in [t, t + T)$.

Proof   The assertion of Lemma 6.2 follows from Lemma 6.1, Corollary 6.1 and the equality

$$\epsilon D_\epsilon(t, s) = W_\epsilon(t + T, t)X_\epsilon(t)\left[\frac{1}{\epsilon}(E - W_\epsilon(T, 0))\right]^{-1} X_\epsilon^{-1}(t)W_\epsilon(t, s). \quad \square$$

Theorem 6.5   Let the following conditions hold.

1. Conditions (H3), H4 and H6.1–H6.4 are met.
2. The matrix $Q = \int_0^T A(\tau)\, d\tau + \sum_{0 \leq \tau_k < T} B_k$ is non-singular.
3. The vector $x_0 = -Q^{-1}(\int_0^T g(t)\, dt + \sum_{0 \leq \tau_k < T} h_k)$ has a norm $|x_0| = m < H$.

Then there exists $\epsilon_0 \in (0, \bar{\epsilon})$ such that for $|\epsilon| \leq \epsilon_0$ equation (6.13) has a T-periodic solution $x_\epsilon(t)$ satisfying the inequality

$$|x_\epsilon(t) - x_0| \leq H - m \quad (t \in \mathbb{R})$$

Moreover,

$$\lim_{\epsilon \to 0} x_\epsilon(t) = x_0 \qquad \text{(uniformly on } t \in \mathbb{R}). \tag{6.19}$$

Proof   From Lemma 6.2 it follows that there exist $\epsilon_1 \in (0, \bar{\epsilon})$ and $K > 0$ such that for $0 < |\epsilon| \leq \epsilon_1$ the matrix $D_\epsilon(t, s)$ is defined and

$$|\epsilon D_\epsilon(t, s)| \leq K \qquad (6.20)$$

uniformly with respect to $0 < |\epsilon| \leq \epsilon_1$, $t \in \mathbb{R}$ and $t \leq s < t + T$.

Let $\mathcal{B}$ be the Banach space of T-periodic functions $x(\,\cdot\,) \in PC(\mathbb{R}, \mathbb{R}^n)$ with norm $\|x\| = \sup_{t \in [0,T]} |x(t)|$ and set $r = H - m$, $x_0(t) \equiv x_0$ and $\mathcal{B}(x_0, r) = \{x \in \mathcal{B} : \|x - x_0\| \leq r\}$.

For $0 < |\epsilon| \leq \epsilon_1$ define the operator $U_\epsilon : \mathcal{B}(x_0, r) \to \mathcal{B}$ by the formula

$$
\begin{aligned}
U_\epsilon[x](t) = & \int_t^{t+T} \epsilon D_\epsilon(t, s)[g(s) + f(s, x(s), \epsilon)] \, ds \\
& + \sum_{t \leq \tau_k < t+T} \epsilon D_\epsilon(t, \tau_k^+)[h_k + I_k(x(\tau_k), \epsilon)].
\end{aligned}
\qquad (6.21)
$$

By the properties (4.11) of the matrix $D_\epsilon(t, s)$ it follows that $U_\epsilon[x](t)$ is a T-periodic function if $x \in \mathcal{B}(x_0, r)$. By H6.2, H6.3 and Lemma 6.2 there exists $\epsilon_0 \in (0, \epsilon_1)$ such that

$$\rho = K(T + q) \sup_{|\epsilon| \leq \epsilon_0} \lambda(\epsilon) < 1 \qquad (6.22)$$

and

$$
\begin{aligned}
& K(T + q)\mu(\epsilon) \\
& + \sup_{t \leq s < t+T} |\epsilon D_\epsilon(t, s) + Q^{-1}| \left(\int_0^T |g(t)| \, dt + \sum_{0 \leq \tau_k < T} |h_k| \right) < r(1 - \rho)
\end{aligned}
\qquad (6.23)
$$

for $|\epsilon| \leq \epsilon_0$, $t \in \mathbb{R}$.

From (6.20)–(6.23) it follows that for $0 < |\epsilon| \leq \epsilon_0$ the conditions of Theorem 6.1 hold. Consequently, the operator U_ϵ has a unique fixed point $x_\epsilon = x_\epsilon(t) \in \mathcal{B}(x_0, r)$, that is,

$$
\begin{aligned}
x_\epsilon(t) = & \int_t^{t+T} \epsilon D_\epsilon(t, s)[g(s) + f(s, x_\epsilon(s), \epsilon)] \, ds \\
& + \sum_{t \leq \tau_k < t+T} \epsilon D(t, \tau_k^+)[h_k + I_k(x_\epsilon(\tau_k), \epsilon)].
\end{aligned}
\qquad (6.24)
$$

In view of the properties (4.11) of the matrix $D_\epsilon(t, s)$ we conclude that $x_\epsilon(t)$ is a T-periodic solution of (6.13) and, passing to the limit in (6.24), we obtain (6.19). $\square$

Now consider the singularly perturbed impulsive equation

$$
\begin{aligned}
\epsilon \frac{dx}{dt} &= Bx + f(t, x, \epsilon), \quad t \neq \tau_k, \\
\Delta x &= I_k(x, \epsilon), \qquad\quad\; t = \tau_k,
\end{aligned}
\qquad (6.25)
$$

where $\epsilon \in J = (-\bar{\epsilon}, \bar{\epsilon})$ is a small parameter.

Theorem 6.6   *Let the following conditions hold.*

1. *Conditions H6.1–H6.4 are met.*
2. *The eigenvalues of the matrix $B \in \mathbb{R}^{n \times n}$ have non-zero real parts.*

Then there exists $\epsilon_0 \in (0, \bar{\epsilon})$ such that for $|\epsilon| \leq \epsilon_0$ equation (6.25) has a unique T-periodic solution $x_\epsilon(t)$ satisfying the inequality

$$|x_\epsilon(t)| \leq H.$$

Moreover,

$$\lim_{\epsilon \to 0} x_\epsilon(t) = x_0(t) \equiv 0 \qquad (\text{uniformly on } t \in \mathbb{R}).$$

Proof   Obviously, for $\epsilon = 0$ equation (6.25) has a unique solution $x_0(t) \equiv 0$.

First assume that $0 < \epsilon < \bar{\epsilon}$. From condition 2 of the theorem it follows that the matrices $E - e^{BT/\epsilon}$ and $E - e^{-BT/\epsilon}$ are non-singular for any $\epsilon > 0$ and there exists a constant $K_1 > 0$ such that

$$|(E - e^{BT/\epsilon})^{-1}| \leq K_1,$$
$$|(e^{-BT/\epsilon} - E)^{-1}| \leq K_1, \quad (0 < \epsilon \leq \epsilon_1) \tag{6.26}$$

Moreover, this condition means that the equation $y' = By$ has an exponential dichotomy in $\mathbb{R}$, that is, there exist constants $K_2 \geq 1$, $\alpha > 0$ and a projector P such that

$$|e^{Bt} P e^{-Bs}| \leq K_2 e^{-\alpha(t-s)} \quad (s \leq t),$$
$$|e^{Bt}(E - P)e^{-Bs}| \leq K_2 e^{-\alpha(s-t)} \quad (s \geq t).$$

This implies that

$$|\exp\{B(t+T)/\epsilon\}P\exp\{-Bs/\epsilon\}| \leq K_2\exp\{-\alpha(t+T-s)/\epsilon\} \quad (s \leq t+T),$$
$$|\exp\{B(t)/\epsilon\}(E-P)\exp\{-Bs/\epsilon\}| \leq K_2\exp\{-\alpha(s-t)/\epsilon\} \qquad (s \geq t).$$

$$\tag{6.27}$$

Let $\mathcal{B}$ be the space of T-periodic functions $x(\,\cdot\,) \in PC(\mathbb{R}, \mathbb{R}^n)$ with norm $\|x\| = \sup_{t \in [0,T]} |x(t)|$ and $\mathcal{B}(x_0, H) = \{x \in \mathcal{B} : \|x\| \leq H\}$. For $\epsilon \in (0, \bar{\epsilon})$ define the operator $U_\epsilon : \mathcal{B}(x_0, H) \to \mathcal{B}$ by the formula

$$U_\epsilon[x](t) = \frac{1}{\epsilon} \int_t^{t+T} D_\epsilon(t, s) f(s, x(s), \epsilon)\, ds$$

$$+ \sum_{t \leq \tau_k < t+T} D_\epsilon(t, \tau_k^+) I_k(x(\tau_k), \epsilon)$$

where $D_\epsilon(t, s) = [E - \exp(BT/\epsilon)]^{-1}\exp\{B(t+T-s)/\epsilon\}$.

It is easily verified that

$U_\epsilon[x](t) =$

$$\frac{1}{\epsilon} \int_t^{t+T} [E - \exp(BT/\epsilon)]^{-1} \exp\{B(t+T)/\epsilon\} P \exp(-Bs/\epsilon) f(s, x(s), \epsilon)\, ds$$

$$+ \frac{1}{\epsilon} \int_t^{t+T} [\exp(-BT/\epsilon) - E]^{-1} \exp(Bt/\epsilon)(E - P) \exp(-Bs/\epsilon) f(s, x(s), \epsilon)\, ds$$

$$+ \sum_{t \le \tau_k < t+T} [E - \exp(BT/\epsilon)]^{-1} \exp\{B(t+T)/\epsilon\} P \exp(-B\tau_k/\epsilon) I_k(x(\tau_k), \epsilon)$$

$$+ \sum_{t \le \tau_k < t+T} [\exp(-BT/\epsilon) - E]^{-1} \exp(Bt/\epsilon)(E - P) \exp(-B\tau_k/\epsilon) I_k(x(\tau_k), \epsilon).$$

$$(6.28)$$

Choose $\epsilon_0 \in (0, \bar{\epsilon})$ so that

$$2K_1 K_2 \left(\frac{1}{\alpha} + q \right) \sup_{|\epsilon| \le \epsilon_0} \mu(\epsilon) \le H,$$

$$2K_1 K_2 \left(\frac{1}{\alpha} + q \right) \sup_{|\epsilon| \le \epsilon_0} \lambda(\epsilon) < 1.$$

From (6.28), taking into account (6.26) and (6.27) we find that for all $x, y \in B(x_0, H)$ and $0 < \epsilon \le \epsilon_0$ the following estimates are valid:

$$\|U_\epsilon[x]\| \le 2K_1 K_2 \left(\frac{1}{\alpha} + q \right) \mu(\epsilon) \le H, \qquad (6.29)$$

$$\|U_\epsilon[x] - U_\epsilon[y]\| \le 2K_1 K_2 \left(\frac{1}{\alpha} + q \right) \lambda(\epsilon) \|x - y\|. \qquad (6.30)$$

Since the function $U_\epsilon[x](t)$ is T-periodic as well as the function $x \in B(x_0, H)$, estimates (6.29) and (6.30) mean that the operator U_ϵ maps $B(x_0, H)$ into itself and is a contraction. Consequently, by Theorem 6.1 the operator U_ϵ has a unique fixed point $x_\epsilon \in B(x_0, H)$. It is easily verified that the function $x = x_\epsilon(t)$ is the T-periodic solution of equation (6.25) that we seek.

In the case when $-\bar{\epsilon} < \epsilon < 0$ the arguments are analogous. $\square$

Consider the non-linear T-periodic equations

$$\frac{dx}{dt} = F(t, x), \quad t \ne \tau_k,$$
$$\Delta x = J_k(x), \qquad t = \tau_k, \qquad (6.31)$$

and

$$\frac{dx}{dt} = F(t, x) + g(t) + f(t, x, \epsilon) + r(t, x), \quad t \ne \tau_k,$$
$$\Delta x = J_k(x) + h_k + I_k(x, \epsilon) + r_k(x), \qquad t = \tau_k, \qquad (6.32)$$

where $\tau_{k+q} = \tau_k + T$ $(k \in \mathbb{Z})$.

Assume that equation (6.31) has a T-periodic solution $\varphi(t)$. The question arises: under what conditions has equation (6.32) a T-periodic solution if the additional terms g, f, r, h_k, I_k and r_k are 'small'? We shall answer this question in the non-critical case when the equation in variations

$$\frac{dx}{dt} = \frac{\partial F}{\partial x}(t, \varphi(t))x, \quad t \neq \tau_k,$$

$$\Delta x = \frac{\partial J_k}{\partial x}(\varphi(\tau_k))x, \quad t = \tau_k$$

(6.33)

has no non-trivial T-periodic solution.

Assume that f, I_k, g, h_k satisfy conditions H6.1–H6.4 and H4, and F, J_k, r, r_k satisfy the following conditions:

H6.5 The function $F : \mathbb{R} \times B_H \to \mathbb{R}^n$ is continuously differentiable in the sets $(\tau_k, \tau_{k+1}] \times B_H$ $(k \in \mathbb{Z})$ and for each $k \in \mathbb{Z}$ and $x \in B_H$ there exist the finite limits of $F(t, y)$ and $\frac{\partial F}{\partial x}(t, y)$ as $(t, y) \to (\tau_k, x)$, $t > \tau_k$. Moreover,

$$F(t + T, x) = F(t, x) \quad (t \in \mathbb{R}, x \in B_H).$$

H6.6 The functions $J_k : B_H \to \mathbb{R}^n$ $(k \in \mathbb{Z})$ are continuously differentiable in B_H and

$$J_{k+q}(x) = J_k(x) \quad (k \in \mathbb{Z}, x \in B_H).$$

H6.7 The function $r : \mathbb{R} \times B_H \to \mathbb{R}^n$ is continuous in the sets $(\tau_k, \tau_{k+1}] \times B_H$ and for each $k \in \mathbb{Z}$ and $x \in B_H$ there exists the finite limit of $r(t, y)$ as $(t, y) \to (\tau_k, x)$, $t > \tau_k$.

H6.8 The functions $r_k : B_H \times \mathbb{R}^n$ are continuous in B_H.

H6.9 The following relations are valid:

$$r(t + T, x) = r(t, x), \quad r_{k+q}(x) = r_k(x), \quad (t \in \mathbb{R}, k \in \mathbb{Z}, x \in B_H),$$
$$r(t, \varphi(t)) = 0, \quad r_k(\varphi(\tau_k)) = 0, \quad (t \in \mathbb{R}, k \in \mathbb{Z}),$$

$$|r(t, x) - r(t, y)| \leq l_1(\delta)|x - y| \quad (t \in \mathbb{R}, |x - \varphi(t)| \leq \delta, |y - \varphi(t)| \leq \delta),$$

$$|r_k(x) - r_k(y)| \leq l_2(\delta)|x - y| \quad (k \in \mathbb{Z}, |x - \varphi(\tau_k)| \leq \delta, |y - \varphi(\tau_k)| \leq \delta)$$

where $l_i(\delta) \geq 0$ and $\lim_{\delta \to 0} l_i(\delta) = 0$ $(i = 1, 2)$.

Let $G(t, s)$ be the Green's function for the periodic problem for the linear non-homogeneous T-periodic equation corresponding to equation (6.33).

Introduce the notation:

$$K = \sup\{|G(t, s)| : t, s \in [0, T]\},$$

$$L_1(\delta) = \sup\left\{\left|\frac{\partial F}{\partial x}(t, \varphi(t) + z) - \frac{\partial F}{\partial x}(t, \varphi(t))\right| : t \in [0, T], |z| \leq \delta\right\}$$

$$L_2(\delta) = \sup\left\{\left|\frac{\partial J_k}{\partial x}(\varphi(\tau_k) + z) - \frac{\partial J_k}{\partial x}(\varphi(\tau_k))\right| : k \in \mathbb{Z}, |z| \leq \delta\right\}.$$

We notice that from conditions H6.5 and H6.6 it follows that

$$\lim_{\delta \to 0_+} L_i(\delta) = 0 \quad (i = 1, 2).$$

The following theorem is valid.

Theorem 6.7   *Let the following conditions hold.*

1. *Conditions H4 and H6.1–H6.9 are met.*
2. *Equation (6.31) has a T-periodic solution $\varphi(t)$ for which $|\varphi(t)| \leq m < H$ ($t \in \mathbb{R}$).*
3. *The equation in variations (6.35) has no non-trivial T-periodic solution.*
4. *The constants $\epsilon_0 \in (0, \bar{\epsilon})$ and $\delta_0 \in (0, H - m)$, the functions g and r and the vectors h_k and r_k satisfy the following conditions:*

$$\rho = K[TL_1(\delta_0) + Tl_1(\delta_0) + qL_2(\delta_0) + ql_2(\delta_0) + (T + q) \sup_{|\epsilon| \leq \epsilon_0} \lambda(\epsilon)] < 1,$$

$$\sup_{\substack{t \in [0,T] \\ |\epsilon| \leq \epsilon_0}} \left| \int_0^T G(t, s)[g(s) + f(s, \varphi(s), \epsilon)] \, ds \right.$$

$$\left. + \sum_{0 \leq \tau_k < T} G(t, \tau_k^+)[h_k + I_k(\varphi(\tau_k), \epsilon)] \right| < \delta_0(1 - \rho).$$

Then for each $\epsilon \in (-\epsilon_0, \epsilon_0)$ equation (6.32) has a unique T-periodic solution $x_\epsilon(t)$ satisfying the relations

$$|x_\epsilon(t) - \varphi(t)| \leq \delta_0 \quad (t \in \mathbb{R}), \tag{6.34}$$

$$\lim_{\epsilon \to 0} x_\epsilon(t) = x_0(t) \quad \text{(uniformly on } t \in \mathbb{R}). \tag{6.35}$$

Sketch of the proof   In (6.32) we carry out the change of variables $x = \varphi(t) + z$ and obtain the equation

$$\frac{dz}{dt} = \frac{\partial F}{\partial x}(t, \varphi(t))z + R(t, z) + g(t) + f(t, \varphi(t) + z, \epsilon) + r(t, \varphi(t) + z),$$

$$t \neq \tau_k,$$

$$\Delta z = \frac{\partial J_k}{\partial x}(\varphi(\tau_k))z + R_k(z) + h_k + I_k(\varphi(\tau_k) + z, \epsilon) + r_k(\varphi(\tau_k) + z),$$

$$t = \tau_k$$

\tag{6.36}

where

$$R(t, z) = F(t, \varphi(t) + z) - F(t, \varphi(t)) - \frac{\partial F}{\partial x}(t, \varphi(t))z,$$

$$R_k(z) = J_k(\varphi(\tau_k) + z) - J_k(\varphi(\tau_k)) - \frac{\partial J_k}{\partial x}(\varphi(\tau_k))z.$$

Let $\mathcal{B} = \{x \in PC(\mathbb{R}, \mathbb{R}^n) : x(t + T) = x(t), t \in \mathbb{R}\}$ and in the ball $\mathcal{B}(\delta_0) = \{z \in \mathcal{B} : \|z\| \le \delta_0\}$ define the operator $U_\epsilon : \mathcal{B}(\delta_0) \to \mathcal{B}$ by the formula

$$U_\epsilon[z](t) = \int_0^T G(t, s)[R(s, z(s)) + g(s) + f(s, \varphi(s) + z(s), \epsilon) + r(s, \varphi(s) + z(s))]\, ds$$

$$+ \sum_{0 \le \tau_k < T} G(t, \tau_k^+)[R_k(z(\tau_k)) + h_k + I_k(\varphi(\tau_k) + z(\tau_k), \epsilon) + r_k(\varphi(\tau_k) + z(\tau_k))].$$

In view of the conditions of the theorem we conclude that the operator U_ϵ is a contraction and has a unique fixed point $z_\epsilon = z_\epsilon(t) \in B(\delta_0)$. Since $z_\epsilon(t)$ is a solution of the T-periodic equation (6.36), then $x_\epsilon(t) = \varphi(t) + z_\epsilon(t)$ is a T-periodic solution of equation (6.32). Estimate (6.34) follows from the fact that $z_\epsilon \in B(\delta_0)$, and the limit relation (6.35) is proved as in Theorem 6.3. $\square$

Note that if the equation in variations (6.33) is exponentially stable then the T-periodic solution $\varphi(t)$ of (6.33) is also exponentially stable [82]. In the case when the equation in variations (6.33) has an exponential dichotomy, the following more general result is valid.

Theorem 6.8   *Let the following conditions hold.*

1. *Conditions H6.5 and H6.6 are valid.*
2. *Equation (6.31) has a T-periodic solution $\varphi(t)$ for which $|\varphi(t)| \le \sigma < H$ $(t \in \mathbb{R})$.*
3. *The equation in variations (6.33) has exponential dichotomy.*

Then there exist real numbers $\delta > 0$, $K \ge 1$, $\alpha > 0$ and an integer m $(0 \le m \le n)$ such that:

1. *in a δ-neighbourhood of $\varphi(0)$ manifolds S_m^+ and S_{n-m}^- are defined such that*

$$S_m^+ \cap S_{n-m}^- = \varphi(0), \qquad \dim S_m^+ = m, \qquad \dim S_{n-m}^- = n - m;$$

2. *for any solution $x(t)$ of (6.31) the relation*

$$|x(t) - \varphi(t)| \le K e^{-\alpha t} \quad (t \in \mathbb{R}_+) \tag{6.37}$$

is valid whenever $x(0) \in S_m^+$, $|x(0) - \varphi(0)| < \delta$, or

$$|x(t) - \varphi(t)| \le K e^{\alpha t} \quad (t \in \mathbb{R}_-) \tag{6.38}$$

whenever $x(0) \in S_{n-m}^-$, $|x(0) - \varphi(0)| < \delta$.

Proof   From conditions H6.5 and H6.6 it follows that the functions

$$R(t, x) = F(t, x) - F(t, \varphi(t)) - \frac{\partial F}{\partial x}(t, \varphi(t))[x - \varphi(t)],$$

and

$$R_k(x) = J_k(x) - J_k(\varphi(\tau_k)) - \frac{\partial J_k}{\partial x}(\varphi(\tau_k))[x - \varphi(\tau_k)]$$

satisfy the following relations

$$R(t, \varphi(t)) \equiv 0, \qquad \frac{\partial R}{\partial x}(t, \varphi(t)) \equiv 0, \qquad R_k(\varphi(\tau_k)) \equiv 0, \qquad \frac{\partial R_k}{\partial x}(\varphi(\tau_k)) \equiv 0,$$

$$|R(t, x) - R(t, y)| \le L(\Delta)|x - y|, \qquad R(t, x) = o(|x - \varphi(t)|),$$

$$|R_k(u) - R_k(v)| \le L(\Delta)|u - v|, \qquad R_k(u) = o(|u - \varphi(\tau_k)|)$$

for all $t \in \mathbb{R}$, $k \in \mathbb{Z}$, $x, y \in \mathbb{R}^n$, $|x - \varphi(t)| < \Delta$, $|y - \varphi(t)| < \Delta$ and $u, v \in \mathbb{R}^n$, $|u - \varphi(\tau_k)| < \Delta$, $|v - \varphi(\tau_k)| < \Delta$, where $L(\Delta) \to 0$ as $\Delta \to 0$ uniformly on $t \in \mathbb{R}$ and $k \in \mathbb{Z}$.

Let $X(t)$ be the fundamental matrix of equation (6.33) normalized at $t = 0$. Since equation (6.33) has an exponential dichotomy, there exist constants $K \ge 1$, $\alpha > 0$ and a projector P such that the functions

$$X_1(t, s) = X(t)PX^{-1}(s),$$

$$X_2(t, s) = X(t)(E - P)X^{-1}(s)$$

satisfy the inequalities

$$|X_1(t, s)| \le Ke^{-2\alpha(t-s)} \quad (t \ge s),$$

$$|X_2(t, s)| \le Ke^{-2\alpha(s-t)} \quad (s \ge t).$$

Choose $\Delta > 0$ so that $\Delta < H - \sigma$ and

$$4KL(\Delta)\left[\frac{1}{\alpha} + \frac{1}{1 - e^{-\alpha\theta}}\right] < 1$$

where $\theta = \min_{k \in \mathbb{Z}}(\tau_{k+1} - \tau_k) > 0$.

Let $m = \operatorname{rank} P$, $L_m^+ = P(\mathbb{R}^n)$ and $L_{n-m}^- = (E - P)(\mathbb{R}^n)$. For any $a \in L_m^+$ with $|a| < \frac{1}{2}\Delta/K = \delta_0$ consider the equation

$$x(t) = \varphi(t) + X(t)a$$

$$+ \int_0^t X_1(t, s)R(s, x(s))\, ds - \int_t^\infty X_2(t, s)R(s, x(s))\, ds \qquad (6.39)$$

$$+ \sum_{0 \le \tau_k < t} X_1(t, \tau_k^+)R_k(x(\tau_k)) - \sum_{t \le \tau_k} X_2(t, \tau_k^+)R_k(x(\tau_k)).$$

We shall prove that the solution $x(t) = x(t, a)$ of (6.39) is defined for $t \in \mathbb{R}_+$, and also satisfies equation (6.31) and the estimate

$$|x(t, a) - \varphi(t)| \leq 2K e^{-\alpha t} \quad (t \in \mathbb{R}_+). \tag{6.40}$$

For this purpose define the sequence of functions $x_\nu = x_\nu(t, a)$ so that $x_0 = \varphi(t)$ and

$$x_\nu(t) = \varphi(t) + X(t)a$$
$$+ \int_0^t X_1(t, s) R(s, x_{\nu-1}(s)) \, ds - \int_t^\infty X_2(t, s) R(s, x_{\nu-1}(s)) \, ds \tag{6.41}$$
$$+ \sum_{0 \leq \tau_k < t} X_1(t, \tau_k^+) R_k(x_{\nu-1}(\tau_k)) - \sum_{t \leq \tau_k} X_2(t, \tau_k^+) R_k(x_{\nu-1}(\tau_k))$$

for $\nu \in \mathbb{N}$ and $t \in \mathbb{R}_+$.

We claim that for any $\nu \in \mathbb{N}$ and $t \in \mathbb{R}_+$ the following estimates are valid:

$$|x_\nu(t, a) - x_{\nu-1}(t, a)| \leq K |a| 2^{1-\nu} e^{-\alpha t}, \tag{6.42}$$
$$|x_\nu(t, a) - \varphi(t)| \leq 2K |a| e^{-\alpha t} < \Delta. \tag{6.43}$$

Indeed, for $\nu = 1$ estimates (6.42) and (6.43) are valid since

$$|x_1(t) - x_0(t)| = |x_1(t) - \varphi(t)| = |X(t)a| = |X(t) P X^{-1}(0)a|$$
$$\leq K e^{-2\alpha t} |a| < 2K |a| e^{-\alpha t} < \Delta.$$

Let (6.42) and (6.43) be valid for $\nu = 1, \ldots, \mu$. Then

$$|x_{\mu+1}(t) - x_\mu(t)| \leq \int_0^t K e^{-2\alpha(t-s)} L K |a| 2^{1-\mu} e^{-\alpha s} \, ds$$
$$+ \int_t^\infty K e^{-2\alpha(s-t)} L K |a| 2^{1-\mu} e^{-\alpha s} \, ds$$
$$+ \sum_{0 \leq \tau_k < t} K e^{-2\alpha(t-\tau_k)} L K |a| 2^{1-\mu} e^{-\alpha \tau_k}$$
$$+ \sum_{t \leq \tau_k} K e^{-2\alpha(\tau_k - t)} L K |a| 2^{1-\mu} e^{-\alpha \tau_k}.$$

In view of

$$\int_0^t e^{-2\alpha(t-s)} e^{-\alpha s} \, ds \leq \frac{e^{-\alpha t}}{\alpha},$$

$$\int_t^\infty e^{-2\alpha(s-t)} e^{-\alpha s} \, ds \leq \frac{e^{-\alpha t}}{\alpha},$$

$$\sum_{0 \leq \tau_k < t} e^{-2\alpha(t-\tau_k)} e^{-\alpha \tau_k} \leq \frac{e^{-\alpha t}}{1 - e^{-\alpha \theta}},$$

$$\sum_{t \leq \tau_k} e^{-2\alpha(\tau_k - t)} e^{-\alpha \tau_k} \leq \frac{e^{-\alpha t}}{1 - e^{-\alpha \theta}},$$

we obtain successively that

$$|x_{\mu+1}(t) - x_\mu(t)| \le K^2 L |a| 2^{1-\mu} \left[\frac{2}{\alpha} + \frac{2}{1 - e^{-\alpha\theta}} \right] e^{-\alpha t},$$

$$\le K |a| 2^{-\mu} e^{-\alpha t},$$

and

$$|x_{\mu+1}(t) - \varphi(t)| \le \sum_{j=0}^{\mu} |x_{j+1}(t) - x_j(t)|$$

$$\le K |a| e^{-\alpha t} [2^{-\mu} + 2^{-\mu+1} + \cdots + 2^{-1} + 1]$$

$$\le 2K |a| e^{-\alpha t} \le \Delta.$$

Hence by induction estimates (6.42) and (6.43) are valid for each $\nu \in \mathbb{N}$. This implies that the functions $x_\nu(t)$ are defined for each $\nu \ge 1$ and there exists the limit

$$\lim_{\nu \to \infty} x_\nu(t, a) = x(t, a)$$

uniformly on $t \in \mathbb{R}_+$ and $a \in L_m^+$, $|a| < \delta_0$.

Passing to the limit in (6.41) and (6.43), we conclude that $x(t, a)$ is a solution of (6.39) satisfying estimate (6.40). A straightforward verification shows that $x(t, a)$ is also a solution of equation (6.31).

Consider the set S_m^+ of the initial values $\xi = x(0, a)$ of the solutions $x(t, a)$. We have the relation

$$\xi = \varphi(0) + a + \psi^+(a) \quad (a \in L_m^+) \tag{6.44}$$

where the function

$$\psi^+(a) = - \int_0^\infty (E - P) X^{-1}(s) R(s, x(s, a))\, ds$$

$$- \sum_{0 \le \tau_k} (E - P) X^{-1}(\tau_k^+) R(x(\tau_k^+, a))$$

is such that $\psi^+(a) = o(|a|)$ as $|a| \to 0$.

Consequently, S_m^+ is an m-dimensional manifold defined in a neighbourhood of the point $\varphi(0)$ and each solution $x(t)$ of (6.31) satisfies (6.37) if $x(0) \in S_m^+$, $|x(0) - \varphi(0)| < \delta_1$ and $\delta_1 \in (0, \delta_0)$ is small enough.

Analogously, we can construct an $(n-m)$-dimensional manifold S_{n-m}^- such that for any solution $x(t)$ of (6.31) the estimate (6.38) is valid if $x(0) \in S_{n-m}^-$, $|x(0) - \varphi(0)| < \delta_2$ and $\delta_2 > 0$ is small enough. Moreover, S_{n-m}^- is given by a function of the form

$$\eta = \varphi(0) + b + \psi^-(b) \quad (b \in L_{n-m}^-) \tag{6.45}$$

where $\psi^-(b) = o(|b|)$ as $|b| \to 0$. Thus from (6.44) and (6.45) it follows that $S_m^+ \cap S_{n-m}^- = \varphi(0)$ if $\delta \le \min(\delta_1, \delta_2)$ is small enough. $\square$

Example 6.1   Consider the logistic impulsive equation

$$\begin{aligned}\frac{dx}{dt} &= rx - \frac{r}{K}x^2, \quad t \neq nT, \\ \Delta x &= \varphi(x), \qquad t = nT\end{aligned} \tag{6.46}$$

where $r > 0$, $K > 0$, $T > 0$, the function $\varphi : \mathbb{R}_+ \to \mathbb{R}$ is differentiable in $\mathbb{R}_+$ and $\varphi(x) > -x$ for $x \in \mathbb{R}_+$.

We seek a positive T-periodic solution $x(t)$ of (6.46) with $x(0+) = x_0 > 0$. For $t \in (0, T]$ we have

$$x(t) = \frac{x_0}{e(t) + b(t)x_0}$$

where

$$e(t) = e^{-rt},$$

$$b(t) = \frac{1 - e^{-rt}}{K}.$$

In view of the dependences

$$x(T) = \frac{x_0}{e(T) + b(T)x_0},$$

$$x_0 = x(T) + \varphi(x(T)),$$

we conclude that the condition of T-periodicity $x(T+) = x_0$ has the form

$$g(x(T)) \equiv \frac{x(T)[e(T) - 1 + b(T)x(T)]}{1 - b(T)x(T)} = \varphi(x(T)). \tag{6.47}$$

Since $g(x) \geq -x$ for $x \in [0, 1/b(T))$ and $\varphi(x) > -x$ for $x \geq 0$, equation (6.47) can have a positive solution only in the interval $(0, 1/b(T))$ (Fig. 6.1). Suppose that equation (6.47) has a positive solution $x(T)$ and denote by $\pi(t)$ the corresponding positive T-periodic solution of equation (6.46) for which

$$\pi(0^+) = x_0 = \frac{e(T)x(T)}{1 - b(T)x(T)}, \quad \pi(T) = x(T),$$

and

$$\pi(t) = \frac{x_0}{e(t) + b(t)x_0}, \quad (t \in (0, T]).$$

The equation in variations corresponding to $\pi(t)$ is

$$\begin{aligned}\frac{dz}{dt} &= (r - \frac{2r}{K}\pi(t))z, \quad t \neq nT, \\ \Delta z &= \varphi'(\pi(T))z, \qquad t = nT,\end{aligned} \tag{6.48}$$

and the multiplier of this equation is

$$\mu = \{1 + \varphi'(x(T))\}\exp\left\{\int_0^T \left(r - \frac{2r}{K}\pi(t)\right) dt\right\}.$$

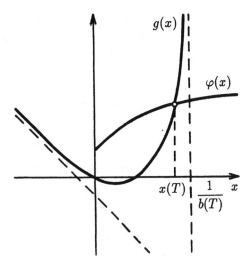

Fig. 6.1

Taking into account that

$$\int_0^T \left(r - \frac{2r}{K}\pi(t)\right) dt = \int_0^T \left(\frac{2\dot{\pi}(t)}{\pi(t)} - r\right) dt = \ln\left|\frac{\pi(T)}{\pi(0+)}\right|^2 - rT,$$

$$\frac{\pi(T)}{\pi(0+)} = \frac{x(T)}{x_0} = \frac{1 - b(T)x(T)}{e(T)},$$

we conclude that

$$\mu = \{1 + \varphi'(x(T))\}\frac{(1 - b(T)x(T))^2}{e(T)}. \tag{6.49}$$

Consider the case when $\varphi(x) = cx$ and $c > -1$. Then

$$x(T) = \frac{1 + c - e(T)}{(1 + c)b(T)} \quad \text{and} \quad \mu = \frac{e(T)}{1 + c},$$

and since $0 < x(T) < 1/b(T)$, we have

$$0 < \mu = \frac{e(T)}{1 + c} < 1.$$

Consequently, in this case the condition for the existence of the positive T-periodic solution of (6.46) is

$$\beta \equiv 1 + c - e^{-rT} > 0$$

and this solution is exponentially stable and unique (compare with Example 4.1).

Example 6.2   Let $u(t)$ and $v(t)$ denote the population densities (or biomasses) of two species at time t in a habitat competing for a common pool of resources. The classical Lotka–Volterra system of ordinary differential equations describing $u(t)$ and $v(t)$ is

$$\frac{du}{dt} = u(r_1 - a_1 u - b_1 v),$$
$$\frac{dv}{dt} = v(r_2 - a_2 u - b_2 v), \tag{6.50}$$

where a_i, b_i, r_i $(i = 1, 2)$ are positive constants. We shall assume that

$$\frac{a_1}{a_2} > \frac{r_1}{r_2} > \frac{b_1}{b_2}. \tag{6.51}$$

From condition (6.51) it follows that system (6.50) has a global attractor (u^*, v^*) in the positive quadrant, where

$$u^* = \frac{r_1 b_2 - r_2 b_1}{a_1 b_2 - a_2 b_1}, \qquad v^* = \frac{r_2 a_1 - r_1 a_2}{a_1 b_2 - a_2 b_1}.$$

The global asymptotic stability of (u^*, v^*) implies that the system in variations of (6.50) with respect to (u^*, v^*) has multipliers with moduli less than 1.

Consider the following T-periodic system with an impulse effect:

$$\frac{du}{dt} = u(r_1 - a_1 u - b_1 v) + \epsilon r_1(t, u, v), \quad t \neq \tau_k,$$
$$\frac{dv}{dt} = v(r_2 - a_2 u - b_2 v) + \epsilon r_2(t, u, v), \quad t \neq \tau_k, \tag{6.52}$$

$$\Delta u = g_{1,k} + \epsilon r_{1,k}(u, v), \quad t = \tau_k, k \in \mathbb{Z},$$
$$\Delta v = g_{2,k} + \epsilon r_{2,k}(u, v), \quad t = \tau_k, k \in \mathbb{Z} \tag{6.53}$$

where ϵ is a small parameter, the functions r_i, $r_{i,k}$ $(i = 1, 2; k \in \mathbb{Z})$ are continuously differentiable for $(t, u, v) \in \mathbb{R}^3$ and there exists a real number $T > 0$ and an integer $q > 0$ such that

$$r_i(t + T, u, v) = r_i(t, u, v)$$

$$r_{i,k+q}(u, v) = r_{i,k}(u, v)$$

$$\tau_{k+q} = \tau_k + T, \quad g_{i,k+q} = g_{i,k},$$

for $t \in \mathbb{R}$, $u \in \mathbb{R}$, $v \in \mathbb{R}$, $k \in \mathbb{Z}$ and $i = 1, 2$.

With conditions (6.53) we can take into account the possible exterior effects under which the population densities change very rapidly. For instance, impulsive reduction of the population density of a given species is possible after its partial destruction by catching or by poisoning with chemicals used in agriculture. An impulsive increase of the density is possible by artificial breeding of the species.

By introducing periodic perturbations ϵr_i $(i = 1, 2)$, small perturbations of the growth and interaction coefficients can be taken into account.

For instance, if in system (6.50) instead of the coefficients r_i, a_i, b_i $(i = 1, 2)$ we set the functions $r_i + \epsilon R_i(u, v)\sin(\omega t + \rho_i)$, $a_i + \epsilon A_i(u, v)\sin(\omega t + \alpha_i)$, $b_i + \epsilon B_i(u, v)\sin(\omega t + \beta_i)$, respectively, where $\omega = 2\pi/T$, we obtain a system of the form (6.52) in which

$$r_1(t, u, v) = uR_1(u, v)\sin(\omega t + \rho_1)$$
$$- u^2 A_1(u, v)\sin(\omega t + \alpha_1)$$
$$- uv B_1(u, v)\sin(\omega t + \beta_1),$$
$$r_2(t, u, v) = vR_2(u, v)\sin(\omega t + \rho_2)$$
$$- uv A_2(u, v)\sin(\omega t + \alpha_2)$$
$$- v^2 B_2(u, v)\sin(\omega t + \beta_2).$$

From the given assumptions about system (6.52), (6.53), Theorem 6.7 implies the existence of numbers $\epsilon_1 > 0$ and $\delta_1 > 0$ such that if $|\epsilon| \leq \epsilon_1$ and $|g_{i,k}| \leq \delta_1$ $(i = 1, 2; k = 1, \dots, q)$, system (6.52), (6.53) has a T-periodic solution $u = u_\epsilon(t)$, $v = v_\epsilon(t)$ which is exponentially stable.

7. Non-linear equations with unfixed moments of an impulse effect

Consider the T-periodic impulsive equation with unfixed moments of an impulse effect

$$\frac{dx}{dt} = f(t, x, \epsilon), \quad t \neq \tau_k(x, \epsilon),$$
$$\Delta x = I_k(x, \epsilon), \quad t = \tau_k(x, \epsilon) \tag{7.1}$$

where $\epsilon \in J = (-\bar{\epsilon}, \bar{\epsilon})$ is a small parameter and $\tau_k(x, \epsilon) < \tau_{k+1}(x, \epsilon)$ for $x \in \mathbb{R}^n$, $\epsilon \in J$.

Suppose that for $\epsilon = 0$ equation (7.1) has a T-periodic solution $x = \varphi(t)$ with moments of an impulse effect τ_k.

We associate with the solution $\varphi(t)$ the variational equation

$$\frac{dz}{dt} = \frac{\partial f}{\partial x}(t, \varphi(t), 0)z, \quad t \neq \tau_k,$$
$$\Delta z = L_k z, \quad t = \tau_k \tag{7.2}$$

where

$$L_k = \frac{\partial I_k}{\partial x} + \left[\frac{\partial I_k}{\partial x} f_k + f_k - f_k^+\right] \frac{\frac{\partial \tau_k}{\partial x}}{1 - \frac{\partial \tau_k}{\partial x} f_k}$$

and

$$f_k = f(\tau_k, \varphi(\tau_k), 0), \quad f_k^+ = f(\tau_k^+, \varphi(\tau_k^+), 0),$$
$$\frac{\partial I_k}{\partial x} = \frac{\partial I_k}{\partial x}(\varphi(\tau_k), 0), \quad \frac{\partial \tau_k}{\partial x} = \frac{\partial \tau_k}{\partial x}(\varphi(\tau_k), 0).$$

In the non-critical case for equation (7.2) we shall prove that for small ϵ equation (7.1) also has a T-periodic solution $x_\epsilon(t)$ which is 'close' to $\varphi(t)$.

Without loss of generality we assume that $\tau_0(\varphi(0), 0) < 0 < \tau_1(\varphi(0), 0)$ and denote by $x(t; x_0, \epsilon)$ the solution of equation (7.1) for which $x(0; x_0, \epsilon) = x_0$.

Theorem 7.1   *Let the following conditions hold:*

1. *Equation (7.1) is T-periodic, that is, there exists a $q \in \mathbb{N}$ such that*

$$f(t + T, x, \epsilon) = f(t, x, \epsilon), \qquad (t \in \mathbb{R}, x \in \mathbb{R}^n, \epsilon \in J),$$
$$I_{k+q}(x, \epsilon) = I_k(x, \epsilon), \qquad (k \in \mathbb{Z}, x \in \mathbb{R}^n, \epsilon \in J),$$
$$\tau_{k+q}(x, \epsilon) = \tau_k(x, \epsilon) + T, \quad (k \in \mathbb{Z}, x \in \mathbb{R}^n, \epsilon \in J).$$

2. *For $\epsilon = 0$ equation (7.1) has a T-periodic solution $x = \varphi(t)$ with moments of an impulse effect τ_k.*
3. *For each $k = 1, \ldots, q$ the function $\tau_k(x, \epsilon)$ is differentiable in some neighbourhood of the point $(\varphi(\tau_k), 0)$ and*

$$1 - \frac{\partial \tau_k}{\partial x}(\varphi(\tau_k), 0) f(\tau_k, \varphi(\tau_k), 0) \neq 0.$$

4. *There exists a $\delta > 0$ such that, for any $\epsilon \in (-\delta, \delta)$ and $x_0 \in \mathbb{R}^n$, $|x_0 - \varphi(0)| < \delta$, the solution $x(t; x_0, \epsilon)$ of equation (7.1) is defined for $t \in [0, T]$, and for the function $x(t; x_0, \epsilon)$ Theorems 2.9 and 2.10 describing continuity on (t, x_0, ϵ) and differentiability on x_0 are valid in some neighbourhood of the point $(T, \varphi(0), 0)$.*
5. *The variational equation (7.2) has no non-trivial T-periodic solutions.*

Then there exists $\epsilon_0 \in (0, \bar{\epsilon})$ such that for $|\epsilon| \leq \epsilon_0$ equation (7.1) has a unique T-periodic solution $x_\epsilon(t)$ with moments of the impulse effect $t_k(\epsilon)$ such that for $t \in [0, T]$

$$x_\epsilon(t) \overset{B}{\to} \varphi(t) \qquad \text{and} \qquad t_k(\epsilon) \to \tau_k \qquad \text{as} \qquad \epsilon \to 0. \tag{7.3}$$

Proof   Since $\tau_0(\varphi(0), 0) < 0 < \tau_1(\varphi(0), 0)$ and the solution $\varphi(t)$ is T-periodic, then $\tau_q(\varphi(T), 0) < T < \tau_{q+1}(\varphi(T), 0)$. Hence by Theorem 2.9 on the continuity of the solution $x(t; x_0, \epsilon)$ it follows that there exists $\delta_1 \in (0, \delta)$ such that for $|x_0 - \varphi(0)| < \delta_1$ and $|\epsilon| < \delta_1$ we have

$$\tau_0(x_0, \epsilon) < 0 < \tau_1(x_0, \epsilon),$$
$$\tau_q(x(T; x_0, \epsilon), \epsilon) < 0 < \tau_1(x(T; x_0, \epsilon), \epsilon).$$

In the domain $D = \{(x_0, \epsilon) \in \mathbb{R}^n \times J : |x_0 - \varphi(0)| < \delta_1, |\epsilon| < \delta_1\}$ define the function

$$\psi(x_0, \epsilon) = x_0 - x(T; x_0, \epsilon).$$

The solution $x(t; x_0, \epsilon)$ of equation (7.1) is T-periodic if and only if $x(T; x_0, \epsilon) = x_0$, that is, if

$$\psi(x_0, \epsilon) = 0. \tag{7.4}$$

Obviously, $\psi(\varphi(0), 0) = 0$ and by condition 4 of the theorem the function $\psi(x_0, \epsilon)$ is continuous on (x_0, ϵ) and differentiable on x_0 in the domain D. Moreover,

$$\frac{\partial \psi}{\partial x_0}(\varphi(0), 0) = E - \frac{\partial x}{\partial x_0}(T; \varphi(0), 0).$$

But since the function $z(t) = \frac{\partial x}{\partial x_0}(t; \varphi(0), 0)$ satisfies equation (7.2) with initial value $z(0) = E$, then $z(T) = \frac{\partial x}{\partial x_0}(T; \varphi(0), 0)$ is a monodromy matrix of the T-periodic equation (7.2). By condition 5 the matrix $E - \frac{\partial \psi}{\partial x_0}(\varphi(0), 0) = E - z(T)$ is non-singular and by the implicit function theorem there exists $\epsilon_0 \in (0, \bar{\epsilon}) \cap (0, \delta_1)$ such that for any $\epsilon \in (-\epsilon_0, \epsilon_0)$ equation (7.4) has a unique continuous solution $x_0 = x_0(\epsilon)$ such that $x_0(0) = \varphi(0)$. Then to the initial value $x_0(\epsilon)$ there corresponds a T-periodic solution $x_\epsilon(t) = x(t; x_0(\epsilon), \epsilon)$ of equation (7.1) which, by Theorem 2.9 on continuity, satisfies the limit relations (7.3). $\square$

Note [82] that if the variational equation (7.2) is exponentially stable then the T-periodic solution $\varphi(t)$ of equation (7.1) is also exponentially stable. In the more general case, when equation (7.2) has an exponential dichotomy then the solution $\varphi(t)$ is conditionally stable. We shall formulate more precisely this result for the equation

$$\begin{aligned}
\frac{dx}{dt} &= f(t, x), \quad t \neq \tau_k(x), \\
\Delta x &= I_k(x), \quad t = \tau_k(x)
\end{aligned} \tag{7.5}$$

where $\tau_k(x) < \tau_{k+1}(x)$ $(k \in \mathbb{Z}, x \in \mathbb{R}^n)$.

Define the sets

$$\begin{aligned}
\sigma_k &= \{(t, x) \in \mathbb{R} \times \mathbb{R}^n : t = \tau_k(x)\} & (k \in \mathbb{Z}), \\
G_k &= \{(t, x) \in \mathbb{R} \times \mathbb{R}^n : \tau_{k-1}(x) < t \leq \tau_k(x)\} & (k \in \mathbb{Z})
\end{aligned}$$

and introduce the following conditions.

H7.1 Equation (7.5) is T-periodic, that is, there exists a $q \in \mathbb{N}$ such that

$$\begin{aligned}
f(t + T, x) &= f(t, x) & (t \in \mathbb{R}, x \in \mathbb{R}^n), \\
I_{k+q}(x) &= I_k(x) & (k \in \mathbb{Z}, x \in \mathbb{R}^n), \\
\tau_{k+q}(x) &= \tau_k(x) + T & (k \in \mathbb{Z}, x \in \mathbb{R}^n).
\end{aligned}$$

H7.2 Equation (7.5) has a T-periodic solution $x = \varphi(t)$ with moments of the impulse effect τ_k.

H7.3 For $k = 1, \ldots, q$ the functions $I_k(x)$ and $\tau_k(x)$ are differentiable in some neighbourhood of the point $\varphi(\tau_k)$ and

$$1 - \frac{\partial \tau_k}{\partial x}(\varphi(\tau_k)) f(\tau_k, \varphi(\tau_k)) \neq 0.$$

H7.4 In the set G_k $(k = 1, \ldots, q)$ the function $f(t, x)$ coincides with a function $f_k(t, x)$ which is differentiable in some ρ-neighbourhood of the set

$$\Gamma_k = \{(t, x) \in \mathbb{R} \times \mathbb{R}^n : x = \varphi(t), \tau_{k-1} < t \leq \tau_k\}.$$

H7.5 There exists a $\delta > 0$ such that, for each $x_0 \in \mathbb{R}^n$, $|x_0 - \varphi(0)| < \delta$ and $t \in [0, T]$, the solution $x(t; x_0)$ of equation (7.5) is defined for which $x(0; x_0) = x_0$. Moreover, for $t \in [0, T]$ the point $(t, x(t; x_0))$ meets each hypersurface σ_k $(k = 1, \ldots, q)$ just once.

Let the following variational equation correspond to the solution $x = \varphi(t)$:

$$\begin{aligned}
\frac{dz}{dt} &= \frac{\partial f}{\partial x}(t, \varphi(t))z, \quad t \neq \tau_k, \\
\Delta z &= L_k z, \qquad\qquad t = \tau_k
\end{aligned} \tag{7.6}$$

where

$$L_k = \frac{\partial I_k}{\partial x} + \left[\frac{\partial I_k}{\partial x} f_k + f_k - f_k^+\right] \frac{\dfrac{\partial \tau_k}{\partial x}}{1 - \dfrac{\partial \tau_k}{\partial x} f_k} \tag{7.7}$$

and

$$\begin{aligned}
f_k &= f_k(\tau_k, \varphi(\tau_k)), \quad f_k^+ = f_{k+1}(\tau_k, \varphi(\tau_k^+)), \\
\frac{\partial I_k}{\partial x} &= \frac{\partial I_k}{\partial x}(\varphi(\tau_k)), \quad \frac{\partial \tau_k}{\partial x} = \frac{\partial \tau_k}{\partial x}(\varphi(\tau_k)).
\end{aligned}$$

Theorem 7.2 *Let conditions H7.1–H7.5 hold and let the variational equation (7.6) have an exponential dichotomy. Then there exist real numbers $\delta_0 > 0$, $K \geq 1$, $\alpha > 0$ and an integer m $(0 \leq m \leq n)$ such that:*

1. *in a δ_0-neighbourhood of the point $\varphi(0)$ an m-dimensional manifold S_m^+ is defined and $\varphi(0) \in S_m^+$;*
2. *for each $x_0 \in S_m^+$, $|x_0 - \varphi(0)| < \delta_0$ the solution $x(t; x_0) = x(t)$ of equation (7.5) is defined for $t \in \mathbb{R}_+$ and the following relations are valid*

$$\begin{aligned}
|t_k - \tau_k| &\leq K e^{-\alpha \tau_k} = \delta_k \quad (k \in \mathbb{N}), \\
|x(t) - \varphi(t)| &\leq K e^{-\alpha t} \qquad (t \in \mathbb{R}_+, |t - \tau_k| \geq \delta_k),
\end{aligned} \tag{7.8}$$

where t_k are the moments of the impulse effect for $x(t)$.

Sketch of the proof The proof is similar to the proof of Theorem 6.8 with a modification taking account of the fact that $x(t)$ and $\varphi(t)$ have different moments of the impulse effect.

Let $X(t)$ be the fundamental matrix of equation (7.6) normalized at $t = 0$. Since (7.6) has an exponential dichotomy there exist constants $K_0 \geq 1$, $\alpha > 0$ and a projector P such that for the functions

$$X_1(t, s) = X(t)PX^{-1}(s),$$

$$X_2(t, s) = X(t)(E - P)X^{-1}(s)$$

the following estimates are valid

$$|X_1(t, s)| \leq K_0 e^{-2\alpha(t-s)} \quad (t \geq s),$$
$$|X_2(t, s)| \leq K_0 e^{-2\alpha(s-t)} \quad (s \geq t). \tag{7.9}$$

Define the function

$$\tilde{f}(t, x) = f_k(t, x) \quad (\tau_{k-1} < t \leq \tau_k, x \in \mathbb{R}^n, |x - \varphi(t)| < \rho)$$

and denote by $u_k(t; t_0, x_0)$ the solution of the initial value problem

$$\frac{du}{dt} = f_k(t, u), \quad u(t_0) = x_0.$$

Let $m = \operatorname{rank} P$ and $L_m = P(\mathbb{R}^n)$ be the range of the projector P. For each $a \in L_m$ with $|a| < \delta$ construct the sequences

$$y_\nu(t) = y_\nu(t, a), \quad t_k^\nu = t_k^\nu(a) \quad (t \in \mathbb{R}_+, k \in \mathbb{N})$$

setting

$$y_0(t) = \varphi(t), \quad t_k^0 = \tau_k$$

after which successively define

$$y_{\nu+1}(t) = \varphi(t) + X(t)a + \int_0^t X_1(t, s)R(s, y_\nu(s)) \, ds - \int_t^\infty X_2(t, s)R(s, y_\nu(s)) \, ds$$

$$+ \sum_{0 \leq \tau_k < t} X_1(t, \tau_k^+)R_k(y_\nu(\tau_k)) - \sum_{t \leq \tau_k} X_2(t, \tau_k^+)R_k(y_\nu(\tau_k))$$

$$\tag{7.10}$$

where

$$R(t, y) = \tilde{f}(t, y) - \tilde{f}(t, \varphi(t)) - \frac{\partial \tilde{f}}{\partial x}(t, \varphi(t))(y - \varphi(t)),$$

$$R_k(y_\nu(\tau_k)) = I_k(u_k(t_k^\nu; \tau_k, y_\nu(\tau_k))) - I_k(\varphi(\tau_k)) + y_\nu(\tau_k^+) - y_\nu(\tau_k)$$

$$+ u_k(t_k^\nu; \tau_k, y_\nu(\tau_k)) - u_{k+1}(t_k^\nu; \tau_k, y_\nu(\tau_k^+))$$

$$- [y_\nu(\tau_k) - \varphi(\tau_k)].$$

and $t_k^{\nu+1}$ is obtained as a solution with respect to t of the equation

$$t = \tau_k(u_k(t; \tau_k, y_{\nu+1}(\tau_k))). \tag{7.11}$$

By Lemma 2.1 equation (7.11) has a unique solution $t = T_k(y_{\nu+1}(\tau_k))$ if $|y_{\nu+1}(\tau_k) - \varphi(\tau_k)| < \delta$ and $\delta > 0$ is small enough. Moreover, for the function $T_k(w)$ the following relations are valid

$$T_k(\varphi(\tau_k)) = \tau_k, \tag{7.12}$$

$$\begin{aligned}
\frac{\partial T_k}{\partial w}(w) = &\frac{\partial \tau_k}{\partial x}(u_k(T_k(w); \tau_k, w)) \\
&\times \left[f(T_k(w), u_k(T_k(w); \tau_k, w)) \frac{\partial T_k}{\partial w}(w) + \frac{\partial u_k}{\partial w}(T_k(w); \tau_k, w) \right],
\end{aligned} \tag{7.13}$$

$$|T_k(w) - T_k(v)| \leq L|w - v| \tag{7.14}$$

for $|w - \varphi(\tau_k)| < \delta$ and $|v - \varphi(\tau_k)| < \delta$, where $L > 0$ is a constant.

In view of (7.7), (7.12) and (7.13) we conclude that

$$|R(t, u) - R(t, v)| \leq L(\Delta)|u - v|, \tag{7.15}$$

$$|R_k(y_{\nu+1}(\tau_k)) - R_k(y_\nu(\tau_k))| \leq L(\Delta)|y_{\nu+1}(\tau_k) - y_\nu(\tau_k)| \tag{7.16}$$

if $|u - \varphi(t)| < \Delta$, $|v - \varphi(t)| < \Delta$, $|y_{\nu+1}(\tau_k) - \varphi(\tau_k)| < \Delta$, $|y_\nu(\tau_k) - \varphi(\tau_k)| < \Delta$ where $L(\Delta) > 0$ and $\lim_{\Delta \to 0+} L(\Delta) = 0$.

Taking into account (7.9), (7.15), (7.16) and (7.14) we prove inductively the inequalities

$$\begin{aligned}
|y_\nu(t) - y_{\nu-1}(t)| &\leq K_0|a|2^{-n}e^{-\alpha t}, \\
|y_\nu(t) - \varphi(t)| &\leq K_0|a|e^{-\alpha t}, \\
|t_k^\nu - t_k^{\nu-1}| &\leq K_0 L|a|2^{-n}e^{-\alpha \tau_k}, \\
|t_k^\nu - \tau_k| &\leq K_0 L|a|e^{-\alpha \tau_k},
\end{aligned} \tag{7.17}$$

for $t \in \mathbb{R}_+$, $k \in \mathbb{N}$, $\nu \in \mathbb{N}$.

From (7.17) it follows that the sequences $y_\nu(t, a)$ and $t_k^\nu(a)$ are convergent and

$$\lim_{\nu \to \infty} t_k^\nu(a) = t_k(a) \quad \text{(uniformly on } k \in \mathbb{N}),$$

$$\lim_{\nu \to \infty} y_\nu(t, a) = y(t, a) \quad \text{(uniformly on } t \in \mathbb{R}_+).$$

Passing to the limit in (7.10) and (7.11) we find that the function $y(t) = y(t, a)$ and the moments $t_k = t_k(a)$ satisfy the equations

$$y(t) = \varphi(t) + X(t)a + \int_0^t X_1(t, s)R(s, y(s)) \, ds - \int_t^\infty X_2(t, s)R(s, y(s)) \, ds$$

$$+ \sum_{0 \leq \tau_k < t} X_1(t, \tau_k^+)R_k(y(\tau_k)) - \sum_{t \leq \tau_k} X_2(t, \tau_k^+)R_k(y(\tau_k))$$

and

$$t_k = \tau_k(u_k(t_k; \tau_k, y(\tau_k))).$$

It is easy to verify that

$$\frac{dy}{dt} = \tilde{f}(t, y(t)), \quad t \neq \tau_k$$

and

$$u_{k+1}(t_k; \tau_k, y(\tau_k^+)) = u_k(t_k; \tau_k, y(\tau_k)) + I_k(u_k(t_k; \tau_k, y(\tau_k))).$$

Define the function $x(t, a)$ by the formula

$$x(t, a) = x(t) = \begin{cases} y(t, a) & \text{if } t \notin [\tau_k; t_k], \\ u_{k+1}(t; \tau_k, y(\tau_k^+)) & \text{if } t_k < t \leq \tau_k, \\ u_k(t; \tau_k, y(\tau_k)) & \text{if } \tau_k \leq t < t_k, \end{cases}$$

$$x(t_k) = x(t_k^-).$$

A straightforward verification shows that

$$t_k = \tau_k(x(t_k)),$$

$$\frac{dx}{dt}(t) = f(t, x(t)), \quad t \neq t_k,$$

$$\Delta x(t_k) = I_k(x(t_k)).$$

Consequently, $x(t, a)$ is a solution of equation (7.5). Moreover, from estimates (7.17) it follows that

$$|t_k(a) - \tau_k| \leq LK_0|a|e^{-\alpha\tau_k} = \delta_k \quad (k \in \mathbb{N})$$
$$|x(t, a) - \varphi(t)| \leq K_0|a|e^{-\alpha t} \qquad (t \in \mathbb{R}_+, |t - \tau_k| \geq \delta_k).$$

Finally, we show, as in the proof of Theorem 6.8, that the set

$$S_m^+ = \{x(0, a) : a \in L_m, |a| < \delta_0\}$$

is the m-dimensional manifold we seek. $\square$

Assertions analogous to the assertions of Theorems 7.1 and 7.2 can also be proved for the T-periodic equation

$$\begin{align} \frac{dx}{dt} &= f(t, x, \epsilon), \quad \phi(t, x, \epsilon) \neq 0, \\ \Delta x &= I(t, x, \epsilon), \quad \phi(t, x, \epsilon) = 0, \end{align} \tag{7.18}$$

where $\epsilon \in J = (-\bar{\epsilon}, \bar{\epsilon})$ is a small parameter and the equation $\phi(t, x, \epsilon) = 0$, for each $\epsilon \in J$, defines in $\mathbb{R} \times \mathbb{R}^n$ a hypersurface $\sigma(\epsilon)$.

For $\epsilon = 0$ let equation (7.18) have a T-periodic solution $x = \varphi(t)$ with moments of an impulse effect τ_k, and let the corresponding variational equation be

$$\frac{dz}{dt} = \frac{\partial f}{\partial x}(t, \varphi(t), 0)z, \quad t \neq \tau_k,$$

$$\Delta z = M_k z, \qquad\qquad t = \tau_k,$$

(7.19)

where

$$M_k = \frac{\partial I}{\partial x} + \left[f^+ - f - \frac{\partial I}{\partial t} - \frac{\partial I}{\partial x} f \right] \frac{\frac{\partial \phi}{\partial x}}{\frac{\partial \phi}{\partial x} f + \frac{\partial \phi}{\partial t}}$$

and

$$f = f(\tau_k, \varphi(\tau_k), 0), \qquad f^+ = f(\tau_k^+, \varphi(\tau_k^+), 0),$$

$$\frac{\partial I}{\partial x} = \frac{\partial I}{\partial x}(\tau_k, \varphi(\tau_k), 0), \qquad \frac{\partial I}{\partial t} = \frac{\partial I}{\partial t}(\tau_k, \varphi(\tau_k), 0),$$

$$\frac{\partial \phi}{\partial x} = \frac{\partial \phi}{\partial x}(\tau_k, \varphi(\tau_k), 0), \qquad \frac{\partial \phi}{\partial t} = \frac{\partial \phi}{\partial t}(\tau_k, \varphi(\tau_k), 0).$$

Denote by $x(t; x_0, \epsilon)$ the solution of equation (7.18) for which $x(0; x_0, \epsilon) = x_0$, and without loss of generality assume $\phi(0, \varphi(0), 0) \neq 0$.

The following theorems are valid.

Theorem 7.3   *Let the following conditions hold.*

1. *Equation (7.18) is T-periodic, that is,*

$$f(t+T, x, \epsilon) = f(t, x, \epsilon), \qquad I(t+T, x, \epsilon) = I(t, x, \epsilon), \qquad \phi(t+T, x, \epsilon) = \phi(t, x, \epsilon)$$

for $t \in \mathbb{R}$, $x \in \mathbb{R}^n$, $\epsilon \in J$.

2. *For $\epsilon = 0$ equation (7.18) has a T-periodic solution $x = \varphi(t)$ with moments of an impulse effect τ_k, and $\tau_{k+q} = \tau_k + T$ $(k \in \mathbb{Z})$ for some $q \in \mathbb{N}$.*

3. *For each $k = 1, \ldots, q$ the function $\phi(t, x, \epsilon)$ is differentiable in some neighbourhood of the point $(\tau_k, \varphi(\tau_k), 0)$ and*

$$\frac{\partial \phi}{\partial t}(\tau_k, \varphi(\tau_k), 0) + \frac{\partial \phi}{\partial x}(\tau_k, \varphi(\tau_k), 0)f(\tau_k, \varphi(\tau_k), 0) \neq 0.$$

4. *There exists a $\delta > 0$ such that, for each $\epsilon \in (-\delta, \delta)$ and $x_0 \in \mathbb{R}^n$, $|x_0 - \varphi(0)| < \delta$, the solution $x(t; x_0, \epsilon)$ of equation (7.18) is defined for $t \in [0, T]$, and for the function $x(t; x_0, \epsilon)$ Theorems 2.9 and 2.10 describing continuity on (t, x_0, ϵ) and differentiability on x_0 are valid in some neighbourhood of the point $(T, \varphi(0), 0)$.*

5. *The variational equation (7.19) has no non-trivial T-periodic solutions.*

Then there exists $\epsilon_0 \in (0, \bar{\epsilon})$ such that for $|\epsilon| \leq \epsilon_0$ equation (7.18) has a unique T-periodic solution $x_\epsilon(t)$ with moments of the impulse effect $t_k(\epsilon)$ such that for $t \in [0, T]$

$$x_\epsilon(t) \xrightarrow{B} \varphi(t) \qquad \text{and} \qquad t_k(\epsilon) \to \tau_k \qquad \text{as} \qquad \epsilon \to 0.$$

Theorem 7.4   *For equation (7.18) with $\epsilon = 0$ let conditions 1–4 of Theorem 7.3 hold, and let the variational equation (7.19) have exponential dichotomy. Then there exist real numbers $\delta_0 > 0$, $K \geq 1$, $\alpha > 0$ and an integer m $(0 \leq m \leq n)$ such that:*

1. *in a δ_0-neighbourhood of the point $\varphi(0)$ an m-dimensional manifold S_m^+ is defined and $\varphi(0) \in S_m^+$;*
2. *for each $x_0 \in S_m^+$, $|x_0 - \varphi(0)| < \delta_0$, the solution $x(t) = x(t; x_0, 0)$ of equation (7.18) with $\epsilon = 0$ is defined for $t \in \mathbb{R}_+$ and the following relations are valid*

$$|t_k - \tau_k| \leq Ke^{-\alpha\tau_k} = \delta_k \quad (k \in \mathbb{N}),$$
$$|x(t) - \varphi(t)| \leq Ke^{-\alpha t} \qquad (t \in \mathbb{R}_+, |t_k - \tau_k| \geq \delta_k)$$

where t_k are the moments of the impulse effect for $x(t)$.

Example 7.1   Consider the impulsive equation

$$\dot{x} = y, \qquad \qquad \qquad \text{for } x \geq 0, \qquad (7.20)$$
$$\dot{y} = -p + a\cos(\omega t + \varphi),$$
$$\Delta x = 0, \qquad \qquad \qquad \text{for } x = 0 \qquad (7.21)$$
$$\Delta y = -(1 + R)y,$$

where $a > 0$, $p > 0$, $\omega > 0$, $R \in (0,1)$ and φ are constants.

System (7.20), (7.21) describes the motion of a material body of mass $m = 1$ under the action of the gravitational force p and a periodically changing force $F = a\cos(\omega t + \varphi)$ (Fig. 7.1). Under an impact of the body against the obstacle S (for $x = 0$) the velocity $y = \dot{x}$ changes by a jump with a coefficient of regeneration $R \in (0,1)$.

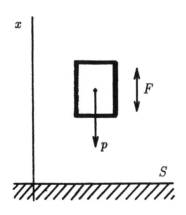

Fig. 7.1

Let $x(t)$, $y(t)$ be a T-periodic solution of system (7.20), (7.21) with one impulse per period and

$$x(0+) = 0, \qquad y(0+) = y_0 > 0.$$

From (7.20) it follows that for $t \in (0, T]$ we have

$$x(t) = -\frac{pt^2}{2} + \left[y_0 - \frac{a}{\omega} \sin \varphi \right] t + \frac{a}{\omega^2} \cos \varphi - \frac{a}{\omega^2} \cos(\omega t + \varphi),$$

$$y(t) = -pt + y_0 - \frac{a}{\omega} \sin \varphi + \frac{a}{\omega} \sin(\omega t + \varphi). \tag{7.22}$$

Since the right-hand sides of equation (7.20) are periodic functions of t with period $2\pi/\omega$, the possible period T is equal to

$$T = \frac{2\pi}{\omega} \nu \quad (\nu \in \mathbb{N}).$$

Thus from (7.22) it follows that, for $t = T$,

$$0 = \left[y_0 - \frac{a}{\omega} \sin \varphi \right] T - \frac{pT^2}{2}, \qquad y(T) = y_0 - pT,$$

and in view of the jump condition $y_0 = -Ry(T)$ we obtain

$$y(T) = -\frac{pT}{1 + R},$$

$$y_0 = \frac{pTR}{1 + R}, \tag{7.23}$$

$$a \sin \varphi = -\frac{p\omega T(1 - R)}{2(1 + R)} \equiv -pB. \tag{7.24}$$

Consequently, if condition (7.24) for the existence of a T-periodic solution is met, then we can determine y_0 from (7.23) and obtain the T-periodic solution of (7.20), (7.21) as a periodic extension of the functions of (7.22).

System (7.20), (7.21) is of the form (1.12) so in order to find the system in variations corresponding to this periodic solution we use formula (2.25). We obtain the system

$$\frac{dz}{dt} = Az, \quad t \neq nT,$$

$$\Delta z = Nz, \quad t = nT, \tag{7.25}$$

where

$$z = \begin{bmatrix} u \\ v \end{bmatrix}, \qquad A = \begin{bmatrix} 0 & 1 \\ 0 & 0 \end{bmatrix}, \qquad N = \begin{bmatrix} -(1 + R) & 0 \\ \dfrac{(1 + R)(-p + a \cos \varphi)}{y(T)} & -(1 + R) \end{bmatrix}.$$

Since $e^{AT} = E + AT$, then the monodromy matrix $M = (E + N)e^{AT}$ has the form

$$M = \begin{bmatrix} -R & -RT \\ b & bT - R \end{bmatrix}$$

where

$$b = \frac{(1 + R)^2(p - a\cos\varphi)}{pT}.$$

Taking into account that $\det M = R^2$ and $\operatorname{Tr} M = -2R + bT$ we obtain the equation for the multipliers

$$\mu^2 + (2R - bT)\mu + R^2 = 0. \tag{7.26}$$

The roots μ_i $(i = 1, 2)$ of equation (7.26) satisfy the condition

$$|\mu_i| < 1 \quad (i = 1, 2) \tag{7.27}$$

if and only if

$$|bT - 2R| < 1 + R^2,$$

or if

$$0 < a\cos\varphi < pC \tag{7.28}$$

where

$$C = 1 + \left(\frac{1 - R}{1 + R}\right)^2.$$

After equivalent transformations of (7.28) making use of (7.24) we conclude that condition (7.27) is met if and only if

$$\frac{a}{\sqrt{B^2 + C^2}} < p < \frac{a}{B},$$

that is

$$a\left[\frac{\omega^2 T^2(1 - R)^2}{4(1 + R)^2} + \left\{1 + \left(\frac{1 - R}{1 + R}\right)^2\right\}^2\right]^{-1/2} < p < \frac{2a(1 + R)}{\omega T(1 - R)}. \tag{7.29}$$

Thus, if conditions (7.24) and (7.29) hold, then system (7.20), (7.21) has a unique periodic solution with period $T = (2\pi/\omega)\nu$ $(\nu \in \mathbb{N})$ and this solution is asymptotically stable.

Finally we note that the multipliers of system (7.25) are distinct from 1 if

$$a \neq \frac{p\omega T(1 - R)}{2(1 + R)}.$$

Example 7.2

Let a ball B of mass m fall freely under the action of its weight with acceleration g. Suppose that during its fall it meets a platform S of mass $M \gg m$ which performs periodic oscillations with amplitude a and frequency ω by the law $x = a \sin \omega t$ (Fig. 7.2). Let the regeneration coefficient under the impact be $R \in (0,1)$ and suppose that the platform S does not change its velocity under the impact, since $M \gg m$. Then the motion of this mechanical system is given by the equations

$$\dot{x} = y,$$
$$\dot{y} = -g, \qquad \qquad \text{if } \phi(t,x) > 0, \qquad (7.30)$$
$$\Delta x = 0,$$
$$\Delta y = -(1+R)y + (1+R)a\omega \cos \omega t, \qquad \text{if } \phi(t,x) = 0 \qquad (7.31)$$

where $\phi(t,x) = x - a \sin \omega t$.

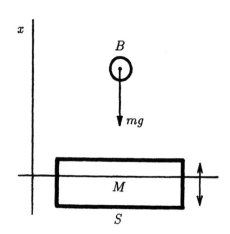

Fig. 7.2

Let $x(t)$, $y(t)$ be a periodic solution of system (7.30), (7.31) with period $T = (2\pi/\omega)\nu$ $(\nu \in \mathbb{N})$ (Fig. 7.3) such that

$$x_T = x(t_0 + T), \quad y_T = y(t_0 + T),$$
$$x_0 = x(t_0^+), \qquad y_0 = y(t_0^+) > 0,$$

$$\phi(t_0, x_0) = x_0 - a \sin \omega t_0 = 0. \qquad (7.32)$$

Then for $t \in (t_0, t_0 + T]$ we have

$$x = -\tfrac{1}{2}g(t - t_0)^2 + y_0(t - t_0) + x_0,$$
$$y = -g(t - t_0) + y_0.$$

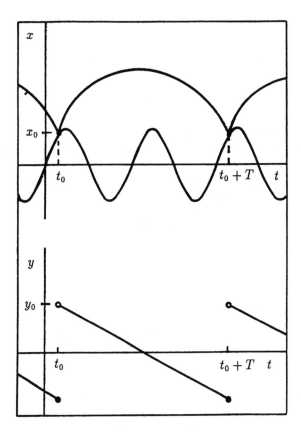

Fig. 7.3

In particular, for $t = t_0 + T$ we obtain

$$x_0 = x_T = -\tfrac{1}{2}gT^2 + y_0 T + x_0,$$
$$y_T = -gT + y_0,$$

whence it follows that

$$y_0 = \tfrac{1}{2}gT, \qquad y_T = -\tfrac{1}{2}gT.$$

In view of (7.31) and the equality $y(t_0 + T+) = y_0$ we find that

$$y_0 - y_T = -(1 + R)y_T + (1 + R)a\omega \cos \omega t_0$$

which implies the condition for the existence of a T-periodic solution

$$\cos \omega t_0 = \frac{B}{\omega^2} \tag{7.33}$$

where

$$B = \frac{\omega g T (1 - R)}{2a(1 + R)}.$$

Then from (7.32) we find x_0:

$$x_0^\pm = \pm a \sqrt{1 - \frac{B^2}{\omega^4}}.$$

Consequently, provided that condition (7.33) holds, system (7.30), (7.31) has two T-periodic solutions if $\omega^2 > B$, one T-periodic solution if $\omega^2 = B$, and no T-periodic solutions if $B > \omega^2$.

Making use of formula (2.25) we obtain the system in variations corresponding to the T-periodic solution of (7.30), (7.31):

$$\frac{dz}{dt} = Az, \quad t \neq nT,$$
$$\Delta z = Nz, \quad t = nT \tag{7.34}$$

where

$$z = \begin{bmatrix} u \\ v \end{bmatrix}, \qquad A = \begin{bmatrix} 0 & 1 \\ 0 & 0 \end{bmatrix}, \qquad N = \begin{bmatrix} -1 - R & 0 \\ \frac{\delta}{c} & -1 - R \end{bmatrix}$$

$$\delta = (1 + R)(a\omega^2 \sin \omega t_0 - g),$$

$$c = -\frac{gT}{1 + R}.$$

Then the monodromy matrix $M = (E + N)e^{AT}$ equals

$$M = \begin{bmatrix} -R + \frac{\delta T}{c} & -RT \\ \frac{\delta}{c} & -R \end{bmatrix}$$

and the equation for the multipliers of (7.34) is

$$\mu^2 + \left(2R - \frac{\delta T}{c}\right)\mu + R^2 = 0.$$

The condition $|\mu_i| < 1$ $(i = 1, 2)$ is equivalent to the inequalities

$$-(1 - R)^2 < \frac{\delta T}{c} < (1 + R)^2$$

which after equivalent transformations take the form

$$\sin \omega t_0 > 0, \tag{7.35}$$

$$\omega^2 \sin \omega t_0 < \frac{g}{a}\left[1 + \left(\frac{1 - R}{1 + R}\right)^2\right] \equiv C. \tag{7.36}$$

Condition (7.35) is met only if $x_0 > 0$, that is, if $x_0 = x_0^+$ in the case when $\omega^2 > B$. In this case condition (7.36) takes the form

$$\omega^2 \sqrt{1 - \cos^2 \omega t_0} < C$$

whence it follows that $\omega^2 < \sqrt{B^2 + C^2}$. Thus if $B < \omega^2 < \sqrt{B^2 + C^2}$, that is, if

$$\frac{\omega g T(1-R)}{2a(1+R)} < \omega^2 < \sqrt{\frac{\omega^2 g^2 T^2 (1-R)^2}{4a^2(1+R)^2} + \frac{g^2}{a^2}\left[1 + \left(\frac{1-R}{1+R}\right)^2\right]^2},$$

then system (7.30), (7.31) has a unique periodic solution with period $T = (2\pi/\omega)\nu$ ($\nu \in \mathbb{N}$) which is exponentially stable.

8. Non-linear autonomous equations

Consider the autonomous impulsive equation

$$
\begin{aligned}
\frac{dx}{dt} &= f(x, \epsilon), \quad x \notin \sigma(\epsilon), \\
\Delta x &= I(x, \epsilon), \quad x \in \sigma(\epsilon)
\end{aligned}
\tag{8.1}
$$

where $\epsilon \in J = (-\bar{\epsilon}, \bar{\epsilon})$ is a small parameter and for each $\epsilon \in J$ the set $\sigma(\epsilon)$ is a hypersurface in $\mathbb{R}^n$.

Suppose that $\sigma(\epsilon)$ consists of q non-intersecting smooth hypersurfaces $\sigma_k(\epsilon)$ which are given by the equations $\phi_k(x, \epsilon) = 0$ $(k = 1, \ldots, q)$. For $\epsilon = 0$ let equation (8.1) have a T_0-periodic solution $x = \varphi(t)$ with moments of an impulse effect τ_k and

$$
\begin{aligned}
\tau_{k+q} &= \tau_k + T_0 \quad (k \in \mathbb{Z}), \\
\phi_k(\varphi(\tau_k)) &= 0 \qquad (k = 1, \ldots, q).
\end{aligned}
$$

Associate with the solution $\varphi(t)$ the variational equation

$$
\begin{aligned}
\frac{dz}{dt} &= \frac{\partial f}{\partial x}(\varphi(t), 0)z, \quad t \neq \tau_k, \\
\Delta z &= N_k z, \qquad\qquad t = \tau_k
\end{aligned}
\tag{8.2}
$$

where

$$N_k = \frac{\partial I}{\partial x} + \left[f^+ - f - \frac{\partial I}{\partial x}f\right]\frac{\frac{\partial \phi}{\partial x}}{\frac{\partial \phi}{\partial x}f}$$

and

$$
\begin{aligned}
f &= f(\varphi(\tau_k), 0), \qquad f^+ = f(\varphi(\tau_k^+), 0), \\
\frac{\partial I}{\partial x} &= \frac{\partial I}{\partial x}(\varphi(\tau_k), 0), \quad \frac{\partial \phi}{\partial x} = \frac{\partial \phi}{\partial x}(\varphi(\tau_k), 0).
\end{aligned}
$$

A peculiarity of the autonomous equation (8.1) is that the derivative $\varphi'(t)$ of the solution $\varphi(t)$ is a solution of the variational equation (8.2). Indeed, for $t \neq \tau_k$ we have that

$$[\varphi'(t)]' = [f(\varphi(t), 0)]' = \frac{\partial f}{\partial x}(\varphi(t), 0)\varphi'(t)$$

and for $t = \tau_k$ we successively obtain

$$\frac{\partial \phi_k}{\partial x}(\varphi(\tau_k), 0)\varphi'(\tau_k) = \frac{\partial \phi_k}{\partial x}(\varphi(\tau_k), 0)f(\varphi(\tau_k), 0),$$

$$N_k \varphi'(\tau_k) = f(\varphi(\tau_k^+), 0) - f(\varphi(\tau_k), 0),$$

$$\Delta\varphi'(\tau_k) = \varphi'(\tau_k^+) - \varphi'(\tau_k) = f(\varphi(\tau_k^+), 0) - f(\varphi(\tau_k), 0) = N_k \varphi'(\tau_k).$$

If $\varphi'(t) \not\equiv 0$, then $\varphi'(t)$ is a non-zero T_0-periodic solution of (8.2) and then equation (8.2) has a multiplier $\mu_1 = 1$.

We shall consider the non-critical case when equation (8.2) has no non-trivial T_0-periodic solutions other than $\varphi'(t)$, that is, when just one of the multipliers of equation (8.2) equals 1. Without loss of generality we assume that $\varphi(0) \notin \sigma(0)$ and denote by $x(t; x_0, \epsilon)$ the solution of equation (8.1) for which $x(0; x_0, \epsilon) = x_0$. Before we show that for small ϵ equation (8.1) has a periodic solution $x_\epsilon(t)$ which for $\epsilon \to 0$ tends to the solution $x_0(t) = \varphi(t)$, we shall offer some heuristic arguments related to the geometric interpretation of the problem.

Let S be a hyperplane in $\mathbb{R}^n$ with normal vector ν, intersecting the trajectory of the solution $\varphi(t)$ at the point $\varphi(0)$ and $(\varphi'(0)|\nu) \neq 0$. Denote by S_0 the disc $\{x \in S : |x - \varphi(0)| < \delta\}$, where $\delta > 0$ is chosen small enough so that S_0 intersects the trajectory of $\varphi(t)$ only at the point $\varphi(0)$.

The solution $x(t)$ of equation (8.1) for small ϵ with initial value $x(0) \in S_0$ remains 'close' to $\varphi(t)$ for t changing in the finite interval $[0, 2T_0]$ if $x(0)$ is close to $\varphi(0)$. Since the point $\varphi(t)$ for $t = T_0$ again meets S_0 $(\varphi(T_0) = \varphi(0))$, then at some moment $t = T$ (close to T_0) the point $x(t)$ will again meet S_0 at a point $x(T)$ which is close to $\varphi(0)$, and the meeting will also be without tangency: $(x'(T)|\nu) \neq 0$. Moreover, the solution $x(t)$ will be periodic if $x(T) = x(0)$. Since equation (8.1) is autonomous the trajectory of the periodic solution $x(t)$ is determined uniquely by the initial point $x(0) \in S_0$.

Thus in order to find a periodic solution of (8.1) for small ϵ we have to determine the initial value $x(0)$ satisfying the equation of the hyperplane S, and the period T satisfying the periodicity condition $x(T) = x(0)$, that is, we have $n + 1$ equations with $n + 1$ unknowns. Since we can choose the coordinate system in $\mathbb{R}^n$ so that the equation of S is $x_1 \equiv 0$, the problem can be reduced to finding the period T and $n - 1$ coordinates $x_i(0)$ $(i = 2, \ldots, n)$ of the initial value $x(0)$.

Theorem 8.1   *Let the following conditions hold.*

1. *For $\epsilon = 0$ equation (8.1) has a T_0-periodic solution $x = \varphi(t)$ with moments of an impulse effect $\tau_k : \tau_{k+q} = \tau_k + T_0$ $(k \in \mathbb{Z})$ and $\varphi'(t) \not\equiv 0$ $(t \in \mathbb{R})$.*
2. *For each $k = 1, \ldots, q$ the function $\phi_k(x, \epsilon)$ is differentiable in some neighbourhood of the point $(\varphi(\tau_k), 0)$ and*

$$\phi_k(\varphi(\tau_k), 0) = 0, \tag{8.3}$$

$$\frac{\partial \phi_k}{\partial x}(\varphi(\tau_k), 0)f(\varphi(\tau_k), 0) \neq 0. \tag{8.4}$$

3. There exists a $\delta > 0$ such that, for each $\epsilon \in (-\delta, \delta)$ and $x_0 \in \mathbb{R}^n$, $|x_0 - \varphi(0)| < \delta$, the solution $x(t; x_0, \epsilon)$ of equation (8.1) is defined for $t \in [0, T_0 + \delta]$ and for the function $x(t; x_0, \epsilon)$ Theorems 2.9 and 2.10 describing continuity on (t, x_0, ϵ) and differentiability on x_0 are valid in some neighbourhood of the point $(T_0, \varphi(0), 0)$.
4. The variational equation (8.2) has no non-trivial T_0-periodic solution other than $\varphi'(t)$.

Then there exists $\epsilon_0 \in (0, \bar{\epsilon})$ such that for $|\epsilon| \leq \epsilon_0$ equation (8.1) has a unique periodic solution $x_\epsilon(t)$ with period $T(\epsilon)$ and moments of the impulse effect $t_k(\epsilon)$ such that for $t \in [0, T_0]$

$$x_\epsilon(t) \xrightarrow{B} \varphi(t), \qquad T(\epsilon) \to T_0 \qquad \text{and} \qquad t_k(\epsilon) \to \tau_k \qquad \text{as } \epsilon \to 0. \qquad (8.5)$$

Proof Since $\varphi'(t) \not\equiv 0$ $(t \in \mathbb{R})$ and equation (8.1) is autonomous, then without loss of generality we assume that $\varphi'(0) \neq 0$. Moreover, we can assume that the following conditions hold

$$\varphi(0) = 0, \qquad \varphi'(0) = e_1 = \text{col}\,(1, 0, \dots, 0), \qquad (8.6)$$

since otherwise we could change the variables $x = Ay + \varphi(0)$ (with $A \in \mathbb{R}^{n \times n}$, $\det A \neq 0$, $A^{-1}\varphi'(0) = e_1$) and obtain a new autonomous system with a periodic solution which would satisfy the conditions of Theorem 8.1 and condition (8.6).

Consider the solution $x(t; a, \epsilon)$ of (8.1) for

$$0 \leq t \leq T_0 + \delta, \qquad |a| < \delta, \qquad |\epsilon| < \delta.$$

By Theorem 2.10 on differentiability with respect to the initial values the function $\frac{\partial x}{\partial a_1}(t; 0, 0)$ is a solution of the variational equation (8.2) and

$$\frac{\partial x}{\partial a_1}(0; 0, 0) = e_1.$$

But since $\varphi'(t)$ is a solution of (8.2) and satisfies the same initial condition $\varphi'(0) = e_1$, we have

$$\frac{\partial x}{\partial a_1}(t; 0, 0) \equiv \varphi'(t),$$

and

$$\frac{\partial x}{\partial a_1}(T_0; 0, 0) = \varphi'(T_0) = \varphi'(0) = e_1.$$

Since $\frac{\partial x}{\partial a}(T_0; 0, 0)$ is a monodromy matrix for equation (8.2) the equation for the multipliers of (8.2) is $\det\left(\frac{\partial x}{\partial a}(T_0; 0, 0) - \mu E\right) = 0$ or, in more detail,

$$\det \begin{bmatrix} 1-\mu & \frac{\partial x_1}{\partial a_2} & \cdots & \frac{\partial x_1}{\partial a_n} \\ 0 & \frac{\partial x_2}{\partial a_2} - \mu & \cdots & \frac{\partial x_2}{\partial a_n} \\ \vdots & \vdots & & \vdots \\ 0 & \frac{\partial x_n}{\partial a_2} & \cdots & \frac{\partial x_n}{\partial a_n} - \mu \end{bmatrix} = (1-\mu)A_{11}(\mu) = 0$$

where $\dfrac{\partial x_i}{\partial a_j} = \dfrac{\partial x_i}{\partial a_j}(T_0; 0, 0)$ and $A_{11}(\mu)$ is the cofactor of the entry $1-\mu$. Consequently,

$$A_{11}(1) \neq 0$$

since the multiplier $\mu = 1$ of equation (8.2) is simple.

Now choose the initial value a of the solution $x(t; a, \epsilon)$ on the hyperplane $a_1 \equiv 0$. The solution $x(t; a, \epsilon)$ is T-periodic if and only if

$$\psi(T, a, \epsilon) = x(T; a, \epsilon) - a = 0. \tag{8.7}$$

Obviously, $\psi(T_0, 0, 0) = 0$. Since $\dfrac{\partial x}{\partial T}(T_0; 0, 0) = \varphi'(T_0) = e_1$, the Jacobi matrix of the function ψ with respect to the variables $\beta = (T, a_2, \ldots, a_n)$ at the point $(T_0, 0, 0)$ is equal to

$$\frac{\partial \psi}{\partial \beta} = \begin{bmatrix} 1 & \dfrac{\partial x_1}{\partial a_2} & \cdots & \dfrac{\partial x_1}{\partial a_n} \\ 0 & \dfrac{\partial x_2}{\partial a_2} - 1 & \cdots & \dfrac{\partial x_2}{\partial a_n} \\ \vdots & \vdots & & \vdots \\ 0 & \dfrac{\partial x_n}{\partial a_2} & \cdots & \dfrac{\partial x_n}{\partial a_n} - 1 \end{bmatrix}.$$

Consequently, $\det \dfrac{\partial \psi}{\partial \beta} = 1 \cdot A_{11}(1) \neq 0$ and by the implicit function theorem there exists $\epsilon_0 \in (0, \bar{\epsilon})$ such that for $|\epsilon| \leq \epsilon_0$ equation (8.7) determines uniquely the continuous functions $T(\epsilon), a_2(\epsilon), \ldots, a_n(\epsilon)$ for which $T(0) = T_0$ and $a_i(0) = 0$ $(i = 2, \ldots, n)$. To the initial value $a(\epsilon) = (0, a_2(\epsilon), \ldots, a_n(\epsilon))$ there corresponds the $T(\epsilon)$-periodic solution $x_\epsilon(t) = x(t; a(\epsilon), \epsilon)$ of equation (8.1) satisfying the limit relations (8.5). $\square$

It is well known [23] that if $\varphi(t)$ is a periodic solution of an autonomous equation without an impulse effect and $\varphi'(t) \not\equiv 0$, then $\varphi(t)$ cannot be asymptotically stable but can be orbitally asymptotically stable. The situation with the autonomous impulsive equations is the same.

Let $x = \varphi(t)$, $t \in \mathbb{R}_+$ be a solution of equation (8.1) with moments of the impulse effect $\tau_k : 0 < \tau_1 < \tau_2 < \cdots$, $\lim_{k \to \infty} \tau_k = \infty$ and $L_+ = \{x \in \mathbb{R}^n : x = \varphi(t), t \in \mathbb{R}_+\}$. Let $x(t; t_0, x_0)$ denote the solution of (8.1) for which $x(t_0^+; t_0, x_0) = x_0$ and let $J^+(t_0, x_0)$ denote the right maximal interval of existence of this solution.

Definition 8.1   The solution $x = \varphi(t)$ of equation (8.1) is said to be:

1. *orbitally stable* if

$$(\forall \rho > 0) \, (\forall \eta > 0) \, (\forall t_0 \in \mathbb{R}_+, |t_0 - \tau_k| > \eta) \, (\exists \delta > 0)$$
$$(\forall x_0 \in \mathbb{R}^n, d(x_0, L_+) < \delta, x_0 \notin \bar{B}_\eta(\varphi(\tau_k)) \cup \bar{B}_\eta(\varphi(\tau_k^+))) \, (\forall t \in J^+(t_0, x_0))$$
$$d(x(t; t_0, x_0), L_+) < \rho;$$

2. *orbitally attractive* if

$$(\forall \eta > 0) \, (\forall t_0 \in \mathbb{R}_+, |t_0 - \tau_k| > \eta) \, (\exists \lambda > 0)$$
$$(\forall x_0 \in \mathbb{R}^n, d(x_0, L_+) < \lambda, x_0 \notin \bar{B}_\eta(\varphi(\tau_k)) \cup \bar{B}_\eta(\varphi(\tau_k^+))) \, (\forall \rho > 0) \, (\exists \sigma > 0)$$
$$t_0 + \sigma \in J^+(t_0, x_0) \, (\forall t \geq t_0 + \sigma, t \in J^+(t_0, x_0))$$

$$d(x(t; t_0, x_0), L_+) < \rho;$$

3. *orbitally asymptotically stable* if it is orbitally stable and orbitally attractive.

Definition 8.2   The solution $x = \varphi(t)$ of equation (8.1) is said to enjoy the property of *asymptotic phase* if

$$(\forall \eta > 0) \ (\forall t_0 \in \mathbb{R}_+, |t_0 - \tau_k| > \eta) \ (\exists \lambda > 0)$$
$$(\forall x_0 \in \mathbb{R}^n, |x_0 - \varphi(t_0)| < \lambda) \ (\exists c \in \mathbb{R}) \ (\forall \rho > 0) \ (\exists \sigma > |c|)$$
$$t_0 + \sigma \in J^+(t_0, x_0) \ (\forall t \ge t_0 + \sigma, t \in J^+(t_0, x_0), |t_0 - \tau_k| > \eta)$$

$$|x(t + c; t_0, x_0) - \varphi(t)| < \rho.$$

The following stability result is valid [84].

Theorem 8.2   For equation (8.1) with $\epsilon = 0$ let conditions 1–3 of Theorem 8.1 hold and let the multipliers μ_j $(j = 1, \ldots, n)$ of equation (8.2) satisfy the condition

$$\mu_1 = 1, \qquad |\mu_j| < 1 \qquad (j = 2, \ldots, n).$$

Then the T_0-periodic solution $x = \varphi(t)$ of equation (8.1) with $\epsilon = 0$ is orbitally asymptotically stable and enjoys the property of asymptotic phase.

In the case $n = 2$ let equation (8.1) have the form

$$\frac{dx}{dt} = P(x, y), \quad \frac{dy}{dt} = Q(x, y), \quad \text{if } \phi(x, y) \ne 0,$$
$$\Delta x = a(x, y), \quad \Delta y = b(x, y), \quad \text{if } \phi(x, y) = 0.$$
$$(8.8)$$

If (8.8) has a T-periodic solution $x = \xi(t)$, $y = \eta(t)$ and the conditions of Theorem 8.2 are met then it can be checked [84] that the corresponding variational system has multipliers $\mu_1 = 1$ and

$$\mu_2 = \prod_{k=1}^{q} \Delta_k \exp\left[\int_0^T \left(\frac{\partial P}{\partial x}(\xi(t), \eta(t)) + \frac{\partial Q}{\partial y}(\xi(t), \eta(t))\right) dt\right] \quad (8.9)$$

where

$$\Delta_k = \frac{P_+\left(\frac{\partial b}{\partial y}\frac{\partial \phi}{\partial x} - \frac{\partial b}{\partial x}\frac{\partial \phi}{\partial y} + \frac{\partial \phi}{\partial x}\right) + Q_+\left(\frac{\partial a}{\partial x}\frac{\partial \phi}{\partial y} - \frac{\partial a}{\partial y}\frac{\partial \phi}{\partial x} + \frac{\partial \phi}{\partial y}\right)}{P\frac{\partial \phi}{\partial x} + Q\frac{\partial \phi}{\partial y}}$$

and $P, Q, \frac{\partial a}{\partial x}, \frac{\partial a}{\partial y}, \frac{\partial b}{\partial x}, \frac{\partial b}{\partial y}, \frac{\partial \phi}{\partial x}, \frac{\partial \phi}{\partial y}$ are calculated at the point $(\xi(\tau_k), \eta(\tau_k))$ and $P_+ = P(\xi(\tau_k^+), \eta(\tau_k^+))$, $Q_+ = Q(\xi(\tau_k^+), \eta(\tau_k^+))$.

Thus for $n = 2$ the following corollary is valid [84].

Corollary 8.1 (Analogue of Poincarés criterion)   The T-periodic solution $x = \xi(t)$, $y = \eta(t)$ of system (8.8) is orbitally asymptotically stable and enjoys the property of asymptotic phase if the multiplier μ_2 calculated by formula (8.9) satisfies the condition $|\mu_2| < 1$.

Example 8.1 (Damping oscillator subject to an impulse effect)   Consider the linear equation

$$\ddot{x} + 2h\dot{x} + \beta^2 x = 0 \quad (0 < h^2 < \beta^2) \tag{8.10}$$

describing the motion of a damping oscillator. Equation (8.10) has no non-trivial periodic solutions. The motion of the mapping point $(x, \dot{x})$ is carried out along a spiral winding around the unique stationary point $(0,0)$ which is a stable focus. If, however, the damping oscillator is subject to an impulse effect, periodic solutions may appear (discontinuous cycles). Such periodic solutions are observed, for instance, in the operation of a clock mechanism.

Let the linear oscillator with equation (8.10) be subject to an impulse effect when the mapping point $P_t = (x, \dot{x})$ meets the straight line $x = 0$ as a result of which the velocity $\dot{x}$ changes by a jump

$$\Delta \dot{x} = I(x, \dot{x})|_{x=0} = I(0, \dot{x}) = b(\dot{x}). \tag{8.11}$$

We shall investigate the question of the existence and orbital stability of non-zero periodic solutions of the impulsive equation (8.10), (8.11).

We write down equations (8.10), (8.11) in the form of a system

$$\begin{aligned} \frac{dx}{dt} &= y, \quad \frac{dy}{dt} = -\beta^2 x - 2hy, \quad \text{if } x \neq 0, \\ \Delta x &= 0, \quad \Delta y = b(y), \qquad\qquad \text{if } x = 0. \end{aligned} \tag{8.12}$$

Suppose that system (8.12) has a non-trivial T-periodic solution $x = \xi(t)$, $y = \eta(t)$ with initial values $\xi(0+) = 0$, $\eta(0+) = \eta_0^+ > 0$ and for $t \in (0, T]$ let the mapping point P_t meet ν times the straight line $x = 0$ at the points $(0, \eta_k)$ for which $\eta_k \neq 0$ and $\eta_k^+ = \eta_k + b(\eta_k)$ $(k = 1, \ldots, \nu)$.

By (8.10) for $t \in (\tau_k, \tau_{k+1}]$ the solution $\xi(t)$, $\eta(t)$ has the form

$$\xi(t) = \frac{\eta_k^+}{\omega} e^{-ht} \sin \omega t, \qquad \eta(t) = \frac{\eta_k^+}{\omega} e^{-ht} (\omega \cos \omega t - h \sin \omega t) \tag{8.13}$$

where $\omega = \sqrt{\beta^2 - h^2}$. From (8.13) it follows that the point P_t is transferred from the position $(0, \eta_k^+)$ to the position $(0, \eta_{k+1})$ for time $\tau = \pi/\omega$ and the following relations are valid:

$$\tau_k = k\tau, \qquad T = \nu\tau, \qquad \eta_{k+1} = -e^{-h\tau} \eta_k^+ \quad (k = 1, \ldots, \nu), \tag{8.14}$$

$$\eta_{k+1}^+ = -e^{-h\tau} \eta_k^+ + b(-e^{-h\tau} \eta_k^+). \tag{8.15}$$

The condition for T-periodicity is $\eta_\nu^+ = \eta_0^+$ and in view of (8.15) we obtain the equation for finding the initial value $y = \eta_0^+$ of the T-periodic solution of system (8.12):

$$U^\nu(y) = y \qquad (8.16)$$

where $U(y) = \lambda y + b(\lambda y)$, $\lambda = -e^{-h\tau}$ and

$$U^\nu = \underbrace{U \circ \cdots \circ U}_{\nu \text{ times}}.$$

In particular, for $\nu = 1$ equation (8.16) has the form

$$U(y) = \lambda y + b(\lambda y) = y, \qquad (8.17)$$

and for $\nu = 2$ it has the form

$$\lambda^2 y + \lambda b(\lambda y) + b(\lambda^2 y + \lambda b(\lambda y)) = y. \qquad (8.18)$$

From formula (8.9) we calculate the multiplier μ_2 of the variational system corresponding to the T-periodic solution $x = \xi(t)$, $y = \eta(t)$:

$$\frac{\partial P}{\partial x} = 0, \qquad \frac{\partial Q}{\partial y} = -2h,$$

$$\int_0^T \left(\frac{\partial P}{\partial x} + \frac{\partial Q}{\partial y} \right) dt = -2h\tau\nu,$$

$$\frac{\partial a}{\partial x} = \frac{\partial a}{\partial y} = 0,$$

$$\frac{\partial b}{\partial x} = 0, \qquad \frac{\partial b}{\partial y} = b'(\eta_k),$$

$$\frac{\partial \phi}{\partial x} = 1, \qquad \frac{\partial \phi}{\partial y} = 0,$$

$$\mu_2 = \frac{\eta_1^+}{\eta_1}(1 + b'(\eta_1)) \cdots \frac{\eta_\nu^+}{\eta_\nu}(1 + b'(\eta_\nu))e^{-2h\tau\nu}.$$

Taking into account (8.14) and the periodicity condition $\eta_\nu^+ = \eta_0^+$ we finally obtain

$$\mu_2 = (1 + b'(\eta_1)) \cdots (1 + b'(\eta_\nu))(-1)^\nu e^{-h\tau\nu}. \qquad (8.19)$$

Suppose system (8.12) is not subject to an impulse effect if $x = 0$ and $\dot{x} = y < 0$, that is,

$$\Delta y = b(y) = \begin{cases} b_1(y) & y > 0, \\ 0 & y \le 0. \end{cases}$$

If the T-periodic solution $x = \xi(t)$, $y = \eta(t)$ has ν moments of the impulse effect in the interval $(0, T]$, then we find analogously that

$$\tau_k = 2\tau k, \qquad T = 2\tau\nu, \qquad \eta_{k+1} = \eta_k^+ e^{-2h\tau},$$

$$\eta_{k+1}^+ = e^{-2h\tau}\eta_k^+ + b_1(e^{-2h\tau}\eta_k^+),$$

$$\mu_2 = (1 + b_1'(\eta_1)) \cdots (1 + b_1'(\eta_\nu))e^{-2h\tau\nu}. \tag{8.20}$$

The equation for the initial value $y = \eta_0^+$ of this solution is

$$U_1^\nu(y) = y,$$

where $U_1(y) = e^{-2h\tau}y + b_1(e^{-2h\tau}y)$. In particular, if in this case the equation

$$ye^{-2h\tau} + b_1(ye^{-2h\tau}) = y$$

has a solution $y = \eta_0^+$, then $\xi(t) = x(t; 0, 0, \eta_0^+)$, $\eta(t) = y(t; 0, 0, \eta_0^+)$ is a 2τ-periodic solution of system (8.12). This solution is orbitally asymptotically stable if the following condition is valid:

$$|1 + b_1'(\eta_0^+ e^{-2h\tau})|e^{-2h\tau} < 1.$$

The following particular cases of system (8.12) are simple models of the operation of a clock mechanism.

Case 1   Let the absolute value of the momentum of the oscillator increase by a constant quantity under the impulse effect. This means that $\Delta \dot{x} = I_0 > 0$ if at the moment of the impulse effect the velocity $\dot{x}$ is positive, and $\Delta \dot{x} = -I_0 < 0$ if $\dot{x} < 0$.

Case 1.1 (unilateral effect)   Suppose that there is no impulse effect if $\dot{x} \leq 0$, that is,

$$\Delta \dot{x} = \Delta y = b(y) = \begin{cases} I_0 & \text{if } y > 0, \\ 0 & \text{if } y \leq 0. \end{cases}$$

Then the periodicity condition $\eta_\nu^+ = \eta_0^+$ becomes

$$\eta_0^+ e^{-2h\tau\nu} + I_0 \sum_{k=0}^{\nu-1} e^{-2h\tau k} = \eta_0^+. \tag{8.21}$$

For each $\nu \in \mathbb{N}$ equation (8.21) has a unique solution $\eta_0^+ = I_0/(1 - e^{-2h\tau})$. Consequently, system (8.12) has a unique non-trivial periodic solution. This solution has a period $T = 2\pi/\omega$ and is orbitally asymptotically stable since $\mu_2 = e^{-2h\tau} \in (0,1)$.

Case 1.2 (bilateral effect)   Let

$$b(y) = \begin{cases} I_0 & \text{if } y > 0, \\ -I_0 & \text{if } y \leq 0. \end{cases}$$

In this case the periodicity condition $\eta_\nu^+ = \eta_0^+$ has the form

$$\eta_0^+ e^{-h\tau\nu} + I_0 \sum_{k=0}^{\nu-1} e^{-h\tau k} = \eta_0^+ \qquad \text{(if } \nu \text{ is even)} \tag{8.22}$$

or

$$-\eta_0^+ e^{-h\tau\nu} - I_0 \sum_{k=0}^{\nu-1} e^{-h\tau k} = \eta_0^+ \qquad \text{(if } \nu \text{ is odd).} \tag{8.23}$$

Equation (8.23) has no positive solution, while equation (8.22) has a unique solution $\eta_0^+ = I_0/(1-e^{-h\tau})$ for any even $\nu > 0$. Consequently, system (8.12) has a unique non-trivial periodic solution. This solution has a period $t = 2\tau = 2\pi/\omega$ and is orbitally asymptotically stable.

Case 2 Let the kinetic energy of the oscillator increase by a constant under the impulse effect. This means that at the moment of the impulse effect

$$\Delta \dot{x}^2 = I_0^2 = \text{const.} > 0$$

or

$$\Delta \dot{x}(\Delta \dot{x} + 2\dot{x}) = I_0^2. \tag{8.24}$$

We solve (8.24) with respect to $\Delta \dot{x}$ and obtain $\Delta \dot{x} = -\dot{x} \pm \sqrt{\dot{x}^2 + I_0^2}$.

Case 2.1 (unilateral effect) Let

$$\Delta y = b(y) = \begin{cases} -y + \sqrt{y^2 + I_0^2} & \text{if } y > 0, \\ 0 & \text{if } y \leq 0. \end{cases}$$

Then the periodicity condition $\eta_\nu^+ = \eta_0^+$ has the form

$$\sqrt{(\eta_0^+)^2 e^{-4h\tau\nu} + I_0^2 \sum_{k=0}^{\nu-1} e^{-4h\tau k}} = \eta_0^+. \tag{8.25}$$

For each $\nu \in \mathbb{N}$ equation (8.25) has a unique solution

$$\eta_0^+ = \frac{I_0}{\sqrt{1 - e^{-4h\tau}}}.$$

Consequently, system (8.12) has a unique non-trivial periodic solution and the period of this solution is $T = 2\tau = 2\pi/\omega$.

Case 2.2 (bilateral effect) Let

$$\Delta y = b(y) = \begin{cases} -y + \sqrt{y^2 + I_0^2} & \text{if } y > 0, \\ -y - \sqrt{y^2 + I_0^2} & \text{if } y \leq 0. \end{cases}$$

Then the periodicity condition $\eta_\nu^+ = \eta_0^+$ has the form

$$\sqrt{(\eta_0^+)^2 e^{-2h\tau\nu} + I_0^2 \sum_{k=0}^{\nu-1} e^{-2h\tau k}} = \begin{cases} \eta_0 & \text{if } \nu \text{ is even} \\ -\eta_0 & \text{if } \nu \text{ is odd} \end{cases} \tag{8.26}$$

Equation (8.26) has no positive solution if ν is odd, while for any even $\nu > 0$ this equation has a unique solution

$$\eta_0^+ = \frac{I_0}{\sqrt{1 - e^{-2h\tau}}}.$$

Consequently, system (8.12) has a unique non-trivial periodic solution and the period of this solution is $T = 2\tau = 2\pi/\omega$.

Since in the cases 2.1 and 2.2 the function $b(y)$ satisfies the inequality $|1 + b'(y)| \le 1$, then $|\mu_2| \le e^{-2h\tau} < 1$, that is, in these two cases the non-trivial periodic solution of system (8.12) is orbitally asymptotically stable.

The orbits corresponding to these solutions are given in Figs 8.1 and 8.2.

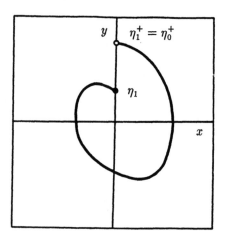

Fig. 8.1

Example 8.2   Let the linear oscillator described by equation (8.10) be subject to an impulse effect when $x = \epsilon$, $\dot{x} > 0$ and as a result of this the velocity $y = \dot{x}$ receives a constant increment $I_0 > 0$. Set

$$\omega = \sqrt{\beta^2 - h^2}, \qquad T_0 = \frac{2\pi}{\omega}, \qquad \sigma(\epsilon) = \{(x, y) \in \mathbb{R}^2 : x = \epsilon, y > 0\},$$

$$z = \begin{bmatrix} x \\ y \end{bmatrix}, \qquad z_0 = \begin{bmatrix} \epsilon \\ y_0 \end{bmatrix}, \qquad A = \begin{bmatrix} 0 & 1 \\ -\beta^2 & -2h \end{bmatrix}, \qquad Q = \begin{bmatrix} 0 \\ I_0 \end{bmatrix}.$$

Then the respective impulsive equation has the form

$$\begin{aligned} \frac{dz}{dt} &= Az, \quad z \notin \sigma(\epsilon), \\ \Delta z &= Q, \quad z \in \sigma(\epsilon). \end{aligned} \tag{8.27}$$

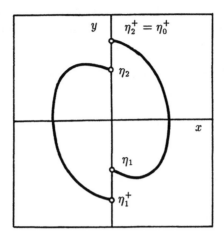

Fig. 8.2

If in (8.27) we fix $\epsilon = 0$, we shall obtain the impulsive system considered in Example 8.1, case 1.1, and this has multipliers $\mu_1 = 1$ and $\mu_2 = e^{-hT_0} \neq 1$. Consequently, by Theorem 8.1 there exists an $\epsilon_0 > 0$ such that for $|\epsilon| \leq \epsilon_0$ equation (8.27) has a unique $T(\epsilon)$-periodic solution $z_\epsilon(t)$ such that for $t \in [0, T_0]$

$$z_\epsilon(t) \xrightarrow{B} z_0(t), \qquad T(\epsilon) \to T_0 \qquad \text{as } \epsilon \to 0, \tag{8.28}$$

where $z_0(t)$ is the unique T_0-periodic solution of the impulsive system of Example 8.1, case 1.1. For comparison we shall prove this fact directly.

If $z(t)$ is a T-periodic solution of (8.27) and $z(0+) = z_0$, then the condition for T-periodicity $z(T+) = z(0+)$ takes the form

$$(E - e^{AT})z_0 = Q \tag{8.29}$$

where

$$E - e^{AT} = \begin{bmatrix} 1 - e^{-hT}\cos\omega T - \frac{h}{\omega}e^{-hT}\sin\omega T & -\frac{1}{\omega}e^{-hT}\sin\omega T \\ \frac{h^2+\omega^2}{\omega}e^{-hT}\sin\omega T & 1 - e^{-hT}\cos\omega T + \frac{h}{\omega}e^{-hT}\sin\omega T \end{bmatrix}.$$

Since $\det(E - e^{AT}) = 1 - 2e^{-hT}\cos\omega T + e^{-2hT} \neq 0$, equation (8.29) is solvable with respect to z_0. We have

$$\epsilon = \frac{I_0 e^{-hT}\sin\omega T}{\omega(1 - 2e^{-hT}\cos\omega T + e^{-2hT})}, \tag{8.30}$$

$$y_0 = \frac{I_0(1 - e^{-hT}\cos\omega T - \frac{h}{\omega}e^{-hT}\sin\omega T)}{1 - 2e^{-hT}\cos\omega T + e^{-2hT}}. \tag{8.31}$$

By the implicit function theorem for sufficiently small ϵ equation (8.30) is solvable with respect to T and the solution $T = T(\epsilon)$ is a continuous function of ϵ with $T(0) = T_0$. Then equation (8.31) defines a continuous function $y_0 = y_0(\epsilon)$ with $y_0(0) = \eta_0 = I_0/(1-e^{-hT_0})$. This means that for sufficiently small ϵ equation (8.27) has a unique periodic solution $z_\epsilon(t)$ with period $T(\epsilon)$ and initial value $z_\epsilon(0+) = \begin{bmatrix} \epsilon \\ y_0(\epsilon) \end{bmatrix}$, and the limit relations (8.28) are valid. Moreover,

$$T = T_0 + \frac{(1 - e^{-hT_0})^2}{I_0 e^{-hT_0}}\epsilon + o(\epsilon), \tag{8.32}$$

$$y_0 = \frac{I_0}{1 - e^{-hT_0}} - 2h\epsilon + o(\epsilon), \quad T_0 = \frac{2\pi}{\omega}. \tag{8.33}$$

Example 8.3 (Vibro-shock system)   Consider the linear oscillator with equation

$$\ddot{x} + \omega^2 x = 0 \tag{8.34}$$

describing the harmonic oscillations of a body B fixed to a spring with rigidity coefficient ω^2. The phase trajectories of (8.34) are ellipses of the form

$$\omega^2 x^2 + \dot{x}^2 = \omega^2 x_0^2 + \dot{x}_0^2. \tag{8.35}$$

Let an obstacle S be situated at a given distance from the body B and suppose that the velocity $\dot{x}$ changes its sign when the body B meets the obstacle S (an absolutely elastic impact). If the impact takes place when the coordinate x takes the value ϵ (Fig. 8.3) then the jump condition becomes

$$\Delta\dot{x} = -2\dot{x}, \quad \text{if } x = \epsilon. \tag{8.36}$$

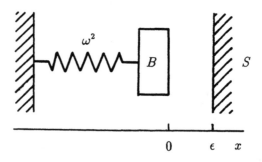

Fig. 8.3

Let $x(t)$ be a solution of the impulsive system (8.34), (8.36) and $x(0+) = x_0$, $\dot{x}(0+) = \dot{x}_0$.

The following cases are possible.

Case 1 $\epsilon > 0$ In this case between the body B situated in an equilibrium position and the obstacle S there is a cleft ϵ and both non-shock and shock periodic regimes are possible.

Case 1.1   Let the energy of the oscillator be small $(\omega^2 x_0^2 + \dot{x}_0^2 \leq \omega^2 \epsilon^2)$ and suppose the body B does not reach the obstacle S or reaches it with velocity $\dot{x} = 0$. Then the oscillator performs linear oscillations with frequency $\Omega = \omega$ and period $T = 2\pi/\omega$, and the phase trajectories are ellipses of the form (8.35) (Fig. 8.4).

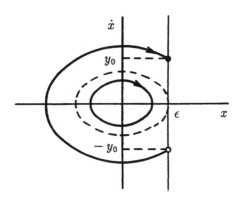

Fig. 8.4

Case 1.2   Let the energy of the oscillator be large enough $(\omega^2 x_0^2 + \dot{x}_0^2 > \omega^2 \epsilon^2)$ and suppose the body B reaches the obstacle S with velocity $\dot{x} = y_0 > 0$. Then the oscillator performs periodic oscillations with one impact per period and for the velocity y_0, the period T, and the frequency Ω, the following relations are valid

$$\omega T = 2\pi - \arccos \frac{\omega^2 \epsilon^2 - y_0^2}{\omega^2 \epsilon^2 + y_0^2}, \qquad \Omega = \frac{2\pi}{T},$$

$$\pi < \omega T < 2\pi, \qquad \omega < \Omega < 2\omega. \tag{8.37}$$

In this case the phase trajectory is that part of the ellipse

$$\omega^2 x^2 + \dot{x}^2 = \omega^2 \epsilon^2 + y_0^2 \tag{8.38}$$

which is situated to the left of the straight line $x = \epsilon$ (Fig. 8.4).

Case 2 $\epsilon \leq 0$ In this case the spring has a tension $|\epsilon|$ (Fig. 8.5). Only a shock periodic regime is possible, provided that the oscillator is given sufficient energy to overcome the force of initial tension $\omega^2 |\epsilon|$. The phase trajectory of such a periodic regime is also the left part of the ellipse (8.38) and is given in Fig. 8.6. For the quantities y_0, T and Ω the following relations are valid

$$\omega T = \arccos \frac{\omega^2 \epsilon^2 - y_0^2}{\omega^2 \epsilon^2 + y_0^2}, \qquad \Omega = \frac{2\pi}{T},$$

$$0 < \omega T \leq \pi, \qquad 2\omega \leq \Omega < \infty.$$

(8.39)

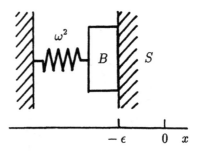

Fig. 8.5

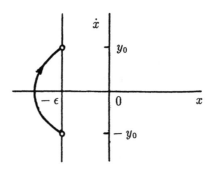

Fig. 8.6

Let the body B be confined to move only between two obstacles S_1 and S_2, and suppose that during this motion the coordinate x changes in the interval $[-\epsilon, \epsilon]$ (Fig. 8.7). If when meeting the obstacles S_1 and S_2 the velocity $\dot{x}$ changes its sign, then the jump condition has the form

$$\Delta \dot{x} = -2\dot{x} \qquad \text{if } |x| = \epsilon > 0.$$

(8.40)

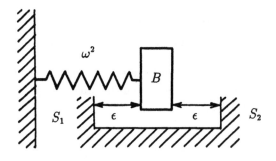

Fig. 8.7

For the impulsive system (8.34), (8.40) both non-shock and shock periodic regimes are possible.

If $\omega^2 x_0^2 + \dot{x}_0^2 \leq \omega^2 \epsilon^2$, then the phase trajectory is the ellipse (8.35).

If $\omega^2 x_0^2 + \dot{x}_0^2 > \omega^2 \epsilon^2$ and $y_0 > 0$ is the velocity with which the body meets the right obstacle S_2, then the phase trajectory is the part of the ellipse $\omega^2 x^2 + y^2 = \omega^2 \epsilon^2 + y_0^2$ which is situated between the straight lines $x = -\epsilon$ and $x = \epsilon$ (Fig. 8.8). For the velocity y_0, the period T, and the frequency Ω, the following relations are valid

$$\omega T = 2 \arccos \frac{-\omega^2 \epsilon^2 + y_0^2}{\omega^2 \epsilon^2 + y_0^2}, \qquad \Omega = \frac{2\pi}{T},$$

$$0 < \omega T < 2\pi, \qquad \omega < \Omega < \infty. \tag{8.41}$$

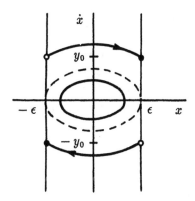

Fig. 8.8

Example 8.4   Consider the impulsive equation

$$\ddot{x} = \alpha\dot{x} - \beta\dot{x}^3 + \epsilon f(x, \dot{x}, \epsilon), \quad \text{if } |x| < \delta(\epsilon),$$
$$\dot{x}^+ = -R(\epsilon)\dot{x}^-, \qquad\qquad \text{if } |x| = \delta(\epsilon), \tag{8.42}$$

where $\epsilon \in J = (-\bar{\epsilon}, \bar{\epsilon})$ is a small parameter, $\alpha > 0$, $\beta > 0$, the function $f(x, y, \epsilon)$ is differentiable for $x \in \mathbb{R}$, $y \in \mathbb{R}$, $\epsilon \in J$, and the functions $R(\epsilon)$ and $\delta(\epsilon)$ are continuous for $\epsilon \in J$ and are such that

$$R(0) = 1, \qquad 0 \le R(\epsilon) \le 1,$$
$$\delta(0) = \delta_0 > 0, \qquad \delta(\epsilon) \ge 0.$$

Equation (8.42) describes the motion of a body under the action of 'cubic friction', which is bounded by two obstacles, and under impact against them changes its velocity by a jump.

We write (8.42) as a system

$$\begin{aligned} \dot{x} &= y, \\ \dot{y} &= \alpha y - \beta y^3 + \epsilon f(x, y, \epsilon), \\ \Delta x &= 0, \\ \Delta y &= -(1 + R(\epsilon))y, \end{aligned} \qquad \begin{aligned} &\text{if } |x| \le \delta(\epsilon), \\[2ex] &\text{if } |x| = \delta(\epsilon). \end{aligned} \tag{8.43}$$

For $\epsilon = 0$ system (8.43) takes the form

$$\begin{aligned} \dot{x} &= y, \\ \dot{y} &= \alpha y - \beta y^3, \\ \Delta x &= 0, \\ \Delta y &= -2y, \end{aligned} \qquad \begin{aligned} &\text{if } |x| \le \delta_0, \\[2ex] &\text{if } |x| = \delta_0. \end{aligned} \tag{8.44}$$

The equation $\dot{y} = \alpha y - \beta y^3$ has stationary solutions

$$y_0 = 0, \qquad y_0^+ = \sqrt{\frac{\alpha}{\beta}}, \qquad y_0^- = -\sqrt{\frac{\alpha}{\beta}}.$$

To $y = y_0$ there corresponds an unstable position of equilibrium. To $y = y_0^+$ there corresponds a uniform rectilinear motion towards the right obstacle, and to $y = y_0^-$ a uniform rectilinear motion towards the left obstacle.

Let $T = 4\delta_0\sqrt{\beta/\alpha}$. Then the T-periodic extensions $\xi(t)$, $\eta(t)$ of the functions

$$x = \sqrt{\frac{\alpha}{\beta}}|t| - \delta_0, \qquad t \in [-\tfrac{1}{2}T, \tfrac{1}{2}T],$$

$$y = \begin{cases} \sqrt{\dfrac{\alpha}{\beta}} & t \in (0, \tfrac{1}{2}T], \\[2ex] -\sqrt{\dfrac{\alpha}{\beta}} & t \in (-\tfrac{1}{2}T, 0] \end{cases}$$

give the T-periodic solution of (8.44) with two impacts per period. The graphs of $\xi(t)$, $\eta(t)$ and the respective phase trajectory are given in Figs 8.9, 8.10 and 8.11. The multiplier μ_2 of the respective variational system is computed by formula (8.9) and is $\mu_2 = e^{-2\alpha T}$. Consequently, by Theorem 8.1 for sufficiently small ϵ system (8.43) has a unique $T(\epsilon)$-periodic solution $p_\epsilon(t) = \text{col}\,(\xi_\epsilon(t), \eta_\epsilon(t))$ and for $t \in [0, T]$

$$p_\epsilon(t) \overset{B}{\to} p_0(t) = \text{col}\,(\xi(t), \eta(t)), \qquad T(\epsilon) \to T \qquad \text{as } \epsilon \to 0.$$

The phase trajectory of this solution is given in Fig. 8.12.

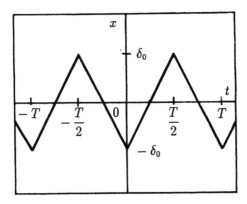

Fig. 8.9

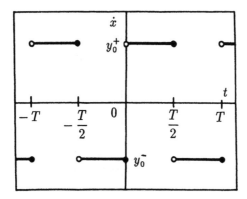

Fig. 8.10

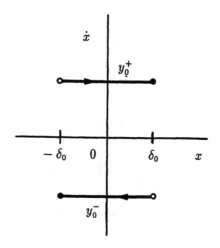

Fig. 8.11

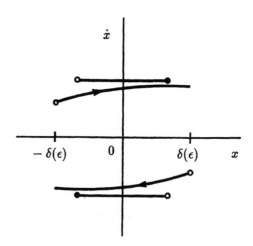

Fig. 8.12

Example 8.5 Consider the system

$$\frac{dx}{dt} = F_1(x,y) + \epsilon g_1(x,y,\epsilon),$$

$$\quad \text{if } |x| \neq \tfrac{1}{2}V_s,$$

$$\frac{dy}{dt} = F_2(x,y) + \epsilon g_2(x,y,\epsilon),$$

$$\Delta x = -2krSx + \epsilon I_1(x,y,\epsilon),$$

$$\quad \text{if } |x| = \tfrac{1}{2}V_s$$

$$\Delta y = -\frac{2rS}{L}x + \epsilon I_2(x,y,\epsilon),$$

$$(8.45)$$

where $\epsilon \in J = (-\bar{\epsilon}, \bar{\epsilon})$ is a small parameter,

$$F_1(x,y) = \begin{cases} \dfrac{kry}{krS - 1} & \text{if } |x| < \tfrac{1}{2}V_s \\ -kry & \text{if } |x| > \tfrac{1}{2}V_s \end{cases}$$

$$F_2(x,y) = \begin{cases} \dfrac{x}{krLC} + \dfrac{ry}{L(krS-1)} & \text{if } |x| < \tfrac{1}{2}V_s \\ \dfrac{x}{krLC} - \dfrac{ry}{L} & \text{if } |x| > \tfrac{1}{2}V_s \end{cases}$$

and k, r, S, L, C, V_s are positive constants.

Suppose that the functions g_j, I_j $(j = 1, 2)$ are differentiable for $(x, y, \epsilon) \in \mathbb{R} \times \mathbb{R} \times J$.

Let us fix $\epsilon = 0$. Then system (8.45) coincides with system (1.9), (1.10) describing the operation of the electronic circuit in Fig. 1.4.

Andronov et al. [6] proved that in the cases when L is small $(L \ll Cr^2/4)$ or L is large $(L \gg Cr^2/4)$ system (1.9), (1.10) has a unique non-zero periodic solution $p(t) = \text{col}\,(\xi(t), \eta(t))$ and the period T of this solution is evaluated as:

$$T \approx \frac{\pi}{\sqrt{\dfrac{1}{LC} + \dfrac{r^2}{4L^2}}} \qquad (\text{for } L \gg Cr^2/4),$$

$$T \approx 2\frac{L}{r}\ln\left(2Skr - 1\right) \quad (\text{for } L \ll Cr^2/4).$$

The phase trajectory of $p(t)$ is given in Fig. 8.13 (for $L \gg Cr^2/4$) and in Fig. 8.14 (for $L \ll Cr^2/4$).

The motion of the mapping point $(\xi(t), \eta(t))$ is carried out in the set defined by the inequality $|x| \geq \tfrac{1}{2}V_s$. This motion is continuous from point A_2 to point A_3 and from point A_4 to point A_1, and it is by jumps from point A_1 to point A_2 and from point A_3 to point A_4 (Figs 8.13 and 8.14). Moreover, $x_1 = -x_3 = \tfrac{1}{2}V_s$, $y_1 = -y_3 > 0$, and $y_2 = -y_4 = y_1 - (2rS/L)x_1$.

Making use of the notation introduced above, we obtain

$$P(x,y) = -kry, \qquad Q(x,y) = \frac{x}{krLC} - \frac{ry}{L},$$

$$a(x,y) = -2krSx, \qquad b(x,y) = -\frac{2rS}{L}x,$$

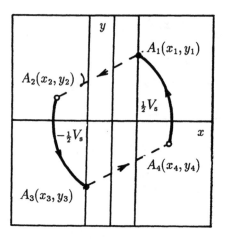

Fig. 8.13

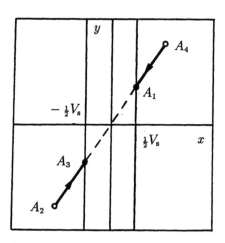

Fig. 8.14

$$\phi(x,y) = |x| - \tfrac{1}{2}V_s, \qquad \frac{\partial \phi}{\partial x}(\pm\tfrac{1}{2}V_s, y) = \pm 1, \qquad \frac{\partial \phi}{\partial y} = 0,$$

$$\frac{\partial P}{\partial x} = 0, \qquad \frac{\partial Q}{\partial y} = -\frac{r}{L},$$

$$\frac{\partial a}{\partial x} = -2krS, \qquad \frac{\partial a}{\partial y} = 0,$$

$$\frac{\partial b}{\partial x} = -\frac{2rS}{L}, \qquad \frac{\partial b}{\partial y} = 0.$$

Then

$$\int_0^T \left(\frac{\partial P}{\partial x} + \frac{\partial Q}{\partial y} \right) dt = -\frac{rT}{L},$$

$$\Delta_1 = \frac{P_+}{P} = \frac{P(x_2, y_2)}{P(x_1, y_1)} = \frac{y_1 - (2rS/L)x_1}{y_1},$$

$$\Delta_2 = \frac{P_+}{P} = \frac{P(x_4, y_4)}{P(x_3, y_3)} = \frac{-y_1 + (2rS/L)x_1}{-y_1} = \Delta_1,$$

$$\mu_2 = \left(\frac{y_1 - (2rS/L)x_1}{y_1} \right)^2 e^{-rT/L}.$$

In the cases when $L \gg Cr^2/4$ or $L \ll Cr^2/4$ the condition $0 < \mu_2 < 1$ is fulfilled. Then the periodic solution of system (1.9), (1.10) is orbitally asymptotically stable. Moreover, by Theorem 8.1 for sufficiently small ϵ system (8.45) has a $T(\epsilon)$-periodic solution $p_\epsilon(t)$ and for $t \in [0, T]$

$$p_\epsilon(t) \overset{B}{\to} p(t), \qquad T(\epsilon) \to T \qquad \text{as } \epsilon \to 0.$$

Example 8.6 Let a body B be suspended by means of a spring so that it could perform oscillations and be subject to impacts when meeting the obstacles S_1 and S_2 (Fig. 8.15). During this motion let the coordinate x vary in the interval $[a, b]$ $(0 < a < b)$. Suppose that during the motion towards the left obstacle S_1 the rigidity coefficient of the spring is w_1^2 and during the motion towards the right obstacle S_2 it is w_2^2. Suppose also that under the impact against S_2 we have $\dot{x}^+ = -\mu \dot{x}^-$ $(0 < \mu < 1)$ and under the impact against S_1, as well as the impact, the body B obtains an additional momentum and its velocity increases by the constant quantity p, that is, $\dot{x}^+ = -\lambda \dot{x}^- + p$ $(0 < \lambda < 1)$. This mechanical system is described by the equations

$$\dot{x} = y,$$

$$\dot{y} = \begin{cases} -w_1^2 x, & \text{if } y \geq 0, \ a \leq x \leq b, \\ -w_2^2 x, & \text{if } y < 0, \ a \leq x \leq b, \end{cases} \tag{8.46}$$

$$\Delta x = 0,$$

$$\Delta y = \begin{cases} -(1 + \mu)y, & \text{if } x = b, \ y \geq 0, \\ -(1 + \lambda)y + p, & \text{if } x = a, \ y < 0. \end{cases} \tag{8.47}$$

The phase trajectory of the periodic solution of system (8.46), (8.47) with two impulses per period is given in Fig. 8.16.

Let us find the value $y_0 > 0$ which determines the initial value (a, y_0) of such a solution. Taking into account that the phase trajectories of the equation $\ddot{x} + w^2 x = 0$ are the ellipses $w^2 x^2 + y^2 = w^2 x_0^2 + y_0^2$, we obtain the relations

$$w_1^2 b^2 + y_1^2 = w_1^2 a^2 + y_0^2, \qquad w_2^2 a^2 + y_2^2 = w_2^2 b^2 + (y_1^+)^2, \tag{8.48}$$

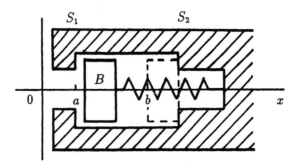

Fig. 8.15

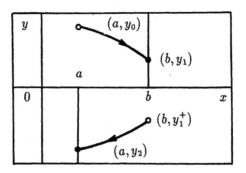

Fig. 8.16

and from the jump conditions (8.47) we have

$$y_1^+ = -\mu y_1, \qquad y_0 = -\lambda y_2 + p. \tag{8.49}$$

From (8.48) and (8.49) we eliminate y_1, y_1^+ and y_2 and obtain the equation for y_0:

$$y_0 = p + \lambda\sqrt{\mu^2 y_0^2 + Q} \equiv F(y_0) \tag{8.50}$$

where $Q = (b^2 - a^2)(\omega_2^2 - \mu^2\omega_1^2)$.

The function $w = F(z)$ is strictly increasing, has an inclined asymptote $w = p + \lambda\mu z$ $(z \geq 0)$ and $F''(z) = \lambda\mu^2 Q(\mu^2 z^2 + Q)^{-3/2}$ has the sign of Q.

The analysis of equation (8.50) shows that:

(i) for $-p^2\mu^2 < Q$ equation (8.50) has a unique positive solution (Figs 8.17 and 8.18);

(ii) for $\dfrac{-p^2\mu^2}{1 - p^2\mu^2} < Q < -p^2\mu^2$ equation (8.50) has two positive solutions (Fig. 8.19);

(iii) for $Q < \dfrac{-p^2\mu^2}{1 - p^2\mu^2}$ equation (8.50) has no positive solutions (Fig. 8.20).

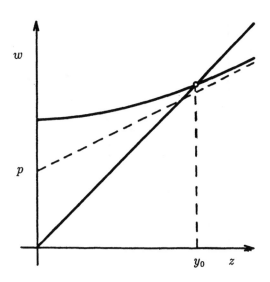

Fig. 8.17

To any positive solution y_0 of equation (8.50) there corresponds a periodic solution with two impulses per period. The multiplier μ_2 computed by formula (8.9) is

$$\mu_2 = \frac{\lambda^2 \mu^2 y_0}{y_0 - p}. \tag{8.51}$$

The condition $|\mu_2| < 1$ is satisfied if y_0 is the unique positive root of (8.50) (case (i)) or y_0 is the greater positive root of (8.50) (case (ii)). In this case the smaller positive root of (8.50) does not satisfy the condition $|\mu_2| < 1$.

Thus for $Q > \frac{-p^2 \mu^2}{1 - p^2 \mu^2}$ system (8.46), (8.47) has a unique orbitally asymptotically stable periodic solution with two impulses per period. The period T of this solution is

$$T = T_1 + T_2$$

where

$$T_1 = \frac{1}{\omega_1} \arccos \frac{\omega_1^2 ab + y_0 y_1}{\omega_1^2 a^2 + y_0^2},$$

$$T_2 = \frac{1}{\omega_2} \arccos \frac{\omega_2^2 ab + y_1^+ y_2}{\omega_2^2 b^2 + (y_1^+)^2},$$

$$y_1 = \sqrt{y_0^2 + \omega_1^2(a^2 - b^2)}, \qquad y_1^+ = -\mu y_1, \qquad y_2 = -\sqrt{\mu^2 y_0^2 + (b^2 - a^2)(\omega_2^2 - \mu^2 \omega_1^2)}.$$

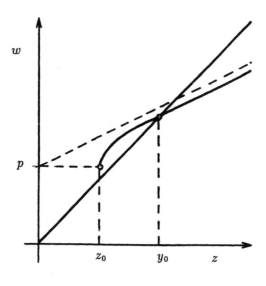

Fig. 8.18

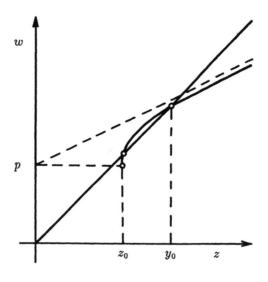

Fig. 8.19

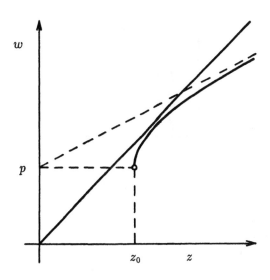

Fig. 8.20

Example 8.7 (Species–food system)   In an isolated medium let a given species A be bred, being fed with food B. We assume that in the absence of the species A the quantity of food B does not change, and in the absence of food B the species A dies out. Such a situation is observed, for instance, in the breeding of a given species of fish in a fish-pond.

If we denote by $x(t)$ and $y(t)$ the absolute or relative quantities of the species A and the food B at the moment t, then the dynamics of the 'species–food' system can be simulated by the system of ordinary differential equations

$$\frac{dx}{dt} = -\gamma xy, \qquad \frac{dy}{dt} = -y(\epsilon - \delta x) \tag{8.52}$$

where $(x,y) \in \mathbb{R}_+^2$ and $\gamma > 0$, $\epsilon > 0$, $\delta > 0$ are constants.

Equations (8.52) are obtained as a limiting case of Volterra's equation (cf. [67] or [88]) describing the dynamics of a 'predator–prey' system. In our example the species bred is a 'predator' and the food is a 'prey' with rate coefficient equal to 0.

If $x(0) = x_0 > 0$ and $y(0) = y_0 > 0$, then the solution of system (8.52) remains in $\mathbb{R}_+^2$ (Fig. 8.21) and

$$\lim_{t \to +\infty} x(t) = x_\infty, \qquad \lim_{t \to +\infty} y(t) = 0.$$

This means that if the food B is not regenerated then the species A dies out. It turns out, however, that under appropriately chosen impulse effects on the 'species–food' system a periodic regime of development of the species A is possible.

Consider the following two cases.

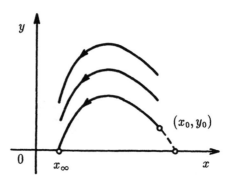

Fig. 8.21

Case 1   At certain moments let the 'species–food' system be subject to an impulse effect such that the quantity of food is increased by the amount $\lambda > 0$ and the population of the species is decreased by αy (for instance, by catching or by exhaustion). Assume that $0 < \alpha < 1$, that is, the whole species A is not exhausted by the 'catch'.

Let these impulse effects occur when the quantity of food reaches a given level $x_1 > 0$, that is, if $x = x_1$. Thus we obtain the following impulsive system

$$\frac{dx}{dt} = -\gamma xy, \quad \frac{dy}{dt} = -y(\epsilon - \delta x), \quad \text{if } x \neq x_1,$$
$$\Delta x = \lambda, \qquad \Delta y = -\alpha y, \qquad \text{if } x = x_1. \tag{8.53}$$

Let us investigate the questions of the existence of a T-periodic solution of system (8.53) with one impulse effect per period and of the stability of this solution.

Let $x = \xi(t)$, $y = \eta(t)$ be such a T-periodic solution. Introduce the notation $\xi_0 = \xi(0+)$, $\eta_0 = \eta(0+)$, $\xi_1 = \xi(T)$, $\eta_1 = \eta(T)$, $\xi_1^+ = \xi(T+)$, $\eta_1^+ = \eta(T+)$. Then, from the condition of T-periodicity, $\xi_1^+ = \xi_0$, $\eta_1^+ = \eta_0$, we obtain

$$x_1 + \lambda = \xi_0, \qquad (1 - \alpha)\eta_1 = \eta_0. \tag{8.54}$$

For $t \in (0, T]$ the solution $x = \xi(t)$, $y = \eta(t)$ of system (8.53) satisfies the relation

$$e^{\gamma(\eta(t)-\eta_0)} = \left(\frac{\xi(t)}{\xi_0}\right)^{\epsilon} e^{-\delta(\xi(t)-\xi_0)}. \tag{8.55}$$

In particular, for $t = T$, we have

$$e^{\gamma(\eta_1 - \eta_0)} = \left(\frac{x_1}{\xi_0}\right)^{\epsilon} e^{-\delta(x_1 - \xi_0)},$$

and in view of (8.54) we obtain

$$\eta_0 = \frac{1 - \alpha}{\alpha\gamma}\left(\epsilon \ln \frac{x_1}{x_1 + \lambda} + \delta\lambda\right).$$

From (8.54) it follows that η_0 is positive if

$$x_1 > x^* = \frac{\lambda\exp(-\delta\lambda/\epsilon)}{1 - \exp(-\delta\lambda/\epsilon)}. \tag{8.56}$$

Thus, if condition (8.56) holds, then system (8.53) has a unique periodic solution with one impulse effect per period. The period T of this solution can be found, taking into account the first equation of (8.53) and (8.55):

$$T = \int_0^T dt = \int_{x_1}^{x_1+\lambda} \frac{dx}{\left(\gamma\eta_0 + \epsilon\ln\frac{x}{x_1+\lambda} - \delta(x - x_1 - \lambda)\right)x}. \tag{8.57}$$

The trajectory of the solution $x = \xi(t)$, $y = \eta(t)$ is given in Fig. 8.22.

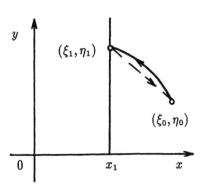

Fig. 8.22

From formula (8.9) we compute the multiplier μ_2 of the system in variations corresponding to the T-periodic solution $x = \xi(t)$, $y = \eta(t)$:

$$\frac{\partial P}{\partial x} = -\gamma y, \qquad \frac{\partial Q}{\partial y} = -\epsilon + \delta x,$$

$$\frac{\partial a}{\partial x} = \frac{\partial a}{\partial y} = 0, \qquad \frac{\partial b}{\partial x} = 0, \qquad \frac{\partial b}{\partial y} = -\alpha,$$

$$\frac{\partial \phi}{\partial x} = 1, \qquad \frac{\partial \phi}{\partial y} = 0,$$

$$\Delta_1 = \frac{P_+(-\alpha+1)}{P} = \frac{-\gamma\xi_0\eta_0(1-\alpha)}{-\gamma\xi_1\eta_1} = \frac{(1-\alpha)\xi_0\eta_0}{\xi_1\eta_1},$$

$$\int_0^T \left(\frac{\partial P}{\partial x} + \frac{\partial Q}{\partial y}\right) dt = \int_0^T (-\gamma\eta(t) - \epsilon + \delta\xi(t)) \, dt$$

$$= \int_0^T \left(\frac{\dot{\xi}(t)}{\xi(t)} + \frac{\dot{\eta}(t)}{\eta(t)}\right) dt = \int_0^T d\ln(\xi(t)\eta(t))$$

$$= \ln\frac{\xi_1\eta_1}{\xi_0\eta_0},$$

$$\mu_2 = \Delta_1 \exp\left\{ \int_0^T \left(\frac{\partial P}{\partial x} + \frac{\partial Q}{\partial y} \right) dt \right\} = (1 - \alpha) \frac{\xi_0 \eta_0}{\xi_1 \eta_1} \cdot \frac{\xi_1 \eta_1}{\xi_0 \eta_0} = 1 - \alpha.$$

Since $\mu_2 = 1 - \alpha \in (0,1)$, then the T-periodic solution $x = \xi(t)$, $y = \eta(t)$ of system (8.53) is orbitally asymptotically stable.

Case 2   Let the 'species–food' system be subject to an impulse effect when the quantity of food reaches the level x_1 satisfying condition (8.56).

Let $n > 0$ be an integer and assume that the quantity of food increases by λ under each impulse effect, while the population of the species decreases by jumps only at the moments of the impulse effect τ_k whose ordinal number k is a multiple of n, that is,

$$\Delta x(\tau_k) = \lambda, \tag{8.58}$$

$$\Delta y(\tau_k) = \begin{cases} 0, & \text{if } k \text{ is not divisible by } n, \\ -\alpha y(\tau_k), & \text{if } k \text{ is divisible by } n. \end{cases} \tag{8.59}$$

First we shall discover under what initial conditions $x(0+) = x_0 > 0$, $y(0+) = y_0 > 0$ the impulsive system (8.52), (8.58), (8.59) has a periodic solution with period $T = \tau_n$. Let $x_k = x(\tau_k)$, $x_k^+ = x(\tau_k^+)$, $y_k = y(\tau_k)$, $y_k^+ = y(\tau_k^+)$ and $\tau_0 = 0$.

In view of (8.58) and (8.59) the condition of T-periodicity $x_n^+ = x_0$, $y_n^+ = y_0$ becomes

$$x_1 + \lambda = x_0, \qquad (1 - \alpha)y_n = y_0. \tag{8.60}$$

From equation (8.55) it follows that

$$e^{\gamma(y_k - y_{k-1}^+)} = \left(\frac{x_k}{x_{k-1}^+} \right)^{\epsilon} e^{-\delta(x_k - x_{k-1}^+)} \qquad (k = 1, \dots, n)$$

and in view of (8.58), (8.59) and (8.60) we find successively

$$y_k - y_{k-1} = \frac{1}{\gamma} \left(\epsilon \ln \frac{x_1}{x_1 + \lambda} + \delta\lambda \right) \equiv \mu \quad (k = 1, \dots, n),$$

$$y_k = y_0 + k\mu \quad (k = 1, \dots, n),$$

$$y_0 = \frac{(1 - \alpha)n\mu}{\alpha}.$$

From (8.56) it follows that y_0 is positive. Hence the solution $x = \xi(t)$, $y = \eta(t)$ of system (8.52), (8.58), (8.59) satisfying the initial conditions $\xi(0+) = x_0$, $\eta(0+) = y_0$ is the unique non-trivial T-periodic solution of this system with n impulse effects per period.

Let us choose a level $\eta^* = \frac{1}{2}(y_{n-1} + y_n) = \mu(\frac{n}{\alpha} - \frac{1}{2})$ of the numbers of the species. Then $x = \xi(t)$, $y = \eta(t)$ is a τ_n-periodic solution of the following autonomous impulsive system

$$\frac{dx}{dt} = -\gamma xy, \quad \frac{dy}{dt} = -y(\epsilon - \delta x), \qquad \text{if } x \neq x_1,$$

$$\Delta x = \lambda, \qquad \Delta y = \begin{cases} 0 & \text{if } y < \eta^* \\ -\alpha y & \text{if } y \geq \eta^* \end{cases} \quad \text{if } x = x_1. \tag{8.61}$$

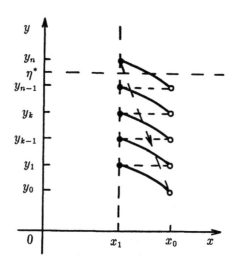

Fig. 8.23

The trajectory of this solution is represented in Fig. 8.23.

From formula (8.9) we compute the multiplier μ_2 of the system in variations corresponding to the τ_n-periodic solution $x = \xi(t)$, $y = \eta(t)$ of system (8.61):

$$\frac{\partial P}{\partial x} = -\gamma y, \qquad \frac{\partial Q}{\partial y} = -\epsilon + \delta x,$$

$$\frac{\partial a}{\partial x} = \frac{\partial a}{\partial y} = 0, \qquad \frac{\partial \phi}{\partial x} = 1, \qquad \frac{\partial \phi}{\partial y} = 0,$$

$$\frac{\partial b}{\partial x} = 0, \qquad \frac{\partial b}{\partial y} = \begin{cases} 0 & \text{if } y < \eta^* \\ -\alpha & \text{if } y \geq \eta^*, \end{cases}$$

$$\Delta_k = \frac{P(x_k^+, y_k^+)}{P(x_k, y_k)} = \frac{-\gamma x_k^+ y_k^+}{-\gamma x_k y_k} = \frac{x_0}{x_1} \qquad (k = 1, \dots, n-1),$$

$$\Delta_n = \frac{P(x_n^+, y_n^+)(-\alpha + 1)}{P(x_n, y_n)} = (1 - \alpha) \frac{-\gamma x_n^+ y_n^+}{-\gamma x_n y_n} = (1 - \alpha) \frac{x_0 y_0}{x_1 y_n},$$

$$\exp\left\{ \int_{\tau_{k-1}}^{\tau_k} \left(\frac{\partial P}{\partial x} + \frac{\partial Q}{\partial y} \right) dt \right\} = \frac{x_k y_k}{x_{k-1}^+ y_{k-1}^+} = \frac{x_1 y_k}{x_0 y_{k-1}} \qquad (k = 1, \dots, n),$$

$$\mu_2 = \prod_{k=1}^{n} \Delta_k \exp\left\{ \int_{\tau_{k-1}}^{\tau_k} \left(\frac{\partial P}{\partial x} + \frac{\partial Q}{\partial y} \right) dt \right\}$$

$$= \frac{y_1}{y_0} \frac{y_2}{y_1} \dots \frac{y_{n-1}}{y_{n-2}} \frac{y_n}{y_{n-1}} \frac{y_0}{y_n} (1 - \alpha)$$

$$= 1 - \alpha.$$

Hence the solution $x = \xi(t)$, $y = \eta(t)$ of system (8.61) is orbitally asymptotically stable. The period $T = \tau_n$ of this solution can be calculated from the formula

$$T = \sum_{k=1}^{n} \int_{x_1}^{x_1+\lambda} \frac{dx}{\left(\gamma y_{k-1} + \epsilon \ln \frac{x}{x_1+\lambda} - \delta(x - x_1 - \lambda)\right)x}. \tag{8.62}$$

Example 8.8 (Cultivation of bacteria)   Consider the following mathematical model of the process of cultivation of bacteria

$$\dot{x} = \mu(y)x, \qquad \dot{y} = \alpha\mu^2(y)x, \tag{8.63}$$

where $x(t) \geq 0$, $y(t) \geq 0$ are the concentrations of the micro-organisms and of the product of the biosynthesis, respectively; $\mu(y) = \beta y/(\gamma + y^2)$ and $\alpha > 0$, $\beta > 0$, $\gamma > 0$ are numerical characteristics of the model.

Assume that at the instants when the concentration $y(t)$ of the product reaches a given level $y_1 > 0$ the system is subject to an instantaneous 'dilution' under which the quantities $x(t)$ and $y(t)$ decrease by $\delta x(t)$ and $\delta y(t)$, respectively, that is,

$$\Delta x = -\delta x, \qquad \Delta y = -\delta y \qquad \text{if } y = y_1. \tag{8.64}$$

We shall look for a T-periodic solution $x = \xi(t)$, $y = \eta(t)$ of the impulsive system (8.63), (8.64) having one jump per period.

Let $\xi_0 = \xi(0+)$, $\eta_0 = \eta(0+)$, $\xi_1 = \xi(T)$, $\eta_1 = \eta(T)$, $\xi_1^+ = \xi(T+)$, $\eta_1^+ = \eta(T+)$. Then the condition for T-periodicity $\xi_1^+ = \xi_0$, $\eta_1^+ = \eta_0$ becomes

$$(1 - \delta)\xi_1 = \xi_0, \qquad (1 - \delta)y_1 = \eta_0. \tag{8.65}$$

From (8.63) it follows that for $t \in (0, T]$ the relation

$$\alpha\beta(x - \xi_0) = \gamma \ln \frac{y}{\eta_0} + \frac{y^2 - \eta_0^2}{2}$$

holds and in view of (8.65) we obtain

$$\alpha\beta\delta\xi_1 = -\gamma \ln(1 - \delta) + \frac{2\delta - \delta^2}{2}y_1^2$$

after which, from (8.65), we uniquely determine ξ_0 and η_0.

Hence system (8.63), (8.64) has a unique T-periodic solution with one jump per period and

$$T = \int_0^T dt = \int_{\eta_0}^{\eta_1} \frac{dy}{\alpha\mu^2(y)\left(\xi_0 + \frac{\gamma}{\alpha\beta}\ln \frac{y}{\eta_0} + \frac{y^2 - \eta_0^2}{2\alpha\beta}\right)}. \tag{8.66}$$

The trajectory is represented in Fig. 8.24. We compute the multiplier μ_2 from formula (8.9):

$$\frac{\partial P}{\partial x} = \mu(y), \qquad \frac{\partial Q}{\partial y} = \alpha x \frac{d(\mu^2)}{dy},$$

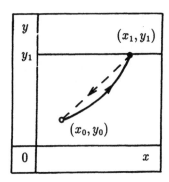

Fig. 8.24

$$\int_0^T \frac{\partial P}{\partial x} \, dt = \int_0^T \mu(\eta(t)) \, dt = \int_0^T \frac{\dot{\xi}(t)}{\xi(t)} \, dt = \ln \frac{\xi(T)}{\xi(0+)} = \ln \frac{\xi_1}{\xi_0},$$

$$\int_0^T \frac{\partial Q}{\partial y} \, dt = \int_0^T \alpha x \frac{d(\mu^2(y))}{dy} \, dt = \int_0^T \frac{d(\mu^2(\eta(t)))}{\mu^2(\eta(t))} = \ln \frac{\mu^2(\eta_1)}{\mu^2(\eta_0)},$$

$$\frac{\partial a}{\partial x} = -\delta, \qquad \frac{\partial a}{\partial y} = 0,$$

$$\frac{\partial b}{\partial x} = 0, \qquad \frac{\partial b}{\partial y} = -\delta,$$

$$\frac{\partial \phi}{\partial x} = 0, \qquad \frac{\partial \phi}{\partial y} = 1,$$

$$\Delta_1 = \frac{Q_+(1-\delta)}{Q} = (1-\delta)\frac{\alpha^2\mu^2(\eta_0)\xi_0}{\alpha^2\mu^2(\eta_1)\xi_1} = (1-\delta)\frac{\xi_0\mu^2(\eta_0)}{\xi_1\mu^2(\eta_1)},$$

$$\mu_2 = \Delta_1 \exp\left\{\int_0^T \left(\frac{\partial P}{\partial x} + \frac{\partial Q}{\partial y}\right) dt\right\}$$

$$= (1-\delta)\frac{\xi_0\mu^2(\eta_0)}{\xi_1\mu^2(\eta_1)} \cdot \frac{\xi_1}{\xi_0} \cdot \frac{\mu^2(\eta_1)}{\mu^2(\eta_0)} = 1 - \delta.$$

(8.67)

From (8.67) it follows that the T-periodic solution $x = \xi(t)$, $y = \eta(t)$ of system (8.63), (8.64) is orbitally asymptotically stable.

Notes and comments for Chapter III

Theorems 6.3 and 6.4 were adapted from [73]. The remaining results in Section 6 were proved by P. S. Simeonov. For equations without impulses analogous results were obtained in [35]. Equation (6.50) in Example 6.2 was considered by K. Gopalsamy in [28].

Theorems 7.1–7.4 are new and are due to P. S. Simeonov. For other results in this direction see A. M. Samoilenko and N. A. Perestyuk [71] and M. U. Ahmetov and N. A. Perestyuk [1–3]. Examples 7.1 and 7.2 were adapted from [50].

Theorem 8.1 is new and was proved by P. S. Simeonov. Example 8.1 was adapted from [73] while Examples 8.3 and 8.4 were adapted from [10]. Example 8.6 was proposed by T. D. Tagarev.

Chapter IV

Method of the Small Parameter. Critical Case

In this chapter we shall discuss the question of the existence of a periodic solution $x(t, \epsilon)$ which depends continuously on the small parameter ϵ entering the equation, and which tends to the generating solution $\varphi(t) = x(t, 0)$. The considerations will be carried out in the critical case when the variational equation corresponding to the generating solution $\varphi(t)$ has $m \geq 1$ linearly independent periodic solutions if the equation is non-autonomous and $m \geq 2$ linearly independent periodic solutions if the equation is autonomous.

9. Quasilinear equations with fixed moments of an impulse effect

Consider the quasilinear T-periodic equation with fixed moments of an impulse effect

$$
\begin{aligned}
\tfrac{dx}{dt} &= A(t)x + g(t) + \epsilon f(t, x, \epsilon), \quad t \neq \tau_k, \\
\Delta x &= B_k x + h_k + \epsilon I_k(x, \epsilon), \qquad t = \tau_k,
\end{aligned}
\tag{9.1}
$$

where $\epsilon \in J = (-\bar{\epsilon}, \bar{\epsilon})$ is a small parameter.
   Together with (9.1) consider the equations

$$
\begin{aligned}
\tfrac{dx}{dt} &= A(t)x + g(t), \quad t \neq \tau_k, \\
\Delta x &= B_k x + h_k, \qquad t = \tau_k;
\end{aligned}
\tag{9.2}
$$

$$
\begin{aligned}
\tfrac{dx}{dt} &= A(t)x, \quad t \neq \tau_k, \\
\Delta x &= B_k x, \qquad t = \tau_k;
\end{aligned}
\tag{9.3}
$$

and

$$
\begin{aligned}
\tfrac{dy}{dt} &= -A^*(t)y, \qquad t \neq \tau_k, \\
\Delta y &= -(E + B_k^*)^{-1} B_k^* y, \quad t = \tau_k.
\end{aligned}
\tag{9.4}
$$

   Assume that the following conditions (H9) hold:

H9.1 $A(\,\cdot\,) \in PC(\mathbb{R}, \mathbb{R}^{n \times n})$, $g(\,\cdot\,) \in PC(\mathbb{R}, \mathbb{R}^n)$, $B_k \in \mathbb{R}^{n \times n}$, $h_k \in \mathbb{R}^n$.
H9.2 The function $f : \mathbb{R} \times \mathbb{R}^n \times J \to \mathbb{R}^n$ is differentiable in the sets $(\tau_k, \tau_{k+1}] \times \mathbb{R}^n \times J$ $(k \in \mathbb{Z})$ and for any $x \in \mathbb{R}^n$, $\epsilon \in J$ and $k \in \mathbb{Z}$ there exist the finite limits of $f(t, y, \mu)$ and $\frac{\partial f}{\partial x}(t, y, \mu)$ as $(t, y, \mu) \to (\tau_k, x, \epsilon)$, $t > \tau_k$.
H9.3 The functions $I_k : \mathbb{R}^n \times J \to \mathbb{R}^n$ $(k \in \mathbb{Z})$ are differentiable in $\mathbb{R}^n \times J$.

H9.4 Equation (9.1) is T-periodic, that is, there exists a $q \in \mathbb{N}$ such that

$$A(t+T) = A(t), \qquad g(t+T) = g(t), \qquad f(t+T, x, \epsilon) = f(t, x, \epsilon),$$

$$\tau_{k+q} = \tau_k + T, \qquad B_{k+q} = B_k, \qquad h_{k+q} = h_k, \qquad I_{k+q}(x, \epsilon) = I_k(x, \epsilon)$$

for $t \in \mathbb{R}$, $k \in \mathbb{Z}$, $x \in \mathbb{R}^n$ and $\epsilon \in J$.

H9.5 Equation (9.3) has m $(1 \le m \le n)$ linearly independent T-periodic solutions $\varphi_1(t), \ldots, \varphi_m(t)$.

H9.6 The compatibility condition

$$\int_0^T \Psi_1^*(t) g(s)\, ds + \sum_{k=1}^q \Psi_1^*(\tau_k^+) h_k = 0 \tag{9.5}$$

is satisfied, where $\Psi_1(t) = [\psi_1(t), \ldots, \psi_m(t)]$ and $\psi_1(t), \ldots, \psi_m(t)$ are the linearly independent T-periodic solutions of equation (9.4).

   From conditions H9.5, H9.6 and Theorem 4.2 it follows that equation (9.2) has an m-parametric family of T-periodic solutions

$$\bar{x}(t, c) = \Phi_1(t) c + \tilde{x}_0(t) \tag{9.6}$$

where $c = \mathrm{col}\,(c_1, \ldots, c_m) \in \mathbb{R}^m$, $\Phi_1(t) = [\varphi_1(t), \ldots, \varphi_m(t)]$ and $\tilde{x}_0(t)$ is a particular T-periodic solution of (9.2).

   Fix $c = c_0 = \mathrm{col}\,(c_1^0, \ldots, c_m^0)$ in (9.6) and obtain the solution

$$p(t) = \Phi_1(t) c_0 + \tilde{x}_0(t)$$

which will be called the *generating solution*. For this solution we formulate the following problem.

Problem A   *Under what conditions does there exist a neighbourhood $J_0 = (-\epsilon_0, \epsilon_0)$ $\subset J$ such that for any $\epsilon \in J_0$ equation (9.1) has a T-periodic solution $x(t, \epsilon)$ for which*

$$\lim_{\epsilon \to 0} x(t, \epsilon) = x(t, 0) = p(t)? \tag{9.7}$$

   It is easily verified that equation (9.3) coincides with the variational equation corresponding to the generating solution $p(t)$. But from condition H9.5 it follows that the monodromy matrix M of equation (9.3) is such that $\det\,(E - M) = 0$; that is why in order to solve Problem A we cannot apply Theorem 6.7.

   Suppose that equation (9.1) has a T-periodic solution $x(t)$. Then $x(t)$ can be considered as a solution of a linear non-homogeneous equation obtained by adding to the right-hand side of the homogeneous equation (9.3) the free terms $g(t) + \epsilon f(t, x(t), \epsilon)$ and $h_k + \epsilon I_k(x(\tau_k), \epsilon)$, respectively. Consequently, the solution $x(t)$ has the form

$$x(t) = x(t, c, \epsilon) = \Phi_1(t) c + \tilde{x}_\epsilon(t)$$

where $\tilde{x}_\epsilon(t)$ is a particular T-periodic solution of (9.1).

By Theorem 4.2 these free terms have to satisfy the compatibility condition (4.19) and in view of (9.5) we find that the T-periodic solution $x(t, c, \epsilon)$ has to satisfy the condition

$$\Delta(c, \epsilon) = \int_0^T \Psi_1^*(s) f(s, \Phi_1(s)c + \tilde{x}_\epsilon(s), \epsilon) \, ds$$

$$+ \sum_{k=1}^q \Psi_1^*(\tau_k^+) I_k(\Phi_1(\tau_k)c + \tilde{x}_\epsilon(\tau_k), \epsilon)$$

$$= 0$$

for $\epsilon \neq 0$. If we suppose that $x(t, c, \epsilon)$ is a solution of Problem A, then letting $\epsilon \to 0$ we obtain the following necessary condition for the solvability of this problem:

$$\Delta(c_0, 0) = \int_0^T \Psi_1^*(s) f(s, p(s), 0) \, ds + \sum_{k=1}^q \Psi_1^*(\tau_k^+) I_k(p(\tau_k^+), 0) = 0. \qquad (9.8)$$

The following assertion is obvious.

Theorem 9.1   *Let conditions (H9) hold. Then Problem A is solvable if and only if there exists $\epsilon_0 \in (0, \bar{\epsilon})$ and m functions $c_1(\epsilon), \ldots, c_m(\epsilon)$ which are defined and continuous in the interval $J_0 = (-\epsilon_0, \epsilon_0)$ and such that the vector-valued function $c(\epsilon) = \mathrm{col}\,(c_1(\epsilon), \ldots, c_m(\epsilon))$ satisfies the conditions*

$$c(0) = c_0 \quad \text{and} \quad \Delta(c(\epsilon), \epsilon) \equiv 0 \quad (\epsilon \in J_0).$$

It is of greater practical interest to find easily verifiable sufficient conditions for the solvability of Problem A. Such a condition is given by the following theorem.

Theorem 9.2   *Let conditions (H9) hold as well as the following conditions*

$$\Delta(c_0, 0) = 0, \quad \det \frac{\partial \Delta}{\partial c}(c_0, 0) \neq 0. \qquad (9.9)$$

Then there exists $\epsilon_0 \in (0, \bar{\epsilon})$ such that for $|\epsilon| \leq \epsilon_0$ equation (9.1) has a unique T-periodic solution $x(t, \epsilon)$ which satisfies the limit relation (9.7).

Theorem 9.2 is an immediate consequence of the implicit function theorem applied to the equation $\Delta(c, \epsilon) = 0$.

In view of $\frac{\partial x}{\partial c}(t, c_0, 0) = \Phi_1(t)$ we find that

$$Q = \frac{\partial \Delta}{\partial c}(c_0, 0) = \int_0^T \Psi_1^*(s) \frac{\partial f}{\partial x}(s, p(s), 0) \Phi_1(s) \, ds$$

$$+ \sum_{k=1}^q \Psi_1^*(\tau_k^+) \frac{\partial I_k}{\partial x}(p(\tau_k), 0) \Phi_1(\tau_k). \qquad (9.10)$$

From (9.8) and (9.10) it is seen that in order to check conditions (9.9) it suffices to know just the generating solution $p(t)$ and the solutions $\varphi_1(t), \ldots, \varphi_m(t)$ and $\psi_1(t), \ldots, \psi_m(t)$ of equations (9.3) and (9.4).

We also note that if $\det \frac{\partial \Delta}{\partial c}(c_0, 0) = 0$ then it is possible, for $\epsilon \neq 0$, that several T-periodic solutions exist satisfying the limit relation (9.7), that is, there is a branching of the generating solution $p(t)$. In this case more detailed investigations are necessary in order to determine the number of these solutions.

Let us discuss the question of the approximate finding of the T-periodic solution $x(t)$ of equation (9.1) under the assumption that conditions (9.9) are met and that ϵ is small enough.

A sequence of successive approximations to the solution $x(t)$ has to be constructed so that after each iteration the free terms newly found satisfy the compatibility condition (4.19) and so that at each step of the iteration process formula (4.21) could be applied. To this end we shall transform the equation so as to be able to use the fact that the matrix Q is non-singular.

We seek the T-periodic solution of (9.1) in the form

$$x(t) = p(t) + z(t)$$

where $z(t)$ is a T-periodic solution of the equation

$$\begin{aligned}
\frac{dz}{dt} &= A(t)z + \epsilon f(t, p(t) + z, \epsilon), \quad t \neq \tau_k, \\
\Delta z &= B_k z + \epsilon I_k(p(\tau_k) + z, \epsilon), \quad t = \tau_k.
\end{aligned} \tag{9.11}$$

As above we conclude that this solution has the form

$$z(t) = \Phi_1(t)a + u(t)$$

where $a \in \mathbb{R}^m$ and $u(t)$ is a particular T-periodic solution of (9.11). The following additional condition is imposed on the function $u(t)$:

$$\Phi_1^*(0)u(0) = 0. \tag{9.12}$$

From conditions H9.2 and H9.3 it follows that the functions f and I_k have the form

$$f(t, p(t) + z, \epsilon) = f(t, p(t), 0) + \frac{\partial f}{\partial x}(t, p(t), 0)z + R(t, z, \epsilon),$$

$$I_k(p(\tau_k) + z, \epsilon) = I_k(p(\tau_k), 0) + \frac{\partial I_k}{\partial x}(p(\tau_k), 0)z + R_k(z, \epsilon)$$

where the functions R, R_k enjoy the properties

$$\begin{aligned}
R(t, 0, 0) &\equiv 0, \quad \frac{\partial R}{\partial z}(t, 0, 0) \equiv 0 \quad (t \in \mathbb{R}), \\
R_k(0, 0) &\equiv 0, \quad \frac{\partial R_k}{\partial z}(0, 0) \equiv 0 \quad (k \in \mathbb{Z}).
\end{aligned} \tag{9.13}$$

Then equation (9.11) can be transformed to the form

$$\frac{du}{dt} = A(t)u$$
$$+\epsilon\left[f(t,p(t),0) + \frac{\partial f}{\partial x}(t,p(t),0)z(t) + R(t,z(t),\epsilon)\right], \quad t \neq \tau_k,$$
$$\Delta u = B_k u$$
$$+\epsilon\left[I_k(p(\tau_k),0) + \frac{\partial I_k}{\partial x}(p(\tau_k),0)z(\tau_k) + R_k(z(\tau_k),\epsilon)\right], \quad t = \tau_k, \tag{9.14}$$

where $z(t) = \Phi_1(t)a + u$.

The successive approximations to $x(t)$ can be obtained, defining the successive approximations to the vector a and the functions $u(t)$ and $z(t)$. As initial approximations choose

$$a_0 = 0, \qquad u_0(t) \equiv 0, \qquad z_0(t) \equiv 0.$$

The next approximations are determined successively from the system

$$Qa_{j+1} = -\int_0^T \Psi_1^*(s)\left[\frac{\partial f}{\partial x}(s,p(s),0)u_j(s) + R(s,z_j(s),\epsilon)\right]ds$$
$$-\sum_{k=1}^q \Psi_1^*(\tau_k^+)\left[\frac{\partial I_k}{\partial x}(p(\tau_k),0)u_j(\tau_k) + R_k(z_j(\tau_k),\epsilon)\right]; \tag{9.15}$$

$$u_{j+1}' = A(t)u_{j+1}$$
$$+\epsilon\left\{f(t,p(t),0) + \frac{\partial f}{\partial x}(t,p(t),0)[\Phi_1(t)a_{j+1} + u_j(t)] + R(t,z_j(t),\epsilon)\right\}, \quad t \neq \tau_k,$$
$$\Delta u_{j+1} = B_k u_{j+1}$$
$$+\epsilon\left\{I_k(p(\tau_k),0) + \frac{\partial I_k}{\partial x}(p(\tau_k),0)[\Phi_1(\tau_k)a_{j+1} + u_j(\tau_k)] + R_k(z_j(\tau_k),\epsilon)\right\}, \quad t = \tau_k; \tag{9.16}$$

$$\Phi_1^*(0)u_{j+1}(0) = 0; \tag{9.17}$$

$$z_{j+1}(t) = \Phi_1(t)a_{j+1} + u_{j+1}(t). \tag{9.18}$$

If the approximations a_j, $u_j(t)$ and $z_j(t)$ have already been determined, then from relations (9.15)–(9.18) we determine uniquely a_{j+1}, $u_{j+1}(t)$ and $z_{j+1}(t)$.

Indeed, from (9.15) we can determine a_{j+1} since $\det Q \neq 0$. In view of (9.8) and (9.15) we conclude that the compatibility condition (4.19) for the free terms of equation (9.16) is met. Since the function $u_{j+1}(t)$ is T-periodic and satisfies system

(9.16), (9.17), then by Theorem 4.2 it is determined uniquely by the formula

$$u_{j+1}(t) = \epsilon \int_0^T \tilde{G}(t,s)$$

$$\times \left\{ f(s,p(s),0) + \frac{\partial f}{\partial x}(s,p(s),0)[\Phi_1(s)a_{j+1} + u_j(s)] + R(s,z_j(s),\epsilon) \right\} ds$$

$$+ \epsilon \sum_{k=1}^q \tilde{G}(t,\tau_k^+)$$

$$\times \left\{ I_k(p(\tau_k),0) + \frac{\partial I_k}{\partial x}(p(\tau_k),0)[\Phi_1(\tau_k)a_{j+1} + u_j(\tau_k)] + R_k(z_j(\tau_k),\epsilon) \right\}$$

$$(9.19)$$

where the function $\tilde{G}(t,s)$ is defined by formula (4.22). Finally, after $u_{j+1}(t)$ has been defined, we find $z_{j+1}(t)$ by formula (9.18).

We shall prove the convergence of the sequences a_j and $u_j(t)$ (respectively $z_j(t)$, $x_j(t)$). Let the real number $h > 0$ be fixed. From (9.13) and conditions H9.2–H9.4 it follows that there exists $\epsilon_1 \in (0,\bar{\epsilon})$ and a non-negative function $\delta(\epsilon)$ with $\lim_{\epsilon\to 0}\delta(\epsilon) = 0$ such that

$$|R(t,z,\epsilon) - R(t,\bar{z},\epsilon)| \leq \delta(\epsilon)|z - \bar{z}|, \quad |R(t,0,\epsilon)| \leq \delta(\epsilon),$$
$$|R_k(z,\epsilon) - R_k(\bar{z},\epsilon)| \leq \delta(\epsilon)|z - \bar{z}|, \quad |R_k(0,\epsilon)| \leq \delta(\epsilon) \qquad (9.20)$$

for $t \in \mathbb{R}$, $k \in \mathbb{Z}$, $|\epsilon| \leq \epsilon_1$ and $z,\bar{z} \in \mathbb{R}^n$, $|z| \leq \epsilon_1$, $|\bar{z}| \leq \epsilon_1$.

For $u \in PC(\mathbb{R},\mathbb{R}^n)$ let $\|u\| = \sup_{t\in\mathbb{R}}|u(t)|$ and denote

$$\Delta v_j = \begin{bmatrix} \|u_{j+1} - u_j\| \\ |a_{j+1} - a_j| \end{bmatrix}.$$

From (9.15) and (9.19) it follows that there exists a non-negative function $\omega(\epsilon)$ with $\lim_{\epsilon\to 0}\omega(\epsilon) = 0$ such that

$$\Delta v_0 = \begin{bmatrix} \|u_1\| \\ |a_1| \end{bmatrix} \leq \begin{bmatrix} \omega(\epsilon) \\ \omega(\epsilon) \end{bmatrix}.$$

Taking into account (9.15), (9.18), (9.19) and (9.20), we conclude that there exists a $K > 0$, independent of ϵ, and such that

$$\Delta v_j \leq S\Delta v_{j-1},$$

where

$$S = \begin{bmatrix} \epsilon K & \epsilon\delta(\epsilon)K \\ K & \delta(\epsilon)K \end{bmatrix}. \qquad (9.21)$$

Then $\Delta v_j \leq S^j \Delta v_0$ and

$$\begin{bmatrix} \|u_{j+1}\| \\ |a_{j+1}| \end{bmatrix} \leq \Delta v_j + \cdots + \Delta v_0 \leq (E - S)^{-1} \begin{bmatrix} \omega(\epsilon) \\ \omega(\epsilon) \end{bmatrix}. \qquad (9.22)$$

From (9.21) and (9.22) it follows that there exists $\epsilon_0 \in (0, \epsilon_1)$ such that for $|\epsilon| \le \epsilon_0$ the functions $u_j(t)$ and the eigenvalues $\lambda_1(\epsilon)$, $\lambda_2(\epsilon)$ of the matrix S satisfy the inequalities

$$\|u_j\| \le h, \qquad \sup_{|\epsilon| \le \epsilon_0} |\lambda_i(\epsilon)| < 1 \qquad (j \in \mathbb{N}, i = 1, 2).$$

Finally, from the estimate

$$\begin{bmatrix} \|u_{j+\nu} - u_j\| \\ |a_{j+\nu} - a_j| \end{bmatrix} \le S^j (E - S)^{-1} \begin{bmatrix} \omega(\epsilon) \\ \omega(\epsilon) \end{bmatrix}$$

we conclude that the sequence a_j is convergent in $\mathbb{R}^n$ and the sequence $u_j(t)$ is convergent in $PC(\mathbb{R}, \mathbb{R}^n)$.

Let $a \in \mathbb{R}^n$ and $u(\cdot) \in PC(\mathbb{R}, \mathbb{R}^n)$ be the limits of these sequences. Then, passing to the limit in (9.19), we find that the vector a and the T-periodic function $u(t)$ satisfy equation (9.14) and condition (9.12) which immediately implies that the T-periodic function $z(t) = \Phi_1(t)a + u(t)$ is a solution of (9.11), and the function $x(t) = p(t) + z(t)$ is the T-periodic solution of equation (9.1) we seek.

Example 9.1   Consider the impulsive equation

$$\ddot{x} + \omega^2 x = \epsilon(a\dot{x} - b\dot{x}^3), \qquad t \ne nT,$$
$$\Delta x = 0, \qquad \Delta \dot{x} = \epsilon K, \qquad t = nT \tag{9.23}$$

where $\epsilon \in J = (-\bar{\epsilon}, \bar{\epsilon})$ is a small parameter and a, b, ω, T, K are positive constants.
In matrix form (9.23) has the form

$$\frac{dz}{dt} = Az + \epsilon g(z), \quad t \ne nT,$$
$$\Delta z = \epsilon I(z), \qquad t = nT, \tag{9.24}$$

where

$$z = \begin{bmatrix} x \\ y \end{bmatrix}, \qquad A = \begin{bmatrix} 0 & 1 \\ -\omega^2 & 0 \end{bmatrix}, \qquad g(z) = \begin{bmatrix} 0 \\ ay - by^3 \end{bmatrix}, \qquad I(z) = \begin{bmatrix} 0 \\ K \end{bmatrix}.$$

For $\epsilon = 0$ equation (9.24) is linear homogeneous without impulses and has $m = 2$ linearly independent periodic solutions with period $T_0 = 2\pi/\omega$.

Let us consider the question of the existence of T_0-periodic solutions of equation (9.24) for small ϵ in the case when $T = T_0$. This case is critical with $m = 2$. For the sake of simplicity in the calculations we assumed that $\omega = 1$, $T = T_0 = 2\pi$. Then equation (9.24) for $\epsilon = 0$ has the following two-parametric family $p(t) = \mathrm{col}\,(\xi(t), \eta(t))$ of 2π-periodic solutions

$$\xi(t) = -c_1 \cos t + c_2 \sin t,$$
$$\eta(t) = c_1 \sin t + c_2 \cos t. \tag{9.25}$$

Let $\Phi(t)$ and $\Psi(t)$ be respectively the fundamental matrices of the equations $\dot{z} = Az$ and $\dot{w} = -A^*w$ and $\Phi(0) = \Psi(0) = E$. Then

$$\Phi(t) = \begin{bmatrix} \cos t & \sin t \\ -\sin t & \cos t \end{bmatrix}, \qquad \Psi^*(t) = \begin{bmatrix} \cos t & -\sin t \\ \sin t & \cos t \end{bmatrix}.$$

Taking into account that

$$\Psi^*(t)g(p(t)) = \begin{bmatrix} (-a\eta(t) + b\eta^3(t))\sin t \\ (a\eta(t) - b\eta^3(t))\cos t \end{bmatrix}, \qquad \Psi^*(0)I(p(0)) = \begin{bmatrix} 0 \\ K \end{bmatrix},$$

$$\Psi^*(t)\frac{\partial g}{\partial z}(p(t))\Phi(t) = \begin{bmatrix} (a - 3b\eta^2(t))\sin^2 t & -(a - 3b\eta^2(t))\sin t \cos t \\ -(a - 3b\eta^2(t))\sin t \cos t & (a - 3b\eta^2(t))\cos^2 t \end{bmatrix},$$

$$\frac{\partial I}{\partial z} = \begin{bmatrix} 0 & 0 \\ 0 & 0 \end{bmatrix}, \qquad \Psi^*(0)\frac{\partial I}{\partial z}(p(0))\Phi(0) = \begin{bmatrix} 0 & 0 \\ 0 & 0 \end{bmatrix},$$

$$\int_0^{2\pi} \eta(t)\sin t \; dt = c_1\pi, \qquad \int_0^{2\pi} \eta(t)\cos t \; dt = c_2\pi,$$

$$\int_0^{2\pi} \eta^3(t)\cos t \; dt = \frac{3\pi}{4}(c_1^2 c_2 + c_2^3),$$

$$\int_0^{2\pi} \eta^3(t)\sin t \; dt = \frac{3\pi}{4}(c_1^3 + c_1 c_2^2),$$

$$\int_0^{2\pi} \eta^2(t)\sin^2 t \; dt = \frac{\pi}{4}(3c_1^2 + c_2^2),$$

$$\int_0^{2\pi} \eta^2(t)\cos^2 t \; dt = \frac{\pi}{4}(c_1^2 + 3c_2^2),$$

$$\int_0^{2\pi} \eta^2(t)\sin t \cos t \; dt = \frac{\pi}{2}c_1 c_2,$$

we find that the compatibility condition (4.19) has the form

$$\begin{aligned} -4ac_1 + 3b(c_1^3 + c_1 c_2^2) &= 0, \\ 4ac_2 - 3b(c_1^2 c_2 + c_2^3) + 4K &= 0, \end{aligned} \tag{9.26}$$

and the matrix Q of (9.10) is

$$Q = \begin{bmatrix} \pi(a - \frac{3}{4}b(3c_1^2 + c_2^2)) & \pi\frac{3}{2}bc_1 c_2 \\ \pi\frac{3}{2}bc_1 c_2 & \pi(a - \frac{3}{4}b(c_1^2 + 3c_2^2)) \end{bmatrix}.$$

Since $K > 0$ the solutions of system (9.26) are

$$c_1 = 0, \qquad c_2 = c$$

where c is some of the roots of the equation

$$f(c) \equiv 3bc^3 - 4ac - 4K = 0. \tag{9.27}$$

An analysis of the function $f(c)$ shows that:

(i) for $K > K_0 = (4a\sqrt{a})/(9\sqrt{b})$ equation (9.27) has one solution;
(ii) for $K = K_0$ equation (9.27) has two solutions;
(iii) for $0 < K < K_0$ equation (9.27) has three solutions.

In these cases system (9.26) has respectively one, two and three solutions of the form (9.25). If for such a solution we have that

$$\det Q(0, c) = \pi^2 \left(a - \frac{3}{4} bc^2 \right) \left(a - \frac{9}{4} bc^2 \right) \neq 0,$$

then by Theorem 9.2 there exists $\epsilon_0 \in (0, \bar{\epsilon})$ such that for $|\epsilon| \leq \epsilon_0$ equation (9.24) has a unique 2π-periodic solution $z_\epsilon(t) = \mathrm{col}\,(x_\epsilon, y_\epsilon)$ for which

$$\begin{aligned} x_\epsilon(t) &\to x_0(t) \\ y_\epsilon(t) &\to y_0(t) \end{aligned} \qquad \text{as } \epsilon \to 0$$

uniformly on $t \in \mathbb{R}$ with

$$x_0(t) = c \sin t, \qquad y_0(t) = c \cos t.$$

10. Quasilinear equations with unfixed moments of an impulse effect

Consider the quasilinear T-periodic equation with unfixed moments of an impulse effect:

$$\begin{aligned} \frac{dx}{dt} &= A(t)x + g(t) + \epsilon f(t, x, \epsilon), & t \neq \tau_k(x, \epsilon), \\ \Delta x &= B_k x + h_k + \epsilon I_k(x, \epsilon), & t = \tau_k(x, \epsilon) \end{aligned} \qquad (10.1)$$

where $\tau_k(x, \epsilon) = \tau_k + \epsilon S_k(x, \epsilon)$ and $\epsilon \in J = (-\bar{\epsilon}, \bar{\epsilon})$ is a small parameter.

Assume that the following conditions (H10) hold:

H10.1 $A(\cdot) \in C(\mathbb{R}, \mathbb{R}^{n \times n})$, $g(\cdot) \in C(\mathbb{R}, \mathbb{R}^n)$, $B_k \in \mathbb{R}^{n \times n}$, $h_k \in \mathbb{R}^n$.
H10.2 The functions $f : \mathbb{R} \times \mathbb{R}^n \times J \to \mathbb{R}^n$, $I_k : \mathbb{R}^n \times J \to \mathbb{R}^n$ and $S_k : \mathbb{R}^n \times J \to \mathbb{R}$ are differentiable in their domains of definition.
H10.3 Equation (10.1) is T-periodic, that is, there exists a $q \in \mathbb{N}$ such that:

$$A(t + T) = A(t), \qquad g(t + T) = g(t), \qquad f(t + T, x, \epsilon) = f(t, x, \epsilon),$$

$$\tau_{k+q} = \tau_k + T, \qquad B_{k+q} = B_k, \qquad h_{k+q} = h_k,$$

$$I_{k+q}(x, \epsilon) = I_k(x, \epsilon), \qquad S_{k+q}(x, \epsilon) = S_k(x, \epsilon)$$

for $t \in \mathbb{R}$, $k \in \mathbb{Z}$, $x \in \mathbb{R}^n$ and $\epsilon \in J$.
H10.4 For the solutions of equation (10.1) the phenomenon of 'beating' is absent.
H10.5 Equation (9.3) has m $(1 \leq m \leq n)$ linearly independent T-periodic solutions $\varphi_1(t), \ldots, \varphi_m(t)$.

H10.6 The compatibility condition is met:

$$\int_0^T \Psi_1^*(s)g(s) \, ds + \sum_{k=1}^q \Psi_1^*(\tau_k^+)h_k = 0$$

where $\Psi_1(t) = [\psi_1(t), \ldots, \psi_m(t)]$ and $\psi_1(t), \ldots, \psi_m(t)$ are the linearly independent m-periodic solutions of equation (9.4).

By Theorem 4.2 the non-homogeneous equation (9.2) which is obtained from (10.1) for $\epsilon = 0$ has a T-parametric family of T-periodic solutions

$$\bar{x}(t, c) = \Phi_1(t)c + \tilde{x}_0(t)$$

where $c = \text{col}(c_1, \ldots, c_m) \in \mathbb{R}^m$, $\Phi_1(t) = [\varphi_1(t) \ldots, \varphi_m(t)]$ and $\tilde{x}_0(t)$ is a particular T-periodic solution of (9.2). Fix $c = c_0$ and for the generating solution

$$p(t) = \Phi_1(t)c_0 + \tilde{x}_0(t)$$

formulate the following problem.

Problem B   *Under what conditions does there exist a neighbourhood $J_0 = (-\epsilon_0, \epsilon_0)$ $\subset J$ such that for each $\epsilon \in J_0$ equation (10.1) has a T-periodic solution $x(t, \epsilon)$ with moments of an impulse effect $t_k(\epsilon)$ such that*

$$x(t, \epsilon) \xrightarrow{B} x(t, 0) = p(t) \quad \text{and} \quad t_k(\epsilon) \to \tau_k \quad \text{as } \epsilon \to 0? \tag{10.2}$$

The goal of the following considerations is to reduce equation (10.1) to an equation with fixed moments of the impulse effect in order to be able to apply the results of the last section.

Denote by $u(t; t_0, u_0, \epsilon)$ the solution of the initial value problem

$$\frac{du}{dt} = A(t)u + g(t) + \epsilon f(t, u, \epsilon), \quad u(t_0) = u_0$$

and for fixed $k \in \mathbb{Z}$ consider the system

$$t_k = \tau_k + \epsilon S_k(u(t_k; \tau_k, \eta, \epsilon), \epsilon), \tag{10.3}$$

$$
\begin{aligned}
u(t_k; \tau_k, \eta^+, \epsilon) &= u(t_k; \tau_k, \eta, \epsilon) + B_k u(t_k; \tau_k, \eta, \epsilon) \\
&\quad + h_k + \epsilon I_k(u(t_k; \tau_k, \eta, \epsilon), \epsilon)
\end{aligned}
\tag{10.4}
$$

with respect to t_k and η^+ for $\epsilon \in J$ and $\eta \in \mathbb{R}^n : |\eta - p(\tau_k)| < \delta$.

Lemma 10.1   *Let conditions H10.1–H10.3 hold. Then there exists $\epsilon_0 \in (0, \bar{\epsilon})$ such that, for $|\epsilon| \leq \epsilon_0$ and $\eta \in \mathbb{R}^n$, $|\eta - p(\tau_k)| < \epsilon_0$, system (10.3), (10.4) defines unique differentiable functions $t_k = t_k(\eta, \epsilon)$ and $\eta^+ = F_k(\eta, \epsilon)$ such that:*

1. *the following relations are valid:*

$$t_k(\eta, 0) = \tau_k, \qquad \frac{\partial t_k}{\partial \epsilon}(\eta, 0) = S_k(\eta, 0), \tag{10.5}$$

$$F_k(\eta, 0) = \eta + B_k\eta + h_k; \tag{10.6}$$

2. *the function $F_k(\eta, \epsilon)$ has the form*

$$F_k(\eta, \epsilon) = \eta + B_k\eta + h_k + \epsilon\Omega_k(\eta, \epsilon)$$

where the function $\Omega_k(\eta, \epsilon)$ is differentiable in a neighbourhood of the point $(p(\tau_k), 0)$ and

$$\begin{aligned}\Omega_k(\eta, 0) &= I_k(\eta, 0) \\ &+ \{(B_kA(\tau_k) - A(\tau_k)B_k)\eta + B_kg(\tau_k) - A(\tau_k)h_k\}S_k(\eta, 0),\end{aligned} \tag{10.7}$$

$$\begin{aligned}\frac{\partial\Omega_k}{\partial\eta}(\eta, 0) &= \frac{\partial I_k}{\partial\eta}(\eta, 0) + \{B_kA(\tau_k) - A(\tau_k)B_k\}S_k(\eta, 0) \\ &+ \{(B_kA(\tau_k) - A(\tau_k)B_k)\eta + B_kg(\tau_k) - A(\tau_k)h_k\}\frac{\partial S_k}{\partial\eta}(\eta, 0);\end{aligned} \tag{10.8}$$

3. *if the function $\eta(t)$ is a solution of the equation*

$$\frac{d\eta}{dt} = A(t)\eta + g(t) + \epsilon f(t, \eta, \epsilon), \quad t \neq \tau_k,$$

$$\eta(\tau_k^+) = F_k(\eta(\tau_k), \epsilon)$$

and $t_k = t_k(\eta(\tau_k), \epsilon)$ then the function

$$x(t) = \begin{cases} \eta(t), & t \notin [\tau_k; t_k], \\ u(t; \tau_k, \eta(\tau_k), \epsilon), & \text{if } \tau_k \leq t < t_k, \\ u(t; \tau_k, \eta(\tau_k^+), \epsilon), & \text{if } t_k < t \leq \tau_k, \end{cases} \tag{10.9}$$

$$x(t_k) = x(t_k^-)$$

is a solution of equation (10.1).

Proof   1. For $\epsilon = 0$ system (10.3), (10.4) implies that

$$t_k = \tau_k, \qquad u(\tau_k; \tau_k, \eta, 0) = \eta, \qquad u(\tau_k; \tau_k, \eta^+, 0) = \eta^+, \qquad \eta^+ = \eta + B_k \eta + h_k$$

and then the existence of the functions $t_k(\eta, \epsilon)$ and $F_k(\eta, \epsilon)$ and the validity of (10.5) and (10.6) follow from the implicit function theorem.

2. From (10.4) it follows that

$$\eta^+ = F_k(\eta, \epsilon) = u(\tau_k; t_k, z, \epsilon) \equiv P(t_k, z, \epsilon) \tag{10.10}$$

where

$$z = v + B_k v + h_k + \epsilon I_k(v, \epsilon), \tag{10.11}$$

$$v = u(t_k; \tau_k, \eta, \epsilon), \qquad t_k = t_k(\eta, \epsilon). \tag{10.12}$$

From (10.6) and the differentiability of $F_k(\eta, \epsilon)$ we conclude that

$$F_k(\eta, \epsilon) = \eta + B_k \eta + h_k + \epsilon \Omega_k(\eta, \epsilon)$$

where

$$\Omega_k(\eta, \epsilon) = \frac{1}{\epsilon}(F_k(\eta, \epsilon) - \eta - B_k \eta - h_k) \quad \text{for } \epsilon \neq 0,$$

and

$$\Omega_k(\eta, 0) = \lim_{\epsilon \to 0} \Omega_k(\eta, \epsilon) = \begin{bmatrix} 0 \\ 0 \end{bmatrix} = \frac{\partial F_k}{\partial \epsilon}(\eta, 0). \tag{10.13}$$

From (10.10)–(10.12) we have that

$$\frac{\partial F_k}{\partial \epsilon} = \frac{\partial P}{\partial t_k}\frac{\partial t_k}{\partial \epsilon} + \frac{\partial P}{\partial z}\frac{\partial z}{\partial \epsilon} + \frac{\partial P}{\partial \epsilon}$$

$$= \frac{\partial P}{\partial t_k}\frac{\partial t_k}{\partial \epsilon} + \frac{\partial P}{\partial z}\left[\frac{\partial z}{\partial v}\frac{\partial v}{\partial \epsilon} + \frac{\partial z}{\partial \epsilon}\right] + \frac{\partial P}{\partial \epsilon}.$$

In view of (10.12), (10.5) and the limit relations

$$v \to \eta, \qquad z \to \eta + B_k \eta + h_k, \qquad \frac{\partial P}{\partial z} = \frac{\partial u}{\partial z}(\tau_k; t_k, z, \epsilon) \to E,$$

$$\frac{\partial P}{\partial t_k} = \frac{\partial u}{\partial t_k}(\tau_k; t_k, z, \epsilon)$$

$$= -\frac{\partial P}{\partial z}[A(t_k)z + g(t_k) + \epsilon f(t_k, z, \epsilon)]$$

$$\to -A(\tau_k)(\eta + B_k \eta + h_k) - g(\tau_k),$$

$$\frac{\partial P}{\partial \epsilon} = \frac{\partial u}{\partial \epsilon}(\tau_k; t_k, z, \epsilon) \to 0,$$

$$\frac{\partial v}{\partial \epsilon} = \frac{\partial u}{\partial \epsilon}(\tau_k; t_k, \eta, \epsilon) \to 0,$$

$$\frac{\partial z}{\partial v} = E + B_k + \epsilon \frac{\partial I_k}{\partial v}(v, \epsilon) \to E + B_k,$$

$$\frac{\partial v}{\partial \epsilon} = \frac{\partial u}{\partial t_k}(t_k; \tau_k, \eta, \epsilon)\frac{\partial t_k}{\partial \epsilon} + \frac{\partial u}{\partial \epsilon}(t_k; \tau_k, \eta, \epsilon)$$

$$\to (A(\tau_k)\eta + g(\tau_k))S_k(\eta, 0),$$

$$\frac{\partial z}{\partial \epsilon} = I_k(v, \epsilon) + \epsilon \frac{\partial I_k}{\partial \epsilon}(v, \epsilon) \to I_k(\eta, 0)$$

as $\epsilon \to 0$ we obtain (10.7) which implies (10.8).

3. From (10.9) it follows that

$$x(t_k) = u(t_k; \tau_k, \eta(\tau_k), \epsilon)$$

and

$$x(t_k^+) = u(t_k; \tau_k, \eta(\tau_k^+), \epsilon).$$

Then from (10.3) and (10.4) we obtain that

$$t_k = \tau_k + \epsilon S_k(x(t_k), \epsilon),$$

$$x(t_k^+) = x(t_k) + B_k x(t_k) + h_k + \epsilon I_k(x(t_k), \epsilon),$$

which means that t_k is a moment of the impulse effect and the condition for a jump in (10.1) is met. From (10.9) it also follows that the function $x(t)$ for $t \neq t_k$ satisfies the differential equation in (10.1). □

The following theorems are corollaries of Lemma 10.1.

Theorem 10.1   *Let conditions (H10) hold and let Problem B have a solution. Then*

$$\Delta(c_0, 0) = \int_0^T \Psi_1^*(s)f(s, p(s), 0) \, ds + \sum_{k=1}^q \Psi_1^*(\tau_k^+)\Omega_k(p(\tau_k), 0) = 0 \qquad (10.14)$$

where $\Omega_k(p(\tau_k), 0)$ is given in (10.7).

Theorem 10.2   *Let conditions (H10) hold as well as the conditions*

$$\Delta(c_0, 0) = 0, \qquad \det \frac{\partial \Delta}{\partial c}(c_0, 0) \neq 0$$

where

$$\begin{aligned} Q = \frac{\partial \Delta}{\partial c}(c_0, 0) = \int_0^T \Psi_1^*(s)\frac{\partial f}{\partial x}(s, p(s), 0)\Phi_1(s) \, ds \\ + \sum_{k=1}^q \Psi_1^*(\tau_k^+)\frac{\partial \Omega_k}{\partial x}(p(\tau_k), 0)\Phi_1(\tau_k), \end{aligned} \qquad (10.15)$$

and $\frac{\partial \Omega_k}{\partial x}(p(\tau_k), 0)$ is given in (10.8).

Then there exists $\epsilon_0 \in (0, \bar{\epsilon})$ such that for $|\epsilon| \leq \epsilon_0$ equation (10.1) has a unique T-periodic solution $x(t, \epsilon)$ with moments of an impulse effect $t_k(\epsilon)$ which satisfy the limit relations (10.2).

11. Autonomous quasilinear equations

Consider the autonomous equation

$$
\begin{aligned}
\frac{dx}{dt} &= Ax + \epsilon g(x, \epsilon), & x \notin \sigma(\epsilon), \\
\Delta x &= B(x, \epsilon), & x \in \sigma(\epsilon),
\end{aligned}
\tag{11.1}
$$

where $x \in \mathbb{R}^n$, $n \geq 2$, $\epsilon \in J = (-\bar{\epsilon}, \bar{\epsilon})$ is a small parameter, $\sigma(\epsilon)$ consists of the hypersurfaces $\sigma_k(\epsilon)$, $k = 1, \ldots, q$, with equations

$$
\sigma_k(\epsilon) \equiv b_k + l_k^* x + \epsilon S_k(x, \epsilon) = 0
$$

and $b_k \in \mathbb{R}, l_k \in \mathbb{R}^n$, $S_k : \mathbb{R}^n \times J \to \mathbb{R}$ $(k = 1, \ldots, q)$. Suppose that $B(x, \epsilon) = B_k x + \epsilon I_k(x, \epsilon)$ for $\epsilon \in J$, $k = 1, \ldots, q$, and $x \in \sigma_k(\epsilon)$.

For $\epsilon = 0$ equation (11.1) takes the form

$$
\begin{aligned}
\frac{dx}{dt} &= Ax, & x \notin \sigma(0), \\
\Delta x &= B_k x, & x \in \sigma_k(0).
\end{aligned}
\tag{11.2}
$$

Suppose that equation (11.2) has a T_0-periodic solution $p(t)$ with moments of an impulse effect τ_k such that $\tau_{k+q} = \tau_k + T_0$ $(k \in \mathbb{Z})$. Denote $p_k = p(\tau_k)$ and assume that $p_k \in \sigma_k(0)$, $l_k^* A p_k \neq 0$ $(k = 1, \ldots, q)$. Without loss of generality suppose that $p(0) \notin \sigma(0)$ and $\dot{p}(0) \neq 0$.

The equation in variations corresponding to the solution $p(t)$ is

$$
\begin{aligned}
\frac{dx}{dt} &= Ax, & t \neq \tau_k, \\
\Delta x &= N_k x \equiv B_k x + [AB_k - B_k A] p_k \frac{l_k^* x}{l_k^* A p_k}, & t = \tau_k.
\end{aligned}
\tag{11.3}
$$

We shall consider the critical case when equation (11.3) has $m \geq 2$ linearly independent T_0-periodic solutions $\varphi_1(t), \ldots, \varphi_m(t)$. Then by Theorem 4.2 the equation adjoint to (11.3) also has m linearly independent T_0-periodic solutions $\psi_1(t), \ldots, \psi_m(t)$. Let $\Phi(t) = [\varphi_1(t), \ldots, \varphi_m(t), \ldots, \varphi_n(t)]$ and $\Psi(t) = [\psi_1(t), \ldots, \psi_m(t), \ldots, \psi_n(t)]$ be the fundamental matrices of these equations. Recall that

$$
\Psi^*(t)\Phi(t) \equiv \Psi^*(0)\Phi(0) \quad (t \in \mathbb{R}).
\tag{11.4}
$$

Moreover the monodromy matrix $M = \Phi(T_0)\Phi^{-1}(0)$ has rank $m' = n - m$ and

$$
(E - \Phi(T_0)\Phi^{-1}(0))p(0) = \sum_{k=1}^{q} \Phi(T_0)\Phi^{-1}(\tau_k^+)(B_k - N_k)p_k.
\tag{11.5}
$$

We set the following problem:

Problem C   Under what conditions does equation (11.1) have a periodic solution $x(t, \epsilon)$ with period $T = T(\epsilon)$ and moments of an impulse effect $t_k(\epsilon)$ such that for $t \in [0, T_0]$

$$x(t, \epsilon) \xrightarrow{B} x(t, 0) = p(t) \quad \text{as } \epsilon \to 0, \quad \text{and}$$

$$T(\epsilon) \to T_0, \qquad t_k(\epsilon) \to \tau_k \quad \text{as } \epsilon \to 0?$$

Assume that the following conditions hold.

H11.1 $A \in \mathbb{R}^{n \times n}$, $B_k \in \mathbb{R}^{n \times n}$, $b_k \in \mathbb{R}, l_k \in \mathbb{R}^n$ $(k = 1, \ldots, q)$.

H11.2 The functions $g : \mathbb{R}^n \times J \to \mathbb{R}^n$, $I_k : \mathbb{R}^n \times J \to \mathbb{R}^n$ and $S_k : \mathbb{R}^n \times J \to \mathbb{R}$ $(k = 1, \ldots, q)$ are continuously differentiable in $\mathbb{R}^n \times J$.

H11.3 The function $p : \mathbb{R} \to \mathbb{R}^n$ is a T_0-periodic solution of equation (11.1) with moments of an impulse effect τ_k and

$$\tau_{k+q} = \tau_k + T_0 \quad (k \in \mathbb{Z}),$$

$$p(0) \notin \sigma(0), \qquad \dot{p}(0) \neq 0,$$

$$p_k = p(\tau_k) \in \sigma_k(0), \qquad l_k^* A p_k \neq 0 \quad (k = 1, \ldots, q).$$

Consider the system

$$b_k + l_k^* u(t_k; \tau_k, \eta, \epsilon) + \epsilon S_k(u(t_k; \tau_k, \eta, \epsilon), \epsilon) = 0, \tag{11.6}$$

$$u(t_k; \tau_k, \eta^+, \epsilon) = u(t_k; \tau_k, \eta, \epsilon) + B_k u(t_k; \tau_k, \eta, \epsilon) + \epsilon I_k(u(t_k; \tau_k, \eta, \epsilon), \epsilon) \tag{11.7}$$

in a neighbourhood of the point $(p_k, 0)$ with respect to the unknowns t_k, η^+. Here $u(t; \tau_k, \eta, \epsilon)$ denotes the solution of the initial value problem

$$\frac{du}{dt} = Au + \epsilon g(u, \epsilon), \qquad u(\tau_k) = \eta.$$

We shall need the following lemma which is a corollary of the implicit function theorem and is proved as in Lemma 10.1.

Lemma 11.1   Let conditions H11.1–H11.3 hold. Then:

1. There exists a $\delta \in (0, \bar{\epsilon})$ such that in the neighbourhood

$$U = \{(\eta, \epsilon) \in \mathbb{R}^n \times J : |\eta - p_k| < \delta, |\epsilon| < \delta\}$$

system (11.6), (11.7) defines unique implicit functions $t_k = t_k(\eta, \epsilon)$ and $\eta^+ = F_k(\eta, \epsilon)$ which are continuously differentiable in U and satisfy the relations:

$$t_k(p_k, 0) = \tau_k, \qquad F_k(p_k, 0) = p_k + B_k p_k, \tag{11.8}$$

$$\frac{\partial t_k}{\partial \eta}(p_k, 0) = -\frac{l_k^*}{l_k^* A p_k}, \qquad \frac{\partial t_k}{\partial \epsilon}(p_k, 0) = -\frac{S_k(p_k, 0)}{l_k^* A p_k}, \qquad (11.9)$$

$$\frac{\partial F_k}{\partial \eta}(p_k, 0) = E + B_k + [AB_k - B_k A]p_k \frac{l_k^*}{l_k^* A p_k} = E + N_k, \qquad (11.10)$$

$$\frac{\partial F_k}{\partial \epsilon}(p_k, 0) = I_k(p_k, 0) + [AB_k - B_k A]p_k \frac{S_k(p_k, 0)}{l_k^* A p_k}, \qquad (11.11)$$

2. If $\eta(t)$ is a solution of the equation

$$\begin{aligned} \frac{d\eta}{dt} &= A\eta + \epsilon g(\eta, \epsilon), \quad t \neq \tau_k, \\ \eta^+ &= F_k(\eta, \epsilon), \qquad\quad t = \tau_k, \end{aligned} \qquad (11.12)$$

and $t_k = t_k(\eta(\tau_k), \epsilon)$ then the function

$$x(t) = \begin{cases} \eta(t), & t \notin [\tau_k; t_k], \\ u(t; \tau_k, \eta(\tau_k), \epsilon), & \tau_k \leq t < t_k, \\ u(t; \tau_k, \eta(\tau_k^+), \epsilon), & t_k < t \leq \tau_k, \end{cases}$$

$$x(t_k) = x(t_k^-)$$

is a solution of equation (11.1) with moment of the impulse effect $t_k = t_k(\eta(\tau_k), \epsilon)$.

We shall seek the T-periodic solution of Problem C in the form $x(t) = x(t, a, \epsilon)$, where

$$\begin{aligned} x(0, a, \epsilon) &= x_0 = p(0) + a. \\ x(t, a, 0) &= p(t) \quad (t \in \mathbb{R}, a \in \mathbb{R}^n, |a| < \bar{\epsilon}). \end{aligned} \qquad (11.13)$$

The solution $x(t, a, \epsilon)$ is T-periodic if and only if it satisfies the condition

$$x(T, a, \epsilon) = x(0, a, \epsilon) = x_0. \qquad (11.14)$$

Since equation (11.1) is autonomous we can impose an additional condition of the form $(\dot{p}(0) \,|\, x_0 - p(0)) = 0$ or

$$(\dot{p}(0) \,|\, a) = 0. \qquad (11.15)$$

Set $Q_k(\eta, \epsilon) = F_k(\eta, \epsilon) - \eta - N_k \eta$ and consider equation (11.12) written in the form

$$\begin{aligned} \frac{d\eta}{dt} &= A\eta + \epsilon g(\eta, \epsilon), \quad t \neq \tau_k, \\ \Delta\eta &= N_k\eta + Q_k(\eta, \epsilon), \quad t = \tau_k. \end{aligned} \qquad (11.16)$$

From Lemma 11.1 we conclude that instead of seeking a solution $x(t, a, \epsilon)$ of equation (11.1) satisfying conditions (11.13), (11.14) and (11.15), we can look for a solution $\eta(t, a, \epsilon) = \eta(t)$ of equation (11.16) satisfying these conditions.

For $t \geq 0$ the solution $\eta(t) = \eta(t, a, \epsilon)$ of (11.16) with initial value $\eta(0, a, \epsilon) = x_0 = p(0) + a$ satisfies the relation

$$
\eta(t) = \Phi(t)\Phi^{-1}(0)x_0 + \epsilon \int_0^t \Phi(t)\Phi^{-1}(s)g(\eta(s), \epsilon)\, ds
$$

$$
+ \sum_{0 < \tau_k < t} \Phi(t)\Phi^{-1}(\tau_k^+)Q_k(\eta_k, \epsilon)
$$

(11.17)

where $\eta_k = \eta(\tau_k) = \eta(\tau_k, a, \epsilon)$.

Then the condition $\eta(T, a, \epsilon) = x_0$ takes the form

$$
[\Phi(T)\Phi^{-1}(0) - E](a + p(0)) + \epsilon \int_0^T \Phi(T)\Phi^{-1}(s)g(\eta(s), \epsilon)\, ds
$$

$$
+ \sum_{k=1}^q \Phi(T)\Phi^{-1}(\tau_k^+)Q_k(\eta_k, \epsilon) = 0
$$

or

$$
[\Phi(T_0)\Phi^{-1}(0) - E](a + p(0)) + \epsilon \int_0^T \Phi(T)\Phi^{-1}(s)g(\eta(s), \epsilon)\, ds
$$

$$
+ [\Phi(T) - \Phi(T_0)]\Phi^{-1}(0)(a + p(0))
$$

(11.18)

$$
+ \sum_{k=1}^q \Phi(T)\Phi^{-1}(\tau_k^+)Q_k(\eta_k, \epsilon) = 0.
$$

We multiply (11.18) by $\Psi^*(T_0)$ and obtain the equivalent equation

$$
[\Psi^*(T_0)\Phi(T_0)\Phi^{-1}(0) - \Psi^*(T_0)](a + p(0)) + \epsilon \int_0^T \Psi^*(T_0)\Phi(T)\Phi^{-1}(s)g(\eta(s), \epsilon)\, ds
$$

$$
+ [\Psi^*(T_0)\Phi(T)\Phi^{-1}(0) - \Psi^*(T_0)\Phi(T_0)\Phi^{-1}(0)](a + p(0))
$$

$$
+ \sum_{k=1}^q \Psi^*(T_0)\Phi(T)\Phi^{-1}(\tau_k^+)Q_k(\eta_k, \epsilon) = 0.
$$

(11.19)

From (11.4) it follows that $\Psi^*(T_0)\Phi(T_0)\Phi^{-1}(0) = \Psi^*(0)$. Then we set $N(T) = \Psi^*(T_0)\Phi(T)$ and, taking into account that

$$
\Psi^*(T_0)\Phi(T)\Phi^{-1}(s) = N(T)N^{-1}(T_0)\Psi^*(s),
$$

we write (11.19) in the form

$$
[\Psi^*(0) - \Psi^*(T_0)](a + p(0)) + \epsilon \int_0^T N(T)N^{-1}(T_0)\Psi^*(s)g(\eta(s), \epsilon)\, ds
$$

$$
+ [N(T)N^{-1}(T_0) - E]\Psi^*(0)(a + p(0))
$$

(11.20)

$$
+ \sum_{k=1}^q N(T)N^{-1}(T_0)\Psi^*(\tau_k^+)Q_k(\eta_k, \epsilon) = 0.
$$

Since the first m columns of the matrix $\Psi(t)$ are T_0-periodic vector-valued functions, the first m rows of the matrix $\Psi^*(0) - \Psi^*(T_0)$ are equal to zero, that is,

$$\Psi^*(0) - \Psi^*(T_0) = \left[\begin{array}{c|c} O_{mm} & O_{mm'} \\ \hline R_{21} & R_{22} \end{array} \right] \qquad (11.21)$$

where $R_{21} \in \mathbb{R}^{m' \times m}$, $R_{22} \in \mathbb{R}^{m' \times m'}$.

Without loss of generality we shall assume that $\det R_{22} \neq 0$.

Set

$$a = (\alpha, \beta) \in \mathbb{R}^m \times \mathbb{R}^{m'},$$

$$P_m = [E_m | O_{mm'}] \in \mathbb{R}^{m \times n},$$

$$P_{m'} = [O_{m'm} | E_{m'}] \in \mathbb{R}^{m' \times n}.$$

Then from conditions (11.15), (11.20) and equality (11.21) we conclude that Problem C has a solution if and only if the following system is solvable with respect to a and T:

$$\Delta_0 \equiv (\dot{p}(0)|a) = 0, \qquad (11.22)$$

$$\begin{aligned}
\Delta_1 \equiv{}& \epsilon \int_0^T P_m N(T) N^{-1}(T_0) \Psi^*(s) g(\eta(s), \epsilon)\, ds \\
&+ P_m[N(T)N^{-1}(T_0) - E]\Psi^*(0)(a + p(0)) \\
&+ \sum_{k=1}^q P_m N(T) N^{-1}(T_0) \Psi^*(\tau_k^+) Q_k(\eta_k, \epsilon) = 0,
\end{aligned} \qquad (11.23)$$

$$\begin{aligned}
\Delta_2 \equiv{}& R_{21}\alpha + R_{22}\beta + \epsilon \int_0^T P_{m'} N(T) N^{-1}(T_0) \Psi^*(s) g(\eta(s), \epsilon)\, ds \\
&+ P_{m'}[N(T)N^{-1}(T_0) - E]\Psi^*(0)(a + p(0)) \\
&+ \sum_{k=1}^q P_{m'} N(T) N^{-1}(T_0) \Psi^*(\tau_k^+) Q_k(\eta_k, \epsilon) \\
&+ P_{m'}[\Psi^*(0) - \Psi^*(T_0)]p(0) = 0.
\end{aligned} \qquad (11.24)$$

Making use of formulae (11.10) and (11.5), we find that for $a = 0$, $T = T_0$ and $\epsilon = 0$,

$$\frac{\partial \Delta_0}{\partial a} = \dot{p}^*(0), \qquad \frac{\partial \Delta_0}{\partial T} = 0, \qquad \frac{\partial \Delta_0}{\partial \epsilon} = 0,$$

$$\frac{\partial \Delta_1}{\partial a} = 0, \qquad \frac{\partial \Delta_1}{\partial T} = \Psi_1^*(0)\dot{p}(0),$$

$$\frac{\partial \Delta_1}{\partial \epsilon} = \int_0^{T_0} \Psi_1^*(s) g(p(s), 0)\, ds + \sum_{k=1}^q \Psi_1^*(\tau_k^+) \frac{\partial F_k}{\partial \epsilon}(p_k, 0),$$

$$\frac{\partial \Delta_2}{\partial a} = [R_{21} | R_{22}], \qquad \frac{\partial \Delta_2}{\partial T} = \Psi_2^*(T_0)\dot{p}(0),$$

$$\frac{\partial \Delta_2}{\partial \epsilon} = \int_0^{T_0} \Psi_2^*(s)g(p(s),0)\,ds + \sum_{k=1}^q \Psi_2^*(\tau_k^+)\frac{\partial F_k}{\partial \epsilon}(p_k,0)$$

where $\Psi_1^*(s) = P_m \Psi^*(s)$ and $\Psi_2^*(s) = P_{m'} \Psi^*(s)$.

Then the Jacobi matrix for system (11.22)–(11.24) for $a = 0$, $T = T_0$, $\epsilon = 0$ is equal to

$$J = \frac{D(\Delta_0, \Delta_1, \Delta_2)}{D(a,T)} = \left[\begin{array}{c|c} \dot{p}^*(0) & 0 \\ \hline O_{mn} & \Psi_1^*(0)\dot{p}(0) \\ \hline R_{21}|R_{22} & \Psi_2^*(T_0)\dot{p}(0) \end{array} \right]$$

and its rank satisfies the inequality

$$1 \leq \operatorname{rank} J < n + 1 \qquad \text{for } 2 \leq m \leq n. \tag{11.25}$$

Moreover, $\Delta_i(0, T_0, 0) = 0$ $(i = 0, 1, 2)$ because of (11.5) and (11.8), and

$$\Delta_1(a, T, \epsilon) = \frac{\partial \Delta_1}{\partial T}(0, T_0, 0)(T - T_0) + \frac{\partial \Delta_1}{\partial \epsilon}(0, T_0, 0)\epsilon \tag{11.26}$$
$$+ o_1(a) + o_2(T - T_0) + o_3(\epsilon).$$

Suppose that Problem C has a solution, that is, system (11.22)–(11.24) defines implicitly functions $a(\epsilon)$ and $T(\epsilon)$ such that $a(\epsilon) \to 0$ and $T(\epsilon) \to T_0$ as $\epsilon \to 0$. Then from (11.26) it follows that the limit

$$\lim_{\epsilon \to 0} \frac{T(\epsilon) - T_0}{\epsilon} = \nu_0 \tag{11.27}$$

exists and necessarily the following equality is valid

$$\frac{\partial \Delta_1}{\partial \epsilon}(0, T_0, 0) + \frac{\partial \Delta_1}{\partial T}(0, T_0, 0)\nu_0 = 0. \tag{11.28}$$

Thus we obtain the following necessary condition for the existence of a solution to Problem C.

Theorem 11.1 *Let the following conditions hold.*

1. *Conditions H11.1–H11.3 are met.*
2. *There exist $\epsilon_0 \in (0, \bar{\epsilon})$ and $\delta > 0$ such that, for any $x_0 \in \mathbb{R}^n$, $|x_0 - p(0)| < \delta$, $|\epsilon| \leq \epsilon_0$ and $t \in [0, T_0 + \delta]$, the solution $x(t; x_0, \epsilon)$ of equation (11.1) is defined, for which $x(0; x_0, \epsilon) = x_0$ and for $t \in [0, T_0 + \delta]$ the point $(t, x(t; x_0, \epsilon))$ meets successively each of the hypersurfaces $\sigma_k(\epsilon)$, $k = 1, \ldots, q$, just once.*
3. *Problem C has a solution.*

Then there exists a $\nu_0 \in \mathbb{R}$ for which the following condition is valid

$$
\int_0^{T_0} \Psi_1^*(s) g(p(s), 0) \, ds + \nu_0 \Psi_1^*(0) \dot{p}(0)
$$

$$
+ \sum_{k=1}^q \Psi_1^*(\tau_k^+) \left[I_k(p_k, 0) + [AB_k - B_k A] p_k \frac{S_k(p_k, 0)}{l_k^* A p_k} \right] = 0. \tag{11.29}
$$

Moreover, the period $T(\epsilon)$ satisfies the limit relation (11.27).

Let relation (11.29) be valid and instead of the variable T introduce the variable $\tau : T = T_0 + \nu_0 \epsilon + \tau \epsilon$. Taking into account (11.27), we conclude that Problem C has a solution if and only if system (11.22)–(11.24) has 'small' solutions $a(\epsilon)$, $\tau(\epsilon)$, that is, solutions for which

$$
a(\epsilon) \to a(0) = 0, \qquad \tau(\epsilon) \to \tau(0) = 0, \qquad \text{as } \epsilon \to 0.
$$

By the implicit function theorem, from the equation $\Delta_2(\alpha, \beta, T, \epsilon) = 0$ we can uniquely express $\beta = \beta(\alpha, \tau, \epsilon)$ for α, τ, ϵ small enough. We replace the β obtained in the equations $\Delta_0 = 0$ and $\Delta_1 = 0$ and obtain the new system

$$
\begin{aligned}
\tilde{\Delta}_0 &\equiv \Delta_0(\alpha, \beta(\alpha, \tau, \epsilon)) = 0, \\
\tilde{\Delta}_1 &\equiv \Delta_1(\alpha, \beta(\alpha, \tau, \epsilon), T_0 + \nu_0 \epsilon + \tau \epsilon, \epsilon) = 0.
\end{aligned} \tag{11.30}
$$

In view of (11.25) and Lemma 1.1 of [89] we conclude that the Jacobi matrix

$$
\frac{D(\tilde{\Delta}_0, \tilde{\Delta}_1)}{D(\alpha, \tau)} = \begin{bmatrix} \dfrac{\partial \tilde{\Delta}_0}{\partial \alpha} & \dfrac{\partial \tilde{\Delta}_0}{\partial \tau} \\[2mm] \dfrac{\partial \tilde{\Delta}_1}{\partial \alpha} & \dfrac{\partial \tilde{\Delta}_1}{\partial \tau} \end{bmatrix}
$$

of system (11.30) equals zero for $\alpha = 0$, $\tau = 0$, $\epsilon = 0$. Consequently, system (11.30) is a branching system [89] and may have more than one small solution $\alpha(\epsilon)$, $\tau(\epsilon)$. As a corollary of Theorem 1.5 of [89] we obtain the following assertion:

Theorem 11.2   *Let the conditions of Theorem 11.1 hold and $\nu_0 \in \mathbb{R}$ exist satisfying (11.29). Then the number of solutions of Problem C coincides with the number of small solutions of system (11.30).*

Notes and comments for Chapter IV

The critical case for periodic impulsive equations with fixed moments of an impulse effect was first considered by S. G. Hristova and D. D. Bainov in [39]. Another consideration in this direction was carried out by A. A. Boichuk, N. A. Perestyuk and A. M. Samoilenko in [19]. The iterative procedure (9.15)–(9.18) for finding the periodic solution was justified by P. S. Simeonov. See [29] for different constructive methods of analysis of periodic equations without impulses.

The results in Sections 10 and 11 are new and were proved by P. S. Simeonov. For considerations in this direction see also M. U. Ahmetov and N. A. Perestyuk [4].

Note that in Sections 7, 10 and 11 the impulsive equation being considered is reduced to an equation with fixed moments of an impulse effect. This 'method of correction' was first used by P. S. Simeonov and D. D. Bainov in [84] and [82].

Chapter V

Non-linear Periodic Impulsive Equations

In this chapter we shall consider the question of the existence of periodic solutions of non-linear periodic impulsive equations.

12. The monodromy operator and the existence of periodic solutions

In this section, making use of the monodromy operator and the theorem of Bohl–Brouwer, we shall prove some results on the existence of a T-periodic solution of the non-linear periodic impulsive equation

$$
\begin{aligned}
\frac{\mathrm{d}x}{\mathrm{d}t} &= f(t,x), \quad t \neq \tau_k, \\
\Delta x &= I_k(x), \quad t = \tau_k
\end{aligned}
\tag{12.1}
$$

where $t \in \mathbb{R}$, $k \in \mathbb{Z}$, $x \in \Omega \subset \mathbb{R}^n$.

Introduce the following conditions (H12).

H12.1 Equation (12.1) is T-periodic, that is, there exists a $q \in \mathbb{N}$ such that

$$
f(t+T,x) = f(t,x), \qquad I_{k+q}(x) = I_k(x), \qquad \tau_{k+q} = \tau_k + T
$$

for $t \in \mathbb{R}$, $k \in \mathbb{Z}$, and $x \in \Omega$.

H12.2 The function $f : \mathbb{R} \times \Omega \to \mathbb{R}^n$ is continuous in each of the sets $(\tau_k, \tau_{k+1}] \times \Omega$ $(k \in \mathbb{Z})$ and for each $x \in \Omega$ and $k \in \mathbb{Z}$ there exists the finite limit of $f(t,y)$ as $(t,y) \to (\tau_k, x)$, $t > \tau_k$.

H12.3 The functions $I_k : \Omega \to \mathbb{R}^n$ $(k \in \mathbb{Z})$ are continuous in Ω.

H12.4 The function $f(t,y)$ is locally Lipschitz continuous with respect to x in the domain $\mathbb{R} \times \Omega$.

Without loss of generality we assume that

$$
\tau_0 < 0 < \tau_1 < \cdots < \tau_q < T < \tau_{q+1}.
$$

Before we go on to the proof of the main results, we shall introduce some auxiliary assertions.

Let $t, s \in \mathbb{R}$, $y \in \Omega$ and $x(t; s, y)$ be a solution of equation (12.1) which is defined in the interval $J(s,y)$ and $x(s^+; s, y) = y$.

Definition 12.1   We call the operator $U(t, s) : \Omega \to \mathbb{R}^n$ defined by the formula $U(t, s)y = x(t; s, y)$ a *translation operator* along the trajectories of equation (12.1) (a Poincaré operator).

Definition 12.2   We call the operator $U = U(T, 0)$ a *monodromy operator.*

Remark 12.1   The operator $U(t, s)$ is defined uniquely for those $y \in \Omega$ for which $t \in J(s, y)$. If, for fixed $t, s \in \mathbb{R}$ and any $y \in \Omega$, there exists a unique solution $x(\tau; s, y)$ of equation (12.1) which is defined for $\tau = t$, then the operator $U(t, s)$ is uniquely defined in Ω. In the general case it is possible for the operator $U(t, s)$ to be undefined or to be defined only in some subset of Ω.

Remark 12.2   If the solution $x(t; s, y)$ of equation (12.1) depends continuously on the initial data then the operator $U(t, s)$ is continuous. Note that conditions H12.2–H12.4 ensure the continuity of $U(t, s)$. Moreover, the following relations hold:

$$U(t, t) = E \qquad\qquad (t \in \mathbb{R}, t \neq \tau_k),$$
$$U(\tau_k^+, \tau_k) = E \qquad\qquad (k \in \mathbb{Z}),$$
$$U(\tau_k^+, \tau_k^-)y = y + I_k(y) \quad (k \in \mathbb{Z}, y \in \Omega).$$

Lemma 12.1   *Let conditions (H12) hold. Then equation (12.1) has a T-periodic solution if and only if there exists an $x_0 \in \Omega$ such that*

$$x(T; 0, x_0) = x_0 \qquad \text{and} \qquad T \in J(0, x_0). \tag{12.2}$$

Proof   The necessity of condition (12.2) is obvious.

    Sufficiency. Let the solution $x(t) = x(t; 0, x_0)$ of equation (12.1) be defined for $t \in [0, T]$ and let condition (12.2) hold. Then the T-periodic extension

$$y(t) = \begin{cases} x(t), & t \in (0, T], \\ x(t - jT), & t \in (jT, jT + T] \end{cases}$$

of the function $x(t)$ is a T-periodic solution of (12.1). Indeed, if $t \in (jT, jT + T]$, then $t - jT \in (0, T]$ and

$$\frac{dy}{dt}(t) = \frac{dx}{dt}(t - jT) = f(t - jT, x(t - jT))$$
$$= f(t, x(t - jT))$$
$$= f(t, y(t)).$$

Analogously, if $\tau_k \in (jT, jT + T]$, then $\tau_{k-jq} = \tau_k - jT \in (0, T]$ and

$$\Delta y(\tau_k) = \Delta x(\tau_k - jT) = I_{k-jq}(x(\tau_k - jT))$$
$$= I_k(x(\tau_k - jT))$$
$$= I_k(y(\tau_k)).$$

$\square$

Remark 12.3   Condition (12.2) means that the monodromy operator $U = U(T, 0)$ has a fixed point $x_0 \in \Omega$.

Let G be a domain in $\mathbb{R}^n$ with boundary Γ and closure $D = G \cup \Gamma \subset \Omega$.

Definition 12.3   The set D is said to be *canonical* if:

(i) the domain G is a bounded and convex set;
(ii) the closure D is defined by a finite number of inequalities

$$\Phi_i(x) \le 0 \quad (i = 1, \dots, r)$$

where the functions $\Phi_i : \mathbb{R}^n \to \mathbb{R}$ $(i = 1, \dots, r)$ are smooth;
(iii) if for $x \in \Gamma$ and $i \in \{1, \dots, r\}$ the equality $\Phi_i(x) = 0$ is valid, then $\frac{\partial \Phi_i}{\partial x}(x) \ne 0$.

For any point $x \in \Gamma$ define the set $\alpha(x) = \{i \in \{1, \dots, r\} : \Phi_i(x) = 0\}$. We need the following theorem [5].

Theorem 12.1 (Bohl–Brouwer fixed point theorem)   *Let S be a non-empty compact convex subset of $\mathbb{R}^n$ and let the operator $U : S \to S$ be continuous. Then U has a fixed point $x \in S$.*

Theorem 12.2   *Let the following conditions hold.*

1. *Conditions (H12) are met.*
2. *The set D is canonical.*
3. *The function f satisfies the condition*

$$\frac{\partial \Phi_i}{\partial x}(x) f(t, x) < 0 \qquad (t \in \mathbb{R}, x \in \Gamma, i \in \alpha(x)).$$

4. $\Phi_j(x + I_k(x)) \le 0$ $(x \in D; j = 1, \dots, r; k = 1, \dots, q)$.

Then equation (12.1) has a T-periodic solution $y(t)$ and $y(t) \in D$ for $t \in \mathbb{R}$.

Proof   Let $y \in D$ and consider the solution $x(t) = x(t; 0, y)$ of equation (12.1) defined for $t \in J = [0, \omega)$.

We shall first prove that $x(t) \in D$ for $t \in J$. Suppose that this is not true and that there exists a $\tau \in J$ such that $x(\tau) \notin D$. Let k be the least of the numbers $i = 0, 1, 2, \ldots$ for which there exists $\tau \in (\tau_i, \tau_{i+1}] \cap J$ such that $x(\tau) \notin D$. For this choice of k we have that

$$x(t) \in D \quad (t \in [0, \tau_k] \cap J),$$
$$x(\tau) \notin D \quad (\text{for some } \tau \in (\tau_k, \tau_{k+1}] \cap J).$$

Introduce the set $\mathcal{N} = \{t \in (\tau_k, \tau_{k+1}] \cap J : x(t) \notin D\}$. The set $\mathcal{N}$ is non-empty since $\tau \in \mathcal{N}$.

Let $t_0 \in \mathcal{N}$, that is, $x(t_0) \notin D$. Then there exists $i \in \{1, \ldots, r\}$ such that

$$\Phi_i(x(t_0)) > 0.$$

But since the function $\Phi_i(x(t))$ is continuous at the point $t_0 \in (\tau_k, \tau_{k+1}]$, there exists a $\delta > 0$ such that $\Phi_i(x(t)) > 0$ for $t \in (t_0 - \delta, t_0]$, that is, there exists $t_1 \in \mathcal{N}$ which is smaller than t_0.

Let $\xi = \inf \mathcal{N}$. From the above arguments it follows that $\xi \notin \mathcal{N}$ and $\xi \neq \tau_{k+1}$. Thus $x(t) \in D$ for $t \in [0, \xi]$ and for the point ξ there are two possibilities: either $\xi = \tau_k$ or $\xi \in (\tau_k, \tau_{k+1}) \cap J$.

Let $\xi = \tau_k$. Since $x(\tau_k) \in D$, then by condition 4 of the theorem $x(\tau_k^+) = x(\tau_k) + I_k(x(\tau_k)) \in D$. We shall prove that there exists a right neighbourhood $(\xi, \xi + \delta)$ of the point $\xi = \tau_k$ such that

$$x(t) \in G \quad \text{for } t \in (\xi, \xi + \delta). \tag{12.3}$$

Indeed, let $x(\tau_k^+) \in G$. Then this assertion follows from the fact that G is an open set, $x(t)$ is a continuous function for $t \in (\tau_k, \tau_{k+1}] \cap J$ and $\lim_{t \to \tau_k^+} x(t) = x(\tau_k^+)$.

Let $x(\tau_k^+) \in \Gamma$. Then

$$\Phi_i(x(\tau_k^+)) = 0 \quad \text{for } i \in \alpha(x(\tau_k^+)),$$
$$\Phi_j(x(\tau_k^+)) < 0 \quad \text{for } j \notin \alpha(x(\tau_k^+)). \tag{12.4}$$

From the continuity of the function Φ_j, from the fact that $x(t)$ is a continuous function in some neighbourhood $(\tau_k, \tau_k + \delta^*)$, and since $\lim_{t \to \tau_k^+} x(t) = x(\tau_k^+)$ it follows that there exists a $\delta_1 > 0$ such that for $t \in (\tau_k, \tau_k + \delta_1)$ inequality (12.4) is valid. For $i \in \alpha(x(\tau_k^+))$ we have

$$\frac{d\Phi_i}{dt}(x(\tau_k^+)) = \frac{\partial \Phi_i}{\partial x}(x(\tau_k^+)) f(\tau_k^+, x(\tau_k^+)) < 0.$$

Consequently, there exists a $\delta_2 > 0$ such that for $t \in (\tau_k, \tau_k + \delta_2)$

$$\Phi_i(x(t)) < 0 \quad (i \in \alpha(x(\tau_k^+))). \tag{12.5}$$

Let $\delta = \min(\delta_1, \delta_2)$. Then inequality (12.5) is valid for $t \in (\tau_k, \tau_k + \delta)$ and $i = 1, \ldots, r$. This means that (12.3) is valid also in this case, which contradicts the fact that $\xi = \inf \mathcal{N}$. Hence $\xi \neq \tau_k$.

In the case when $\xi \in (\tau_k, \tau_{k+1}) \cap J$, by similar arguments we arrive at a contradiction.

Thus the assumption is not true and $x(t) \in D$ for each $t \in J$. Since D is a compact set contained in Ω, by Theorem 2.6 the solution $x(t)$ is continuable for each $t > 0$, and $x(t)$ does not leave D for $t > 0$. In particular, $x(T) = x(T; 0, y) \in D$. This means that the operator $U = U(T, 0)$ is defined for each $y \in D$ and $Uy \in D$, that is, $U : D \to D$. By Remark 12.2 the operator U is continuous in D. Consequently, the conditions of the Bohl–Brouwer theorem are met, which claims that the operator U has a fixed point $x_0 \in D$, that is, $U(T, 0)x_0 = x_0$ or $x(T; 0, x_0) = x_0$. Then by Lemma 12.1 equation (12.1) has a T-periodic solution $y(t)$ and since $x(t; 0, x_0) \in D$ for $t \in [0, T]$, then $y(t) \in D$ for $t \in \mathbb{R}$. $\square$

Let

$$\Delta_1 = [0, \tau_1],$$

$$\Delta_k = (\tau_{k-1}, \tau_k] \qquad (k = 2, \ldots, q),$$

$$\Delta_{q+1} = (\tau_q, T].$$

Lemma 12.2   *Let the following conditions hold.*

1. *The functions $f_m(t, x)$ are continuous in the sets $\Delta_k \times D$ $(k = 1, \ldots, q+1)$ and there exists the finite limit*

$$\lim_{m \to \infty} f_m(t, x) = f(t, x)$$

   uniformly with respect to $(t, x) \in [0, T] \times D$.
2. *For $x \in D$ and $k = 1, \ldots, q$ there exists the finite limit*

$$\lim_{\substack{(t,y) \to (\tau_k, x) \\ t > \tau_k}} f_m(t, y) = f_m(\tau_k^+, x) \qquad (m \in \mathbb{N}).$$

3. *The functions $I_k(x)$ $(k = 1, \ldots, q)$ are continuous in D.*
4. *The set D is compact.*
5. *The equation*

$$\frac{\mathrm{d}x}{\mathrm{d}t} = f_m(t, x), \qquad t \neq \tau_k, t \in [0, T],$$
$$\Delta x(\tau_k) = I_k(x(\tau_k)), \qquad k = 1, \ldots, q$$

   has a solution $x_m(t)$ and $x_m(t) \in D$ for $t \in [0, T]$.

Then from the sequence $\{x_m(t)\}$ a subsequence $\{x_{m_j}(t)\}$ can be chosen, which is uniformly convergent in $[0, T]$ and its limit $x(t) = \lim_{j \to \infty} x_{m_j}(t)$ is a solution of equation (12.1). Moreover, $x(t) \in D$ for $t \in [0, T]$.

Proof   From condition 1 of the lemma, it follows that the function f is continuous in the sets $\Delta_k \times D$ $(k = 1, \ldots, q+1)$, and from conditions 1 and 2 it follows that for $x \in D$ and $k = 1, \ldots, q$ there exists the limit of $f(t, y)$ as $(t, y) \to (\tau_k, x)$, $t > \tau_k$. Since D is compact, there exists a constant $M > 0$ such that

$$|f(t, x)| \le M, \qquad |f_m(t, x)| \le M, \qquad |I_k(x)| \le M, \qquad |x_m(0)| \le M$$

for $t \in [0, T]$, $x \in D$, $m \in \mathbb{N}$ and $k = 1, \ldots, q$.

Then from the equality

$$x_m(t) = x_m(0) + \int_0^t f_m(s, x_m(s))\, ds + \sum_{0 < \tau_k < t} I_k(x_m(\tau_k))$$

there follow the estimates

$$|x_m(t)| \le M + \int_0^t M\, ds + \sum_{0 < \tau_k < t} M \le M(1 + T + q),$$

$$|x_m(t_2) - x_m(t_1)| \le \left| \int_{t_1}^{t_2} f_m(s, x_m(s))\, ds \right| \le M(t_2 - t_1)$$

for $t \in [0, T]$, $t_1, t_2 \in \Delta_k$, $t_1 < t_2$ and $m \in \mathbb{N}$, whence we conclude that the family of functions $\{x_m(t)\}$ is uniformly bounded and quasiequicontinuous in $[0, T]$. Then by Lemma 2.4 the sequence $\{x_m(t)\}$ is a relatively compact subset in the space $S = PC([0, T], \mathbb{R}^n)$; that is why from $\{x_m(t)\}$ one can choose a convergent subsequence $\{x_{m_j}(t)\}$, i.e., there exists $x(t) \in S$ such that

$$\lim_{j \to \infty} x_{m_j}(t) = x(t) \qquad \text{(uniformly on } t \in [0, T]).$$

Hence $x(t) \in D$ for $t \in [0, T]$. From the equality

$$x_{m_j}(t) = x_{m_j}(0) + \int_0^t f_{m_j}(s, x_{m_j}(s))\, ds$$

$$+ \sum_{0 < \tau_k < t} I_k(x_{m_j}(\tau_k)) \qquad (t \in [0, T])$$

passing to the limit as $j \to \infty$ we obtain

$$x(t) = x(0) + \int_0^t f(s, x(s))\, ds$$

$$+ \sum_{0 < \tau_k < t} I_k(x(\tau_k)) \qquad (t \in [0, T])$$

which implies that $x(t)$ is a solution of (12.1). $\square$

Theorem 12.3   *Let the following conditions hold.*

1. *Conditions H12.1–H12.3 are valid.*
2. *The set D is canonical.*
3. *The function f satisfies the condition*

$$\frac{\partial \Phi_i}{\partial x}(x)f(t,x) \le 0 \qquad (t \in \mathbb{R}, x \in \Gamma, i \in \alpha(x)).$$

4. $\Phi_j(x + I_k(x)) \le 0 \ (x \in D; j = 1,\ldots,r; k = 1,\ldots,q).$

Then equation (12.1) has a T-periodic solution $y(t)$ and $y(t) \in D$ for $t \in \mathbb{R}$.

Proof   Let the point $y_0 \in G$ be fixed and let $\delta > 0$ be a real number. Consider the equation

$$\frac{dx}{dt} = f(t,x) + \delta(y_0 - x), \quad t \ne \tau_k,$$
$$\Delta x(\tau_k) = I_k(x(\tau_k)), \qquad k = 1,\ldots,q.$$

Since the set D is canonical and the point y_0 is interior, there exists $a > 0$ such that

$$\frac{\partial \Phi_i}{\partial x}(x)(y_0 - x) \le -a \tag{12.6}$$

for $x \in \Gamma$ and $i \in \alpha(x)$. Then from (12.6) and condition 3, it follows that

$$\frac{\partial \Phi_i}{\partial x}(x)f(t,x) + \delta\frac{\partial \Phi_i}{\partial x}(x)(y_0 - x) < -\delta a \tag{12.7}$$

for $x \in \Gamma$ and $i \in \alpha(x)$.

From conditions H12.1–H12.3 and the compactness of D it follows that for any $\epsilon > 0$ there exists a function $f_\epsilon(t,x)$ such that

(i) $f_\epsilon(t,x)$ is continuous in $(\tau_{k-1},\tau_k] \times D$ $(k \in \mathbb{Z})$ and for $x \in D$ and $k \in \mathbb{Z}$ there exists the finite limit of $f_\epsilon(t,y)$ as $(t,y) \to (\tau_k,x)$, $t > \tau_k$;
(ii) $f_\epsilon(t+T,x) = f_\epsilon(t,x)$ $(t \in \mathbb{R}, x \in D)$;
(iii) $|f_\epsilon(t,x) - f(t,x)| < \epsilon$ $(t \in \mathbb{R}, x \in D)$;   (12.8)
(iv) $f_\epsilon(t,x)$ is Lipschitz continuous with respect to x in $\mathbb{R} \times D$.

Then from (12.7), (12.8) and the boundedness of $\Phi_i(x)$ in Γ it follows that to each $\delta > 0$ there corresponds an $\epsilon = \epsilon(\delta) \in (0,\delta)$ such that

$$\frac{\partial \Phi_i}{\partial x}(x)f_\epsilon(t,x) + \delta\frac{\partial \Phi_i}{\partial x}(x)(y_0 - x) \le \frac{-a\delta}{2}$$

for $x \in \Gamma$ and $i \in \alpha(x)$.

Choose the sequences $\delta_m = m^{-1}$, $\epsilon_m = \epsilon(m^{-1})$, $\lim_{m\to\infty} \epsilon(m^{-1}) = 0$ and consider the system

$$\frac{dx}{dt} = f_m(t,x), \quad t \ne \tau_k,$$
$$\Delta x(\tau_k) = I_k(x(\tau_k)), \quad k \in \mathbb{Z} \tag{12.9}$$

where $f_m(t,x) = f_{\epsilon m}(t,x) + \delta_m(y_0 - x)$.

Since equation (12.9) satisfies the conditions of Theorem 12.2 this equation has a T-periodic solution $x_m(t)$ and $x_m(t) \in D$ for $t \in \mathbb{R}$.

But equation (12.9) and the sequence $\{x_m(t)\}$ satisfy the conditions of Lemma 12.2, so from the sequence $\{x_m(t)\}$ one can choose a subsequence $\{x_{m_j}(t)\}$ which is uniformly convergent in $[0,T]$, its limit $x(t) = \lim_{j \to \infty} x_{m_j}(t)$ is a solution of (12.1), and $x(t) \in D$ for $t \in [0,T]$.

Since the functions $x_{m_j}(t)$ are T-periodic the solution $x(t)$ is also T-periodic and $x(t) \in D$ for $t \in \mathbb{R}$. $\square$

Theorem 12.4   Let the following conditions hold.

1. Conditions H12.1–H12.3 are valid.
2. The set D is canonical.
3. The function f satisfies the condition
$$\frac{\partial \Phi_i}{\partial x}(x)f(t,x) \geq 0 \qquad (t \in \mathbb{R}, x \in \Gamma, i \in \alpha(x)).$$
4. The mappings $\Psi_k : D \to \mathbb{R}^n$, $\Psi_k(x) = x + I_k(x)$ $(k = 1,\ldots,q)$ are homeomorphisms and $\Psi_k(D) \supset D$ $(k = 1,\ldots,q)$.

Then equation (12.1) has a T-periodic solution $x(t)$ and $x(t) \in D$ for $t \in \mathbb{R}$.

Proof   Consider the equation
$$\frac{dy}{dt} = -f(-t,y), \qquad t \neq -\tau_k,$$
$$\Delta y(-\tau_k^+) = \Psi_k^{-1}(y(-\tau_k)), \quad k \in \mathbb{Z}. \tag{12.10}$$

A straightforward verification shows that equation (12.10) satisfies the conditions of Theorem 12.3. Thus it has a T-periodic solution $y(t)$ and, moreover, $y(t) \in D$ for $t \in \mathbb{R}$. Then the function
$$x(t) = \begin{cases} y(-t), & t \neq \tau_k, \\ y(-\tau_k^+), & t = \tau_k \end{cases}$$
is a T-periodic solution of equation (12.1) and $x(t) \in D$ for $t \in \mathbb{R}$. $\square$

Consider equation (12.1) in the particular cases which $n = 1$ and $n = 2$.

Theorem 12.5   Let the following conditions hold.

1. Conditions H12.1 and H12.2 are met for $n = 1$.
2. The functions $\Psi_k : \mathbb{R} \to \mathbb{R}$, $\Psi_k(x) = x + I_k(x)$ $(k = 1,\ldots,q)$ are continuous and strictly monotone in $\mathbb{R}$, and p of them are decreasing $(0 \leq p \leq q)$.
3. Each point $(t_0, x_0) \in \mathbb{R}^2$ is a point of uniqueness for the equation $x' = f(t,x)$.

Then each solution of equation (12.1) which is defined and bounded in $\mathbb{R}_+$ $(\mathbb{R}_-)$ either is ω-periodic itself, or tends asymptotically to an ω-periodic solution of (12.1) as $t \to +\infty$ $(t \to -\infty)$. Moreover, $\omega = T$ if p is even, and $\omega = 2T$ if p is odd.

Proof   Denote by $x(t, x_0)$ the solution of (12.1) for which $x(0, x_0) = x_0$. Let B_+ be the set of the initial values $y = x(0, y)$ of all solutions $x(t, y)$ of equation (12.1) which are defined and bounded in $\mathbb{R}_+$. We define the monodromy operator $U : B_+ \to \mathbb{R}$ by the formula $U(y) = x(T, y)$ $(y \in B_+)$.

From the T-periodicity of equation (12.1) it follows that the function $z(t) = x(t+T, y)$ for $y \in B_+$ is also a solution of (12.1) which is defined and bounded in $\mathbb{R}_+$. Then from the relation $U(y) = x(T, y) = z(0) \in B_+$ it follows that $U(B_+) \subset B_+$, that is, $U : B_+ \to B_+$.

From the T-periodicity of equation (12.1) it also follows that the mth power of the operator U satisfies the equality

$$U^m(y) = x(mT, y) \qquad (m \in \mathbb{N}).$$

Since the integral curves of equation (12.1) do not intersect one another, the function U is monotone. Moreover, U is increasing if the number p is even, and U is decreasing if p is odd.

Let $x(t, x_0)$ be a solution of (12.1) which is defined and bounded in $\mathbb{R}_+$. If $U(x_0) = x_0$, then by Lemma 12.1 there exists a T-periodic solution $\varphi(t)$ of (12.1) for which $\varphi(0) = x_0$, and from condition 3 it follows that $x(t, x_0) \equiv \varphi(t)$.

Suppose that $U(x_0) \neq x_0$. If p is even, then the sequence $\{U^m(x_0)\} = \{x(mT, x_0)\}$ is monotone and bounded, hence there exists the limit $\lim_{m \to \infty} x(mT, x_0) = a$.

We shall show that the solution $x(t, a)$ is T-periodic. We note that the functions $y_m(t) = x(t+mT, x_0)$, due to the T-periodicity of equation (12.1), are also its solutions and $y_m(0) \to a$ as $m \to \infty$. Then by Theorem 2.9 on the continuous dependence of the solutions on the initial values we have that

$$x(T, a) = \lim_{m \to \infty} y_m(T) = a.$$

Hence the solution $x(t, a)$ is T-periodic.

Let $\epsilon > 0$ be given. By Theorem 2.9 there exists a $\delta > 0$ such that

$$|x(t, y) - x(t, a)| < \epsilon$$

if $|y - a| < \delta$ and $0 \leq t \leq T$. Choose $m_1 > 0$ so that

$$|x(mT, x_0) - a| < \delta$$

for $m > m_1$. Then $|x(t, x_0) - x(t, a)| < \epsilon$ for $t > mT$ which proves that

$$\lim_{t \to \infty}[x(t, x_0) - x(t, a)] = 0.$$

Let the number p be odd. Then the function $U^2(x) = U(U(x))$ is increasing. If $U^2(x_0) = x_0$, then by Lemma 12.1 the solution $x(t, x_0)$ is $2T$-periodic.

If $U^2(x_0) \neq x_0$, then the sequence $\{x(2mT, x_0)\} = \{U^{2m}(x_0)\}$ is monotone and bounded, hence there exists the limit

$$\lim_{m \to \infty} x(2mT, x_0) = \lim_{m \to \infty} U^{2m}(x_0) = b.$$

Then, as above, it is proved that the function $x(t, b)$ is a $2T$-periodic solution of (12.1) and

$$\lim_{t \to \infty}[x(t, x_0) - x(t, b)] = 0.$$

By a change of the independent variable $t = -\tau$ the validity of the assertion is established in the case when the solution $x(t, x_0)$ of (12.1) is defined and bounded in $\mathbb{R}_-$. $\square$

Remark 12.4   It is not necessary that $\omega = T$ and $\omega = 2T$ be the minimal periods of the ω-periodic solutions in the formulation of Theorem 12.5.

Remark 12.5   When equation (12.1) is without impulses, that is, $I_k(x) \equiv 0$, then $\Psi_k(x) = x$ are increasing functions. Then $p = 0$ and by Theorem 12.5 we have that $\omega = T$.

When equation (12.1) is impulsive, it is possible that a periodic solution with the smallest period $\omega = 2T$ exists. This is seen from the following example.

Example 12.1   The impulsive differential equation

$$\frac{dx}{dt} = \sin x, \quad t \neq k,$$
$$\Delta x = -2x, \quad t = k \, (k \in \mathbb{Z})$$

is periodic with period $T = 1$. If $0 < x_0 < \pi$, then the solution $x(t, x_0)$ as $t \to +\infty$ asymptotically tends to the solution

$$\varphi(t) = \begin{cases} \pi, & 2k - 1 < t \leq 2k \ (k \in \mathbb{Z}) \\ -\pi, & 2k < t \leq 2k + 1 \ (k \in \mathbb{Z}) \end{cases}$$

which has a period $\omega = 2T = 2$. $\square$

Remark 12.6   If the number p is odd and equation (12.1) has at least one solution which is defined and bounded in $\mathbb{R}_+$ ($\mathbb{R}_-$), then equation (12.1) has a unique T-periodic solution. Indeed, in this case the set B_+ (B_- is defined analogously) is non-empty and the function $U : B_+ \to B_+$ ($U : B_- \to B_-$) defined in Theorem 12.5 is decreasing. Since $U(B_\pm) \subset B_\pm$ and the set $B_\pm$ is connected, U has a unique fixed point $x_0 \in B_+$ ($x_0 \in B_-$), that is, equation (12.1) has a unique T-periodic solution.

The isolated periodic solutions of equation (12.1) in the case $n = 1$ have some remarkable properties.

Definition 12.4   The periodic solution $x = \varphi(t)$ of equation (12.1) is said to be *isolated* if there exists an $h > 0$ such that in the domain

$$\{(t, x) \in \mathbb{R}^2 : \varphi(t) - h \leq x \leq \varphi(t) + h\} \tag{12.11}$$

there are no periodic solutions of (12.1) distinct from $\varphi(t)$.

Theorem 12.6   *Let the conditions of Theorem 12.5 hold and let equation (12.1) have a stable isolated periodic solution. Then this solution is asymptotically stable.*

Proof   Let $x = \varphi(t)$ be a stable isolated periodic solution of (12.1) and let the number $h > 0$ define the domain (12.11) in which there are no other periodic solutions of (12.1).

Since $\varphi(t)$ is stable there exists a $\delta > 0$ such that if $|x_0 - \varphi(0)| < \delta$ then the solution $x(t, x_0)$ of (12.1) satisfies the inequality

$$|x(t, x_0) - \varphi(t)| < h \qquad (t \in \mathbb{R}_+).$$

This means that the solution $x(t, x_0)$ cannot be periodic if $0 < |x_0 - \varphi(0)| < \delta$ and thus by Theorem 12.5 $x(t, x_0)$ tends asymptotically to a periodic solution of (12.1) lying in the domain (12.11). But such a periodic solution is only $\varphi(t)$. Consequently, $\varphi(t)$ is asymptotically stable. $\square$

Theorem 12.7   *Let the conditions of Theorem 12.5 hold and let $x = \varphi(t)$ be an isolated periodic solution of (12.1). Then any other solution of (12.1) taking, at $t = 0$, a value close enough to $\varphi(0)$ tends asymptotically to $\varphi(t)$ either as $t \to +\infty$ or as $t \to -\infty$.*

Proof   Since $\varphi(t)$ is a periodic solution of (12.1), then by Theorem 12.5 $\varphi(t)$ has a period $\omega = T$ if p is even, or $\omega = 2T$ if p is odd. Let the number $h > 0$ define the domain (12.11). Choose $\delta > 0$ such that the solution $x(t)$ of (12.1) satisfies the inequality

$$|x(t) - \varphi(t)| < h \qquad (t \in [0, \omega]) \qquad (12.12)$$

whenever $|x(0) - \varphi(0)| < \delta$. Let $x(t)$ be such a solution. Since $\varphi(t)$ is an isolated periodic solution of (12.1), then from (12.12) it follows that $x(t)$ cannot be periodic, so $x(0) \neq x(\omega)$. Four cases are possible.

First assume that $\varphi(0) - \delta < x(0) < x(\omega) < \varphi(0)$. Then the sequence $x(k\omega)$, $k \in \mathbb{N}$, is increasing and

$$\varphi(0) - \delta < x(k\omega) < \varphi(0)$$

for each $k \in \mathbb{N}$. Then from (12.12) and the periodicity of (12.1) it follows that

$$|x(t) - \varphi(t)| < h \qquad (t \in [k\omega, k\omega + \omega], k \in \mathbb{N}),$$

that is, $x(t)$ does not leave the domain (12.11) for $t \in \mathbb{R}_+$. Thus by what was proved in Theorem 12.6 we have

$$\lim_{t \to +\infty} [x(t) - \varphi(t)] = 0. \qquad (12.13)$$

Analogously we prove that (12.13) is valid also in the case when $\varphi(0) < x(\omega) < x(0) < \varphi(0) + \delta$.

In the case when $\varphi(0) < x(0) < x(\omega) < \varphi(0) + \delta$ or $\varphi(0) - \delta < x(\omega) < x(0) < \varphi(0)$ by similar arguments we conclude that

$$\lim_{t \to -\infty} [x(t) - \varphi(t)] = 0. \qquad \square$$

Theorem 12.8   *Let the following conditions hold.*

1. *Conditions H12.1 and H12.2 are met for $n = 2$.*
2. *The mappings $\Psi_k : \mathbb{R}^2 \to \mathbb{R}^2$, $\Psi_k(x) = x + I_k(x)$ $(k = 1, \ldots, q)$ are homeomorphisms and exactly p of them $(0 \le p \le q)$ invert the orientation of $\mathbb{R}^2$.*
3. *Each point $(t_0, x_0) \in \mathbb{R} \times \mathbb{R}^2$ is a point of uniqueness for the equation $x' = f(t, x)$.*
4. *All solutions $x(t, x_0)$ $(x_0 \in \mathbb{R}^2)$ are continuable on the interval $[0, \omega]$, where $\omega = T$ if p is even, or $\omega = 2T$ if p is odd.*

Then, if equation (12.1) has a solution which is defined and bounded in $\mathbb{R}_+$, equation (12.1) has a periodic solution with period T if p is even, or period $2T$ if p is odd.

We omit the proof of Theorem 12.8 which is similar to the proof of Theorem 12.5 and is based on the following theorem of Brouwer [63].

Theorem 12.9   *Let the mapping $U : \mathbb{R}^2 \times \mathbb{R}^2$ be a homeomorphism preserving the orientation of $\mathbb{R}^2$, and let a point $x \in \mathbb{R}^2$ exist such that the sequence $\{U^m(x)\}$ contains a convergent subsequence. Then the mapping U has a fixed point.*

13. Method of upper and lower solutions

In this section we shall discuss the question of the existence of a solution of the periodic value problem

$$\frac{dx}{dt} = f(t, x), \qquad t \ne \tau_k, t \in J, \tag{13.1}$$

$$x(\tau_k^+) = \Psi_k(x(\tau_k)), \qquad k = 1, \ldots, q, \tag{13.2}$$

$$x(0) = x(T) \tag{13.3}$$

where $J = [0, T]$, $\tau_k \in (0, T)$ $(k = 1, \ldots, q)$, $x \in \mathbb{R}^n$.

To this end we shall use the method of upper and lower solutions. First we shall give the following definitions.

Definition 13.1   The function x is said to be a *solution* of the periodic value problem (13.1)–(13.3) if $x \in PC^1(J, \mathbb{R}^n)$ and x satisfies (13.1)–(13.3).

Definition 13.2   The function $v \in PC^1(J, \mathbb{R}^n)$ is said to be a *lower solution* of the periodic value problem (13.1)–(13.3) if

$$v'(t) \le f(t, v(t)), \qquad t \ne \tau_k, t \in J, \tag{13.4}$$

$$v(\tau_k^+) \le \Psi_k(v(\tau_k)), \qquad k = 1, \ldots, q, \tag{13.5}$$

$$v(0) \le v(T). \tag{13.6}$$

Analogously we define an *upper solution* of (13.1)–(13.3) by an inversion of the inequalities in (13.4)–(13.6).

Definition 13.3   A function $f : J \times \mathbb{R}^n \to \mathbb{R}^n$ is said to be *quasimonotone non-decreasing* if $u, v \in \mathbb{R}^n$, $u \leq v$, $u_i = v_i$, for some $1 \leq i \leq n$, implies $f_i(t, u) \leq f_i(t, v)$. Here and in what follows the inequality $u \leq v$ means that $u_i \leq v_i$ for $i = 1, \ldots, n$.

Definition 13.4   A function $I : \mathbb{R}^n \times \mathbb{R}^n$ is said to be *non-decreasing* in $\mathbb{R}^n$ if for $u, v \in \mathbb{R}^n$, $u \leq v$ implies $I(u) \leq I(v)$.

Let
$$\Delta_0 = [0, \tau_1],$$
$$\Delta_k = (\tau_k, \tau_{k+1}] \qquad (k = 1, \ldots, q - 1),$$
$$\Delta_q = (\tau_q, T].$$

Theorem 13.1   *Let the following conditions hold.*

1. *The functions v, w are lower and upper solutions of (13.1)–(13.3) such that $v(t) \leq w(t)$ in J.*
2. *The function $f : J \times \mathbb{R}^n \to \mathbb{R}^n$ is quasimonotone non-decreasing in $J \times \mathbb{R}^n$, continuous in the sets $\Delta_k \times \mathbb{R}^n$ $(k = 0, \ldots, q)$, and for each $k = 1, \ldots, q$ and $x \in \mathbb{R}^n$ there exists the finite limit of $f(t, y)$ as $(t, y) \to (\tau_k, x)$, $t > \tau_k$.*
3. *There exists a function $\lambda \in L^1(J, \mathbb{R}_+)$ such that*
$$\sup_{v(t) \leq x \leq w(t)} |f_i(t, x)| \leq \lambda(t) \quad \text{a.e. on } J \qquad (i = 1, \ldots, n).$$

4. *The functions $\Psi_k : \mathbb{R}^n \to \mathbb{R}^n$ $(k = 1, \ldots, q)$ are continuous and non-decreasing with respect to $x \in \mathbb{R}^n$.*

Then the periodic value problem (13.1)–(13.3) has a solution $x(t)$ such that $v(t) \leq x(t) \leq w(t)$ in J.

Proof   Without loss of generality we can assume that $n = 2$ and $q = 1$. First we consider the impulsive differential equation (13.1), (13.2) with initial condition
$$x(0) = x_0 \tag{13.7}$$
and we will prove that there exists a solution $x(t, x_0)$ of the initial value problem (13.1), (13.2), (13.7) such that $v(t) \leq x(t, x_0) \leq w(t)$ in J whenever $v(0) \leq x_0 \leq w(0)$. For this purpose we consider the initial value problem
$$x' = F(t, x), \qquad x(0) = x_0 \tag{13.8}$$
for $t \in \Delta_0$, where $F : \Delta_0 \times \mathbb{R}^2 \to \mathbb{R}^2$ is defined by
$$F_i(t, x) = \begin{cases} f_i(t, \bar{x}) + \dfrac{w_i(t) - x_i}{1 + |x_i - w_i(t)|}, & \text{if } x_i > w_i(t), \\ f_i(t, \bar{x}), & \text{if } v_i(t) \leq x_i \leq w_i(t), \\ f_i(t, \bar{x}) + \dfrac{v_i(t) - x_i}{1 + |x_i - v_i(t)|}, & \text{if } x_i < v_i(t), \end{cases}$$

and

$$
\bar{x}_i = \begin{cases} w_i(t), & \text{if } x_i > w_i(t), \\ x_i, & \text{if } v_i(t) \leq x_i \leq w_i(t), \\ v_i(t), & \text{if } x_i < v_i(t) \end{cases}
$$

for $i = 1, 2$. Obviously, (13.8) has a solution $x(t)$ in Δ_0 since

$$
\sup_{x \in \mathbb{R}^2} |F_i(t, x)| \leq \lambda(t) + 1 \quad \text{a.e. on } J \quad (i = 1, 2).
$$

We claim that $x(t) \leq w(t)$ in J. Otherwise for some $i = 1, 2$ the function $m_i(t) = x_i(t) - w_i(t)$ would have a positive maximum at some $t_0 \in (0, \tau_1]$. Consequently, there would exist $t^* \in (0, \tau_1)$ such that $m_i(t^*) > 0$ and $m_i'(t^*) \geq 0$, whence it would follow that $\bar{x}_i(t^*) = w_i(t^*)$. Since, moreover, $\bar{x}(t^*) \leq w(t^*)$ and f is quasimonotone non-decreasing, then $f_i(t^*, \bar{x}(t^*)) \leq f_i(t^*, w(t^*))$. Thus we arrive at a contradiction:

$$
m'(t^*) = x_i'(t^*) - w_i'(t^*)
$$

$$
\leq f_i(t^*, \bar{x}(t^*)) - f_i(t^*, w(t^*)) + \frac{w_i(t^*) - x_i(t^*)}{1 + |x_i(t^*) - w_i(t^*)|}
$$

$$
< 0.
$$

Analogously we can prove that $v(t) \leq w(t)$ in J, after which we conclude that $v(t) \leq x(t) \leq w(t)$ in J and also that $x(t)$ is a solution of the initial value problem

$$
x' = f(t, x), \qquad x(0) = x_0 \qquad (t \in \Delta_0).
$$

Since $v(\tau_1) \leq x(\tau_1) \leq w(\tau_1)$ and $\Psi_1(x)$ is non-decreasing in $\mathbb{R}^2$, then

$$
v(\tau_1^+) \leq \Psi_1(v(\tau_1)) \leq \Psi_1(x(\tau_1)) \leq \Psi_1(w(\tau_1)) \leq w(\tau_1^+).
$$

Setting $y_0 = \Psi_1(x(\tau_1))$ and repeating the same arguments we can prove that the initial value problem

$$
y' = f(t, y), \qquad y(\tau_1) = y_0
$$

has a solution $y(t)$ in $[\tau_1, T]$ such that $v(t) \leq y(t) \leq w(t)$ $(t \in [\tau_1, T])$. Then the function

$$
x(t, x_0) = \begin{cases} x(t), & t \in \Delta_0, \\ y(t), & t \in \Delta_1, \end{cases}
$$

is a solution of the initial value problem (13.1), (13.2), (13.7) such that

$$
v(t) \leq x(t, x_0) \leq w(t) \qquad (t \in J). \tag{13.9}
$$

Let $x_0 = (x_{01}, x_{02}) \in \mathbb{R}^2$ and denote by $X_1(x_{01}, x_{02})$ the set of all solutions (13.1), (13.2), (13.7) satisfying (13.9). By what was proved above $X_1(x_{01}, x_{02}) \neq \emptyset$.

The proof that the periodic value problem (13.1)–(13.3) has a solution will be carried out in two steps. First we shall prove that for each $x_{02} \in [v_2(0), w_2(0)]$ there is an $x_{01} = x_{01}(x_{02}) \in [v_1(0), w_1(0)]$ such that there exists an $x(\,\cdot\,, x_{01}, x_{02}) \in X_1(x_{01}, x_{02})$ which satisfies

$$x_1(0, x_{01}, x_{02}) = x_1(T, x_{01}, x_{02}).$$

This is obviously true if $v_1(0) = w_1(0)$ because then we obtain $v_1(0) = v_1(T) = w_1(0) = w_1(T)$. Therefore suppose that $v_1(0) < w_1(0)$ and that the assertion is not true. This means that for any $z \in [v_1(0), w_1(0)]$ and any $x \in (x_1, x_2) \in X_1(z, x_{02})$ we have that

$$x_1(0) \neq x_1(T).$$

From the conditions imposed on $v(t)$ and $w(t)$ it follows that

$$x_1(0) < x_1(T)$$

for each $x \in X_1(v_1(0), x_{02})$ and

$$y_1(0) > y_1(T) \tag{13.10}$$

for each $y \in X_1(w_1(0), x_{02})$ since

$$x_1(0) = v_1(0) \leq v_1(T) \leq x_1(T) \qquad \text{and} \qquad y_1(0) = w_1(0) \geq w_1(T) \geq y_1(T).$$

We claim that there exists a δ with $0 < \delta < w_1(0) - v_1(0)$ such that $y_1(0) > y_1(T)$ for each $y = (y_1, y_2) \in X_1(z, x_{02})$ whenever $0 \leq w_1(0) - z < \delta$. Otherwise there exist the sequences z_n with $0 \leq w_1(0) - z_n < n^{-1}$ and $y^{(n)} = (y_1^{(n)}, y_2^{(n)}) \in X_1(z_n, x_{02})$ such that

$$y_1^{(n)}(0) < y_1^{(n)}(T). \tag{13.11}$$

Obviously, $y^{(n)}$ satisfy the equation equivalent to (13.1), (13.2) and (13.7):

$$y^{(n)}(t) = \begin{cases} y^{(n)}(0) + \int_0^t f(s, y^{(n)}(s))\, ds, & t \in \Delta_0 = [0, \tau_1], \\ \Psi_1(y^{(n)}(\tau_1)) + \int_{\tau_1}^t f(s, y^{(n)}(s))\, ds, & t \in \Delta_1 = (\tau_1, T], \end{cases} \tag{13.12}$$

where $y^{(n)}(0) = (z_n, x_{02})$.

From (13.2) and conditions 2 and 4 of the theorem we conclude that the sequence $\{y^{(n)}\}$ is uniformly bounded and quasiequicontinuous in J. Consequently, by Lemma 2.4 there exists a subsequence $\{y^{(n_k)}\}$ which is convergent to $y = (y_1, y_2)$ uniformly in J. Passing to the limit in (13.11) and (13.12), which is possible by the conditions imposed on f and Ψ_1, we obtain that $y \in X_1(w_1(0), x_{02})$ and $y_1(0) < y_1(T)$, which contradicts (13.10).

Now we define δ^* as the supremum of all such $\delta > 0$. Then it is clear that $0 < \delta^* \leq w_1(0) - v_1(0)$ and

$$y_1(0) > y_1(T) \tag{13.13}$$

if $0 \leq w_1(0) - z < \delta^*$ and $y \in X_1(z, x_{02})$.

By the definition of δ^* there exists a sequence $z_n \in \mathbb{R}$ such that

$$v_1(0) < z_n < w_1(0) - \delta^*, \qquad z_n \to w_1(0) - \delta^* \qquad \text{as } n \to \infty$$

and for some $x^{(n)} = (x_1^{(n)}, x_2^{(n)}) \in X_1(z_n, x_{02})$ we have

$$x_1^{(n)}(0) < x_1^{(n)}(T).$$

Passing to the limit we conclude that there exists an $x^* = (x_1^*, x_2^*) \in X_1(w_1(0) - \delta^*, x_{02})$ such that

$$x_1^*(0) < x_1^*(T).$$

Then the function $x^*(t)$ is a lower solution of (13.1)–(13.3) and $x^*(t) \leq w(t)$ in J. Consequently as above we can prove that for each $z \in [w_1(0) - \delta^*, w_1(0)]$

$$X_1(z, x_{02}) \cap [x^*; w] \neq \emptyset$$

where $[x^*; w] = \{y : J \to \mathbb{R}^2 : x^*(t) \leq y(t) \leq w(t), t \in J\}$.

Choose the sequences z_n and $x^{(n)}$ so that

$$z_n \in (w_1(0) - \delta^*, w_1(0)], \qquad z_n \to w_1(0) - \delta^* \quad \text{as } n \to \infty,$$

$$x^{(n)} = (x_1^{(n)}, x_2^{(n)}) \in X_1(z_n, x_{02}) \cap [x^*; w].$$

Then $x_1^{(n)}(0) > x_1^{(n)}(T)$ for each $n \geq 1$ by virtue of (13.13). Passing to the limit we find that there exists a function

$$x = (x_1, x_2) \in X_1(w_1(0) - \delta^*, x_{02}) \cap [x^*; w]$$

such that $x_1(0) > x_1(T)$. However, this is impossible due to the fact that

$$x^*(t) \leq x(t) \quad \text{in } J, \qquad x_1^*(0) < x_1^*(T) \qquad \text{and} \qquad x_1(0) = x_1^*(0).$$

For each $x_{02} \in [v_2(0), w_2(0)]$ define the set

$$X_2(x_{02}) = \left\{ \begin{array}{l} x \in \mathrm{PC}^1(J, \mathbb{R}^2) : x \text{ satisfies (13.1), (13.2), (13.9)} \\ \text{and } x_2(0) = x_{02}, x_1(0) = x_1(T) \end{array} \right\}.$$

Thus we have just proved that $X_2(x_{02}) \neq \emptyset$.

Finally, if we prove that for some $x_{02} \in [v_2(0), w_2(0)]$ and $x \in X_2(x_{02})$ we have

$$x_2(0) = x_2(T), \tag{13.14}$$

then such a function $x = (x_1, x_2)$ is a solution of the periodic value problem (13.1)–(13.3) in $[v; w]$. This conclusion is obviously true if $v_2(0) = w_2(0)$ since then $x_2(0) = x_2(T)$ for each $x \in X_2(v_2(0))$.

Suppose that $v_2(0) < w_2(0)$ and that the assertion is not true. From the conditions imposed on v and w we obtain that

$$x_2(0) < x_2(T)$$

for each $x \in (x_1, x_2) \in X_2(v_2(0))$ and

$$y_2(0) > y_2(T)$$

for each $y \in (y_1, y_2) \in X_2(w_2(0))$. This time we consider the sets $X_2(z)$ where $z \in [v_2(0), w_2(0)]$ and, repeating the arguments used in order to prove that $X_2(x_{02}) \neq \emptyset$, we obtain a similar contradiction. Thus (13.14) is true. It is clear that the arguments used up to now are also applicable in the general case when $n \geq 2$ and $q \geq 1$. $\square$

14. Periodic value problem for a second-order impulsive equation

In this section, by applying the method of upper and lower solutions, we shall investigate the periodic value problem for the second-order impulsive differential equation

$$-x'' = f(t, x, x'), \qquad t \in J, \, t \neq \tau_k, 0, T, \tag{14.1}$$

$$x(\tau_k^+) = \Psi_k(x(\tau_k)), \tag{14.2}$$

$$x'(\tau_k^+) = N_k(x'(\tau_k)), \tag{14.3}$$

$$x(0) = x(T), \qquad x'(0) = x'(T), \tag{14.4}$$

where $J = [0, T]$, $\tau_k \in (0, T)$ $(k = 1, \ldots, q)$, $f : J \times \mathbb{R}^n \times \mathbb{R}^n \to \mathbb{R}^n$ and $\Psi_k, N_k : \mathbb{R}^n \to \mathbb{R}^n$ $(k = 1, \ldots, q)$.

Suppose that the functions $f(t, x, y)$, $\Psi_k(y)$, $N_k(y)$ are not coupled with respect to $y \in \mathbb{R}^n$, that is, for each $i = 1, \ldots, n$ we have that $f_i = f_i(t, x, y_i)$, $\Psi_{ki} = \Psi_{ki}(y_i)$, $N_{ki} = N_{ki}(y_i)$, $y_i \in \mathbb{R}$.

Definition 14.1   The function x is said to be a *solution* of (14.1)–(14.4) if $x \in PC^2(J, \mathbb{R}^n)$ and x satisfies (14.1)–(14.4).

Definition 14.2   The function $\alpha \in PC^2(J, \mathbb{R}^n)$ is said to be a *lower solution* of (14.1)–(14.4) if

$$-\alpha'' \leq f(t, \alpha, \alpha'), \qquad t \in J, \, t \neq \tau_k, 0, T, \tag{14.5}$$

$$\alpha(\tau_k^+) = \Psi_k(\alpha(\tau_k)), \tag{14.6}$$

$$\alpha'(\tau_k^+) \geq N_k(\alpha'(\tau_k)), \tag{14.7}$$

$$\alpha(0) = \alpha(T), \qquad \alpha'(0) \geq \alpha'(T). \tag{14.8}$$

Analogously we define an *upper solution* of (14.1)–(14.4) by an inversion of the inequalities of (14.5), (14.7) and (14.8).

Let $\alpha, \beta \in C(J, \mathbb{R}^n)$ and $\alpha(t) \leq \beta(t)$ $(t \in J)$.

Definition 14.3   The function $g : J \times \mathbb{R}^n \times \mathbb{R}^n \to \mathbb{R}^n$ is said to satisfy *Nagumo's condition* with respect to (α, β) if

$$|g_i(t, x, y)| \leq h_i(|y_i|) \qquad (i = 1, \ldots, n)$$

for $\alpha(t) \leq x \leq \beta(t)$, $y \in \mathbb{R}^n$, $t \in J$, where $h_i \in C(\mathbb{R}_+, (0, +\infty))$ and

$$\int_\lambda^\infty \frac{s \, ds}{h_i(s)} > \max_J \beta_i(t) - \min_J \alpha_i(t),$$

λ being given by

$$\lambda T = \max(|\alpha(0) - \beta(T)|, |\alpha(T) - \beta(0)|).$$

We shall need the following well-known results.

Lemma 14.1   Suppose that a function $g : J \times \mathbb{R}^n \times \mathbb{R}^n \to \mathbb{R}^n$ is continuous for $t \in (0,T)$ and satisfies Nagumo's condition with respect to (α, β). Then there exists a $c > 0$ depending only on α, β and h_i such that

$$|x'(t)| \le c \quad (t \in J)$$

for any solution $x(t)$ of the equation

$$-x'' = g(t, x, x') \quad (t \in (0,T)) \tag{14.9}$$

for which $\alpha(t) \le x(t) \le \beta(t)$.

Now consider equation (14.9) with the boundary conditions

$$x(0) = A, \qquad x(T) = B \tag{14.10}$$

where $A, B \in \mathbb{R}^n$ and

$$\alpha(0) \le A \le \beta(0), \qquad \alpha(T) \le B \le \beta(T). \tag{14.11}$$

For the boundary value problem (14.9), (14.10) we have the following result.

Lemma 14.2   Suppose that $\alpha, \beta \in C^1(J, \mathbb{R}^n) \cap C^2((0,T), \mathbb{R}^n)$ and $\alpha(t) \le \beta(t)$ in J; $g : J \times \mathbb{R}^n \times \mathbb{R}^n \to \mathbb{R}^n$, $g(t,x,y)$ is continuous for $t \in (0,T)$, quasimonotone non-decreasing in x, satisfies Nagumo's condition with respect to (α, β), and for $t \in (0,T)$

$$-\alpha'' \le g(t, \alpha, \alpha'), \qquad -\beta'' \ge g(t, \beta, \beta').$$

Then for any $A, B \in \mathbb{R}^n$ satisfying (14.11) the problem (14.9), (14.10) has a solution $x \in C^1(J, \mathbb{R}^n) \cap C^2((0,T), \mathbb{R}^n)$ such that $\alpha(t) \le x(t) \le \beta(t)$ and $|x'(t)| \le c$ in J.

Let
$$\begin{aligned}
\Delta_0 &= [0, \tau_1], \\
\Delta_k &= (\tau_k, \tau_{k+1}] \quad (k = 1, \ldots, q-1), \\
\Delta_q &= (\tau_q, T].
\end{aligned}$$

Now we can begin with the proof of the following theorem on the existence of a solution of the periodic value problem (14.1)–(14.4).

Theorem 14.1   Let the following conditions hold.

1. The functions α, β are lower and upper solutions of (14.1)–(14.4) such that $\alpha(t) \le \beta(t)$ in J.

2. The functions $f(t,x,y)$, $\Psi_k(y)$ and $N_k(y)$ $(k = 1, \ldots, q)$ are not coupled with respect to $y \in \mathbb{R}^n$.

3. The function $f(t,x,y)$ is quasimonotone non-decreasing with respect to x; satisfies Nagumo's condition with respect to (α,β); is continuous in the sets $\Delta_k \times \mathbb{R}^n \times \mathbb{R}^n$; and for each $k = 1,\ldots,q$ and $x \in \mathbb{R}^n$, $y \in \mathbb{R}^n$ there exists the finite limit of $f(t,u,v)$ as $(t,u,v) \to (\tau_k, x, y)$, $t > \tau_k$.

4. The functions $\Psi_k, N_k : \mathbb{R}^n \to \mathbb{R}^n$ $(k = 1,\ldots,q)$ are continuous and non-decreasing in $\mathbb{R}^n$.

Then the periodic value problem (14.1)–(14.4) has a solution $x(t)$ such that $\alpha(t) \leq x(t) \leq \beta(t)$ and $|x'(t)| \leq c$ in J.

Proof   For convenience of the presentation we assume that $n = 2$, $q = 1$ and divide the proof into two steps.

     Step 1   First we shall prove that for any $A, B \in \mathbb{R}^n$ such that $\alpha(0) \leq A \leq \beta(0)$, $\alpha(T) \leq B \leq \beta(T)$ the boundary value problem (14.1)–(14.3) and (14.10) has a solution $x(t)$ such that

$$\alpha(t) \leq x(t) \leq \beta(t) \quad \text{and} \quad |x'(t)| \leq c \quad \text{in } J.$$

Choose an arbitrary $C_0 = (a,b) \in \mathbb{R}^2$ satisfying $\alpha(\tau_1) \leq C_0 \leq \beta(\tau_1)$. By Lemma 14.2 there exists an $x \in C^1([0,\tau_1], \mathbb{R}^2) \cap C^2((0,\tau_1), \mathbb{R}^2)$ such that

$$-x'' = f(t,x,x'), \quad t \in (0,\tau_1), \quad x(0) = A, \quad x(\tau_1) = C_0$$

and $\alpha(t) \leq x(t) \leq \beta(t)$, $|x'(t)| \leq c$ in $[0,\tau_1]$. Since Ψ_1 is non-decreasing, then

$$\alpha(\tau_1^+) \leq \Psi_1(C_0) = C^* \leq \beta(\tau_1^+).$$

Again, applying Lemma 14.2, we find $y \in C^1([\tau_1,T], \mathbb{R}^2) \cap C^2((\tau_1,T), \mathbb{R}^2)$ such that

$$-y'' = f(t,y,y'), \quad t \in (\tau_1,T), \quad y(\tau_1) = C^*, \quad y(T) = B,$$

and $\alpha(t) \leq y(t) \leq \beta(t)$, $|y'(t)| \leq c$ in J. Then the function

$$z(t) = \begin{cases} x(t), & t \in [0,\tau_1], \\ y(t), & t \in (\tau_1,T] \end{cases}$$

is a solution of problem (14.1), (14.2), (14.10) such that $\alpha(t) \leq z(t) \leq \beta(t)$, $|z'(t)| \leq c$ in J. Denote by $X_1(a,b)$ the set of all such solutions so that $X_1(a,b) \neq \emptyset$.

     The next thing we claim is that for each $b \in [\alpha_2(\tau_1), \beta_2(\tau_1)]$ there exists $a = a(b) \in [\alpha_1(\tau_1), \beta_1(\tau_1)]$ such that $X_1(a,b)$ contains a solution $x(t)$ which satisfies the relation

$$x_1'(\tau_1^+) = N_{11}(x'(\tau_1)) \tag{14.12}$$

where $N_1 = (N_{11}, N_{12})$. Suppose that this is not true, i.e., that there exists $b \in [\alpha_2(\tau_1), \beta_2(\tau_1)]$ such that for each $a \in [\alpha_1(\tau_1), \beta_1(\tau_1)]$ and $x \in X_1(a,b)$ we have

$$x_1'(\tau_1^+) \neq N_{11}(x_1'(\tau_1)). \tag{14.13}$$

Let $x \in X_1(\alpha_1(\tau_1), b)$. Since $x_1(\tau_1) = \alpha_1(\tau_1)$ and $\Psi_1(x)$ is not coupled with respect to x, then we also have that $x_1(\tau_1^+) = \alpha_1(\tau_1^+)$. Thus from the fact that $\alpha_1(t) \leq x_1(t)$ $(t \in J)$ we conclude that

$$x_1'(\tau_1) \leq \alpha_1'(\tau_1), \qquad \alpha'(\tau_1^+) \leq x'(\tau_1^+).$$

Consequently,

$$N_{11}(x_1'(\tau_1)) \leq N_{11}(\alpha_1'(\tau_1)) \leq \alpha_1'(\tau_1^+) \leq x_1'(\tau_1^+),$$

whence in view of (14.13) it follows that

$$x_1'(\tau_1^+) > N_{11}(x_1'(\tau_1)).$$

Similarly, we can prove that for each $y \in X_1(\beta_1(\tau_1), b)$ we have

$$y_1'(\tau_1^+) < N_{11}(y_1'(\tau_1)). \tag{14.14}$$

We must then have $\beta_1(\tau_1) > \alpha_1(\tau_1)$ and so we claim that there exists a $\delta > 0$, $\delta < \beta_1(\tau_1) - \alpha_1(\tau_1)$, such that (14.14) is valid for each $y \in X_1(z, b)$ whenever $0 \leq \beta_1(\tau_1) - z < \delta$. Otherwise there would exist a sequence $\{z_n\} \subset \mathbb{R}$ satisfying $0 \leq \beta_1(\tau_1) - z_n < n^{-1}$, and a sequence of functions $y_n = (y_{n1}, y_{n2}) \in X_1(z_n, b)$ such that

$$y_{n1}'(\tau_1^+) > N_{11}(y_{n1}'(\tau_1)).$$

Since each function $x \in X_1(a, b)$ has an equivalent integral representation by means of two appropriate Green's functions and the sequence $\{y_n'\}$ is bounded, then by Lemma 2.4 we can find a subsequence $\{y_{n_k}\}$ which tends to $u = (u_1, u_2)$ in $PC^1(J, \mathbb{R}^n)$. Passing to the limit, we conclude that $u \in X_1(\beta_1(\tau_1), b)$ and $u_1'(\tau_1^+) > N_{11}(u_1'(\tau_1))$, which contradicts (14.14).

Now we define δ^* as the supremum of all such $\delta > 0$. It is clear that $0 < \delta^* \leq \beta_1(\tau_1) - \alpha_1(\tau_1)$ and $y_1'(\tau_1^+) < N_{11}(y_1'(\tau_1))$ for each $y = (y_1, y_2) \in X_1(z, b)$ with $0 \leq \beta_1(\tau_1) - z < \delta^*$.

By the definition of δ^* there exists a sequence z_n such that

$$\alpha_1(\tau_1) < z_n < \beta_1(\tau_1) - \delta^*, \qquad z_n \to \beta_1(\tau_1) - \delta^* \qquad \text{as } n \to \infty,$$

and for some function $x_n = (x_{n1}, x_{n2}) \in X_1(z_n, b)$

$$x_{n1}'(\tau_1^+) > N_{11}(x_{n1}'(\tau_1)).$$

Then, passing to the limit we obtain that there exists an $x^0 = (x_1^0, x_2^0) \in X_1(\beta_1(\tau_1) - \delta^*, b)$ such that

$$x_1^{0'}(\tau_1^+) > N_{11}(x_1^{0'}(\tau_1)).$$

Now we apply Lemma 14.2 with respect to (x^0, β) instead of (α, β) and then it is easily seen that

$$X_1(z, b) \cap [x^0; \beta] \neq \emptyset$$

for each $z \in [\beta_1(\tau_1) - \delta^*, \beta_1(\tau_1)]$, where

$$[x^0; \beta] = \{y : J \to \mathbb{R}^2 : x^0(t) \le y(t) \le \beta(t) \text{ in } J\}.$$

Choose a sequence $z_n \in (\beta_1(\tau_1) - \delta^*, \beta_1(\tau_1)]$ such that $z_n \to \beta_1(\tau_1) - \delta^*$ as $n \to \infty$ and choose some $y_n = (y_{n1}, y_{n2}) \in X_1(z_n, b) \cap [x^0; \beta]$. Then $y'_{n1}(\tau_1^+) < N_{11}(y'_{n1}(\tau_1))$ for each $n \ge 1$ by the definition of δ^*. Passing to the limit, we obtain $u = (u_1, u_2) \in X_1(\beta(\tau_1) - \delta^*, b) \cap [x^0; \beta]$ such that

$$u'_1(\tau_1^+) < N_{11}(u'_1(\tau_1)).$$

Since $x_1^0(t) \le u_1(t)$ in J, $u_1(\tau_1) = x_1^0(\tau_1)$ and $u_1(\tau_1^+) = x_1^0(\tau_1^+)$, then we see that

$$u'_1(\tau_1) \le x_1^{0'}(\tau_1) \qquad \text{and} \qquad x_1^{0'}(\tau_1^+) \le u'_1(\tau_1^+). \tag{14.15}$$

Hence we have that

$$u'_1(\tau_1^+) < N_{11}(u'_1(\tau_1)) \le N_{11}(x_1^{0'}(\tau_1)) < x_1^{0'}(\tau_1^+)$$

which contradicts (14.15).

Thus we have proved that for each $b \in [\alpha_2(\tau_1), \beta_2(\tau_1)]$ problem (14.1), (14.2) and (14.10) has a solution $x(t)$ such that $\alpha(t) \le x(t) \le \beta(t)$ in J, and

$$x_2(\tau_1) = b, \qquad x'_1(\tau_1^+) = N_{11}(x'_1(\tau_1)).$$

Let $X_2(b)$ denote the set of all such solutions.

Now we shall show that there exists $b \in [\alpha_2(\tau_1), \beta_2(\tau_1)]$ such that $X_2(b)$ contains $x = (x_1, x_2)$ which satisfies

$$x'_2(\tau_1^+) = N_{12}(x'_2(\tau_1)). \tag{14.16}$$

Then it follows that $x(t)$ is a solution of the boundary value problem (14.1)–(14.3), (14.10) such that $\alpha(t) \le x(t) \le \beta(t)$ and $|x'(t)| \le c$ in J, which completes step 1 of the proof. Suppose, as above, that this conclusion is not true. Then for each $x \in X_2(\alpha_2(\tau_1))$ we have that $\alpha_2(t) \le x_2(t)$ in J and $\alpha_2(\tau_1) = x_2(\tau_1)$. Consequently, $x_2(\tau_1^+) = \alpha_2(\tau_1^+)$, and

$$x'_2(\tau_1) \le \alpha'_2(\tau_1), \qquad x'_2(\tau_1^+) \ge \alpha_2(\tau_1^+)$$

which implies

$$x'_2(\tau_1^+) > N_{12}(x'_2(\tau_1)) \tag{14.17}$$

since N_1 is non-decreasing and α satisfies (14.3). Similarly, we can obtain that

$$y'_2(\tau_1^+) < N_{12}(y'_2(\tau_1)) \tag{14.18}$$

for each $y \in X_2(\beta_2(\tau_1))$. As before, we can find the supremum δ^* with $0 < \delta^* \le \beta_2(\tau_1) - \alpha_2(\tau_1)$ such that (14.18) is valid for each $y = (y_1, y_2) \in X_2(z)$ whenever $0 \le \beta_2(\tau_1) - z < \delta^*$ since $\beta_2(\tau_1) - \alpha_2(\tau_1) > 0$ in view of (14.17) and (14.18). This time we consider the sets $X_2(z)$, where $z \in [\alpha_2(\tau_1), \beta_2(\tau_1)]$ is a parameter, and relation

(14.16) instead of (14.12) repeating the same procedure which we used in order to prove that $X_2(b) \neq \emptyset$, and obtain a similar contradiction. Thus our conclusion (14.16) is true.

Step 2   We shall prove that there exists $C_0 = (a, b)$ with $\alpha_1(0) \leq a \leq \beta_1(0)$ and $\alpha_2(0) \leq b \leq \beta_2(0)$ such that if $A = B = C_0$ then one of the solutions $x(t)$ of (14.1)–(14.3), (14.10) satisfies $\alpha(t) \leq x(t) \leq \beta(t)$ in J and

$$x'(0) = x'(T).$$

Thus it is a solution of problem (14.1)–(14.4).

For each $C_0 = (a, b)$ with $\alpha_1(0) \leq a \leq \beta_1(0)$, $\alpha_2(0) \leq b \leq \beta_2(0)$ let $X_3(a, b)$ denote the set of all solutions of (14.1)–(14.3), (14.10) such that $\alpha(t) \leq x(t) \leq \beta(t)$ in J, where $A = B = C_0$. Then $X_3(a, b) \neq \emptyset$ by what was proved in step 1. Now we claim that for each $b \in [\alpha_2(0), \beta_2(0)]$ there exists $a \in [\alpha_1(0), \beta_1(0)]$ such that $X_3(a, b)$ contains $x = (x_1, x_2)$ which satisfies

$$x_1'(0) = x_1'(T).$$

Suppose that this claim is not true. Then we obviously have that

$$y_1'(0) < y_1'(T) \tag{14.19}$$

for any $y = (y_1, y_2) \in X_3(\beta_1(0), b)$ and

$$x_1'(0) > x_1'(T) \tag{14.20}$$

for any $x = (x_1, x_2) \in X_3(\alpha_1(0), b)$.

Passing to the limit as in step 1 we can show that there exists a supremum δ^* with $0 < \delta^* < \beta_1(0) - \alpha_1(0)$ such that (14.19) is valid for any $y = (y_1, y_2) \in X_3(z, b)$ while $z \in (\beta_1(0) - \delta^*, \beta_1(0)]$. Then by the definition of δ^* there exists a sequence $z_n \in [\alpha_1(0), \beta_1(0) - \delta^*)$ and $x_n = (x_{n1}, x_{n2}) \in X_3(z_n, b)$ such that $z_n \to \beta_1(0) - \delta^*$ and x_{n1} satisfies (14.20). It is not hard to show that the sequence $\{x_n\}$ tends to $x_0 = (x_{01}, x_{02}) \in X_3(\beta_1(0) - \delta^*, b)$ in PC^1 and x_{01} satisfies (14.20). By the arguments we have used up to now we are sure that

$$X_3(z, b) \cap [x_0; \beta] \neq \emptyset$$

for each $z \in [x_{01}(0), \beta_1(0)]$. Thus we choose a sequence $z_n \in (\beta_1(0) - \delta^*, \beta_1(0))$ and a sequence $y_n = (y_{n1}, y_{n2}) \in X_3(z_n, b) \cap [x_0; \beta]$, where $z_n \to \beta_1(0) - \delta^*$ as $n \to \infty$. It is clear that $y_{n1}'(0) < y_{n1}'(T)$ for each $n \geq 1$. Passing to the limit, we obtain $u = (u_1, u_2) \in X_3(\beta_1(0) - \delta^*, b) \cap [x_0; \beta]$ such that $u_1'(0) < u_1'(T)$. This, however, is impossible since $x_{01}(t) \leq u_1(t)$ in J, $x_{01}(0) = x_{01}(T) = u_1(0) = u_1(T)$ and $x_{01}'(0) > x_{01}'(T)$.

Thus we have proved that for each $b \in [\alpha_2(0), \beta_2(0)]$ there exists a solution $x = (x_1, x_2)$ of problem (14.1)–(14.3) such that $\alpha(t) \leq x(t) \leq \beta(t)$ in J, $x(0) = x(T)$, $x_2(0) = x_2(T) = b$ and $x_1'(0) = x_1'(T)$.

Let $X_4(b)$ denote the set of all such solutions so that $X_4(b) \neq \emptyset$. We see that for each $y = (y_1, y_2) \in X_4(\beta_2(0))$ we have

$$y_2'(0) < y_2'(T)$$

while for $x = (x_1, x_2) \in X_4(\alpha_2(0))$

$$x_2'(0) > x_2'(T).$$

Now consider the sets $X_4(z)$, where $z \in [\alpha_2(0), \beta_2(0)]$ is a parameter, and proceed as above to find $z \in [\alpha_2(0), \beta_2(0)]$ and $x = (x_1, x_2) \in X_4(z)$ such that

$$x_2'(0) = x_2'(T).$$

Obviously, this $x(t)$ is a solution of the periodic value problem (14.1)–(14.4) such that $\alpha(t) \leq x(t) \leq \beta(t)$ and $|x'(t)| \leq c$ in J. $\square$

Notes and comments for Chapter V

Theorems 12.2–12.4 were adapted from S. G. Hristova and D. D. Bainov [40]. Theorem 12.5 has been proved by V. I. Guțu in [33], where Theorems 12.6–12.8 are formulated. For results in this direction see S. G. Hristova and D. D. Bainov [44] and L. H. Erbe and Xinzhi Liu [26].

Theorem 13.1 was proved by D. D. Bainov, S. G. Hristova, S. Hu and V. Lakshmikantham in [12]. Theorem 14.1 is due to S. Hu and V. Lakshmikantham [45]. For similar results see E. Liz and J. J. Nieto [56], C. Pierson-Gorez [62] and A. S. Vatsala and Yong Sun [87].

Chapter VI

Approximate Analytical Methods for Finding Periodic Solutions

15. Monotone-iterative technique for finding periodic solutions

The monotone-iterative technique is a constructive method for finding extremal quasisolutions of systems of differential equations. In the present section the monotone-iterative technique will be applied to find the minimal and maximal quasisolutions of the periodic value problem (13.1)–(13.3). First we introduce some preliminary considerations.

For each $i = 1, \ldots, n$ fix two positive integers p_i, q_i such that $p_i + q_i = n - 1$. Let us fix p_i components of the vector $x \in \mathbb{R}^n$ which are distinct from the ith component x_i. We denote this 'part' of the vector x by $[x]_{p_i}$. The remaining components of the vector x which are distinct from x_i will be denoted by $[x]_{q_i}$. Then the vector $x \in \mathbb{R}^n$ can be represented in the form $x = (x_i, [x]_{p_i}, [x]_{q_i})$ and the T-periodic value problem (13.1)–(13.3) becomes

$$
\begin{aligned}
x_i' &= f_i(t, x_i, [x]_{p_i}, [x]_{q_i}), \quad && t \neq \tau_k, \\
\Delta x_i &= I_{ki}(x_i, [x]_{p_i}, [x]_{q_i}), \quad && k = 1, \ldots, q, \\
x_i(0) &= x_i(T), \quad && i = 1, \ldots, n,
\end{aligned}
\tag{15.1}
$$

where $\tau_0 < 0 < \tau_1 < \cdots < \tau_q < T < \tau_{q+1}$.

Definition 15.1   The function f is said to be *mixed quasimonotone* if for each $i = 1, \ldots, n$ the function $f_i(t, x_i, [x]_{p_i}, [x]_{q_i})$ is non-decreasing with respect to $[x]_{p_i}$ and non-decreasing with respect to $[x]_{q_i}$.

Definition 15.2   The functions $\alpha, \beta \in PC^1([0, T], \mathbb{R}^n)$ are said to be *coupled lower and upper quasisolutions* of the periodic value problem (15.1) if

$$
\begin{aligned}
\alpha_i'(t) &\leq f_i(t, \alpha_i, [\alpha]_{p_i}, [\beta]_{q_i}), \quad && t \neq \tau_k, \\
\Delta \alpha_i(\tau_k) &\leq I_{ki}(\alpha_i, [\alpha]_{p_i}, [\beta]_{q_i}), \quad && k = 1, \ldots, q, \\
\alpha_i(0) &\leq \alpha_i(T), \quad && i = 1, \ldots, n,
\end{aligned}
\tag{15.2}
$$

and

$$
\begin{aligned}
\beta_i'(t) &\geq f_i(t, \beta_i, [\beta]_{p_i}, [\alpha]_{q_i}), \quad && t \neq \tau_k, \\
\Delta \beta_i(\tau_k) &\geq I_{ki}(\beta_i, [\beta]_{p_i}, [\alpha]_{q_i}), \quad && k = 1, \ldots, q, \\
\beta_i(0) &\geq \beta_i(T), \quad && i = 1, \ldots, n,
\end{aligned}
\tag{15.3}
$$

Definition 15.3   The functions $\alpha, \beta \in PC^1([0,T], \mathbb{R}^n)$ are said to be *coupled quasisolutions* of the periodic value problem (15.1) if α and β satisfy the relations (15.2) and (15.3) in which the inequality signs are replaced by equalities.

Let $\alpha, \beta \in PC^1([0,T], \mathbb{R}^n)$ and $\alpha(t) \leq \beta(t)$ for $t \in [0,T]$. Introduce the set

$$D[\alpha, \beta] = \{\sigma \in PC^1([0,T], \mathbb{R}^n) : \alpha(t) \leq \sigma(t) \leq \beta(t), t \in [0,T]\}.$$

Definition 15.4   The function $v \in D[\alpha, \beta]$ is said to be a *minimal (maximal) solution* of the periodic value problem (15.1) in $D[\alpha, \beta]$ if v is a solution of (15.1) and for any other solution $x \in D[\alpha, \beta]$ of (15.1) the following inequality holds:

$$v(t) \leq x(t) \qquad (x(t) \leq v(t)) \qquad \text{for } t \in [0,T].$$

Let α, β be coupled quasisolutions for the periodic value problem (15.1).

Definition 15.5   The functions $\rho, r \in D[\alpha, \beta]$ are said to be *coupled minimal and maximal quasisolutions* of the periodic value problem (15.1) in $D[\alpha, \beta]$ if they are coupled quasisolutions of (15.1) and for any coupled quasisolutions $v, w \in D[\alpha, \beta]$ of (15.1) the following inequalities hold:

$$\rho(t) \leq v(t), w(t) \leq r(t) \qquad \text{for } t \in [0,T].$$

Our further considerations are based on the following lemma.

Lemma 15.1   *Let $m \in PC^1([0,T], \mathbb{R})$ and*

$$\begin{aligned} m'(t) &\leq -Mm(t), \quad t \in [0,T], t \neq \tau_k, \\ \Delta m(\tau_k) &\leq -L_k m(\tau_k), \quad k = 1, \ldots, q, \end{aligned} \qquad (15.4)$$

where $L_k \leq 1$ and $\prod_{k=1}^{q}(1 - L_k)e^{-MT} < 1$.
Then if $m(0) \leq m(T)$, then $m(t) \leq 0$ for $t \in [0,T]$.

Proof   From (15.4) by Lemma 2.2 it follows that

$$m(t) \leq m(0) \prod_{0 < \tau_k < t} (1 - L_k)e^{-Mt} \qquad (t \in [0,T])$$

which implies the assertion of the lemma since $m(0) \leq 0$. In fact, if we suppose that this is not true, that is, $m(0) > 0$, then we obtain the contradiction

$$m(0) \leq m(T) \leq m(0) \prod_{k=1}^{q}(1 - L_k)e^{-MT} < m(0). \qquad \square$$

We shall consider separately the scalar case $n = 1$ when the functions $\alpha(t), \beta(t)$ are coupled lower and upper quasisolutions of the periodic value problem (15.1) if and only if $\alpha(t)$ and $\beta(t)$ are respectively a lower and upper solution of (15.1), and α and β are coupled quasisolutions of (15.1) if and only if α and β are solutions of (15.1).

Theorem 15.1   *Let the following conditions hold.*

1. *The functions α and β are respectively a lower and upper solution of the periodic value problem (15.1) and $\alpha(t) \leq \beta(t)$ in $[0, T]$.*
2. *The function $f : [0, T] \times \mathbb{R} \to \mathbb{R}$ is continuous and the inequality*

$$f(t, u) - f(t, v) \geq -M(u - v) \tag{15.5}$$

   holds for $t \in [0, T]$ and $\alpha(t) \leq v \leq u \leq \beta(t)$, where $M > 0$ is a constant.
3. *The functions $I_k : \mathbb{R} \to \mathbb{R}$ are continuous and satisfy the inequalities*

$$I_k(u) - I_k(v) \geq -L_k(u - v) \qquad (k = 1, \ldots, q) \tag{15.6}$$

   where $\alpha(\tau_k) \leq v \leq u \leq \beta(\tau_k)$, $L_k < 1$ and $\prod_{k=1}^{q}(1 - L_k)e^{-MT} < 1$.

Then there exist monotone sequences $\{\alpha_\nu(t)\}$, $\{\beta_\nu(t)\}$ with $\alpha_0(t) = \alpha(t)$, $\beta_0(t) = \beta(t)$ such that

$$\lim_{\nu \to \infty} \alpha_\nu(t) = \rho(t), \qquad \lim_{\nu \to \infty} \beta_\nu(t) = r(t)$$

uniformly and monotonically in $[0, T]$ and ρ, r are respectively the minimal and maximal solutions of the periodic value problem (15.1) in $D[\alpha, \beta]$.

Proof   For each $\eta \in D[\alpha, \beta]$ consider the periodic value problem

$$\begin{aligned}
y' &= f(t, \eta(t)) - M(y - \eta(t)), \quad t \neq \tau_k, \\
\Delta y &= I_k(\eta(\tau_k)) - L_k(y - \eta(\tau_k)), \quad k = 1, \ldots, q, \\
y(0) &= y(T).
\end{aligned} \tag{15.7}$$

The linear periodic value problem (15.7) has a unique solution

$$\begin{aligned}
y(t) = &\prod_{0 < \tau_k < t} (1 - L_k)e^{-Mt}y(0) \\
&+ \int_0^t \prod_{s < \tau_k < t} (1 - L_k)e^{M(s-t)}(f(s, \eta(s)) + M\eta(s)) \, ds \\
&+ \sum_{0 < \tau_k < t} \prod_{\tau_k < \tau_i < t} (1 - L_i)e^{M(\tau_k - t)}(I_k(\eta(\tau_k)) + L_k\eta(\tau_k))
\end{aligned} \tag{15.8}$$

where

$$\begin{aligned}
y(T) = y(0) = &\left(1 - \prod_{k=1}^{q}(1 - L_k)e^{-MT}\right)^{-1} \\
&\times \Bigg\{ \int_0^T \prod_{s < \tau_k < T} (1 - L_k)e^{M(s-T)}(f(s, \eta(s)) + M\eta(s)) \, ds \\
&+ \sum_{k=1}^{q} \prod_{\tau_k < \tau_i < T} (1 - L_i)e^{M(\tau_k - T)}(I_k(\eta(\tau_k)) + L_k\eta(\tau_k)) \Bigg\}.
\end{aligned}$$

We define the operator $Q : D[\alpha, \beta] \to PC^1([0, T], \mathbb{R})$, $Q\eta = y$. It is easily verified that Q enjoys the following properties:

I. If $\eta \in D[\alpha, \beta]$, then $Q\eta \in D[\alpha, \beta]$.

II. $\alpha \leq Q\alpha$, $Q\beta \leq \beta$.

III. Q is a monotone increasing operator in $D[\alpha, \beta]$, that is, $Qv \leq Qw$ whenever $v \leq w$; $v, w \in D[\alpha, \beta]$.

In order to prove property I we set $Q\eta = y$, where y is the solution of (15.7). We set $p = \alpha - y$ and in view of (15.5) and (15.6) obtain

$$p' = \alpha' - y' \leq f(t, \alpha) - f(t, \eta) + M(y - \eta)$$
$$\leq -M(\alpha - \eta) + M(y - \eta)$$
$$= -Mp, \qquad t \neq \tau_k,$$

$$\Delta p(\tau_k) = \Delta\alpha(\tau_k) - \Delta y(\tau_k) \leq I_k(\alpha(\tau_k)) - I_k(\eta(\tau_k)) + L_k(y(\tau_k) - \eta(\tau_k))$$
$$\leq -L_k(\alpha(\tau_k) - \eta(\tau_k)) + L_k(y(\tau_k) - \eta(\tau_k))$$
$$= -L_k p(\tau_k),$$

$$p(0) = \alpha(0) - y(0) \leq \alpha(T) - y(T) = p(T).$$

Then by Lemma 15.1 $p(t) \leq 0$ for $t \in [0, T]$, that is, $\alpha(t) \leq y(t)$ for $t \in [0, T]$. Analogously, it is proved that $y(t) \leq \beta(t)$ for $t \in [0, T]$.

The proof of properties II and III is carried out by similar arguments applying Lemma 15.1.

Define the sequences $\{\alpha_\nu(t)\}$, $\{\beta_\nu(t)\}$:

$$\alpha_0 = \alpha, \qquad \beta_0 = \beta,$$
$$\alpha_{\nu+1} = Q\alpha_\nu, \quad \beta_{\nu+1} = Q\beta_\nu \quad (\nu = 0, 1, 2, \ldots).$$

From property II it follows that

$$\alpha_0 \leq \alpha_1 \leq \cdots \leq \alpha_\nu \leq \beta_\nu \leq \cdots \leq \beta_1 \leq \beta_0$$

in $[0, T]$ and by standard arguments we conclude that

$$\lim_{\nu \to \infty} \alpha_\nu(t) = \rho(t), \qquad \lim_{\nu \to \infty} \beta_\nu(t) = r(t)$$

uniformly and monotonically in $[0, T]$.

It is easy to show that ρ and r are solutions of the periodic value problem (15.1) using the fact that α_ν, β_ν satisfy the relations

$$\alpha'_{\nu+1} = f(t, \alpha_\nu) - M(\alpha_{\nu+1} - \alpha_\nu), \qquad t \neq \tau_k,$$
$$\Delta\alpha_{\nu+1}(\tau_k) = I_k(\alpha_\nu(\tau_k)) - L_k(\alpha_{\nu+1}(\tau_k) - \alpha_\nu(\tau_k)), \quad k = 1, \ldots, q,$$
$$\alpha_{\nu+1}(0) = \alpha_{\nu+1}(T);$$

and

$$\beta'_{\nu+1} = f(t, \beta_\nu) - M(\beta_{\nu+1} - \beta_\nu), \qquad t \neq \tau_k,$$

$$\Delta\beta_{\nu+1}(\tau_k) = I_k(\beta_\nu(\tau_k)) - L_k(\beta_{\nu+1}(\tau_k) - \beta_\nu(\tau_k)), \quad k = 1, \dots, q,$$

$$\beta_{\nu+1}(0) = \beta_{\nu+1}(T).$$

In order to prove that ρ and r are the minimal and maximal solution of the periodic value problem (15.1) in $D[\alpha, \beta]$ we shall show that if $y \in D[\alpha, \beta]$ is a solution of (15.1) then $\alpha \leq \rho \leq y \leq r \leq \beta$ in $[0, T]$. To this end we suppose that for some $\nu \in \mathbb{N}$ we have $\alpha_\nu \leq y \leq \beta_\nu$ in $[0, T]$ and set $p = \alpha_{\nu+1} - y$. Then

$$p' = f(t, \alpha_\nu) - M(\alpha_{\nu+1} - \alpha_\nu) - f(t, y)$$

$$\leq -Mp, \qquad t \neq \tau_k,$$

$$\Delta p(\tau_k) \leq -L_k p(\tau_k), \qquad k = 1, \dots, q,$$

$$p(0) = p(T)$$

and by Lemma 15.1 we obtain that $\alpha_{\nu+1}(t) \leq y(t)$ in $[0, T]$. Analogously, we find that $y(t) \leq \beta_{\nu+1}(t)$ in $[0, T]$. Then from the fact that $\alpha_0 = \alpha \leq y \leq \beta = \beta_0$ it follows that the inequality $\alpha_\nu(t) \leq y(t) \leq \beta_\nu(t)$ is valid for all $\nu \in \mathbb{N}$ and $t \in [0, T]$. Passing to the limit as $\nu \to \infty$ we conclude that $\alpha \leq \rho \leq y \leq r \leq \beta$ in $[0, T]$. $\square$

Let us consider the case when $n > 1$.

Theorem 15.2 *Let the following conditions hold.*

1. *The functions $\alpha(t)$ and $\beta(t)$ are coupled lower and upper quasisolutions of the periodic value problem (15.1) and $\alpha(t) \leq \beta(t)$ ($t \in [0, T]$).*

2. *The function $f : [0, T] \times \mathbb{R}^n \to \mathbb{R}^n$ is continuous and mixed quasimonotone in the set $\{(t, x) \in \mathbb{R} \times \mathbb{R}^n : 0 \leq t \leq T, \alpha(t) \leq x \leq \beta(t)\}$.*

3. *The function $I_k : \mathbb{R}^n \to \mathbb{R}^n$ is continuous and mixed quasimonotone in the set $\Omega_k = \{x \in \mathbb{R}^n : \alpha(\tau_k) \leq x \leq \beta(\tau_k)\}$ ($k = 1, \dots, q$).*

4. *The following inequalities are valid:*

$$f_i(t, w_i, [u]_{p_i}, [u]_{q_i}) - f_i(t, u_i, [u]_{p_i}, [u]_{q_i}) \geq -M_i(w_i - u_i),$$

$$I_{ki}(w_i, [u]_{p_i}, [u]_{q_i}) - I_{ki}(u_i, [u]_{p_i}, [u]_{q_i}) \geq -L_{ki}(w_i - u_i),$$

   where $u, w \in \mathbb{R}^n$, $\alpha(t) \leq u, w \leq \beta(t)$, $\alpha_i(t) \leq u_i \leq w_i \leq \beta_i(t)$, $t \in [0, T]$, $M_i > 0$, $L_{ki} < 1$, $\prod_{k=1}^q (1 - L_{ki})e^{-MT} < 1$, $i = 1, \dots, n$; $k = 1, \dots, q$.

Then:

1. *there exist monotone sequences $\{\alpha_\nu(t)\}$, $\{\beta_\nu(t)\}$ with $\alpha_0 = \alpha$, $\beta_0 = \beta$ such that*

$$\lim_{\nu \to \infty} \alpha_\nu(t) = \rho(t), \qquad \lim_{\nu \to \infty} \beta_\nu(t) = r(t)$$

   uniformly and monotonically in $[0, T]$ and ρ, r are coupled minimal and maximal quasisolutions of the periodic value problem (15.1) in $D[\alpha, \beta]$;

2. *if $x \in D[\alpha, \beta]$ is a solution of the periodic value problem (15.1), then $\rho(t) \leq x(t) \leq r(t)$ in $[0, T]$.*

Proof For any $\eta, \mu \in D[\alpha, \beta]$ consider the periodic value problem

$$
\begin{aligned}
y_i' &= f_i(t, \eta_i, [\eta]_{p_i}, [\mu]_{q_i}) - M_i(y_i - \eta_i), && t \neq \tau_k, \\
\Delta y_i &= I_{ki}(\eta_i, [\eta]_{p_i}, [\mu]_{q_i}) - L_{ki}(y_i - \eta_i), && t = \tau_k, k = 1, \ldots, q, && (15.9) \\
y_i(0) &= y_i(T), && i = 1, \ldots, n.
\end{aligned}
$$

The periodic value problem (15.9) has a unique solution $y_i(t)$ which is represented explicitly by a formula similar to (15.8).

Define the operator $Q : D[\alpha, \beta] \times D[\alpha, \beta] \to PC^1([0, T], \mathbb{R}^n)$, $Q(\eta, \mu) = y$. It is easily checked that:

   I. $\alpha \leq Q(\alpha, \beta)$, $Q(\beta, \alpha) \leq \beta$;

  II. if $\alpha \leq \eta \leq \mu \leq \beta$, then $\alpha \leq Q(\eta, \mu) \leq Q(\mu, \eta) \leq \beta$.

We shall only prove that the inequality $\alpha \leq \eta \leq \mu \leq \beta$ implies the inequality $y \equiv Q(\eta, \mu) \leq Q(\mu, \eta) \equiv z$. The proof of the remaining inequalities in I and II is analogous.

Set $\psi = y - z$. Then

$$
\begin{aligned}
\psi_i' &= f_i(t, \eta_i, [\eta]_{p_i}, [\mu]_{q_i}) - M_i(y_i - \eta_i) - f_i(t, \mu_i, [\mu]_{p_i}, [\eta]_{q_i}) + M_i(z_i - \mu_i) \\
&\leq f_i(t, \eta_i, [\mu]_{p_i}, [\mu]_{q_i}) - M_i(y_i - \eta_i) - f_i(t, \mu_i, [\mu]_{p_i}, [\mu]_{q_i}) - M_i(\mu_i - z_i) \\
&\leq -M_i(\eta_i - \mu_i) - M_i(y_i - \eta_i) - M_i(\mu_i - z_i) \\
&= -M_i \psi_i, \qquad t \neq \tau_k.
\end{aligned}
$$

Analogously,

$$
\begin{aligned}
\Delta \psi_i(\tau_k) &\leq -L_{ki} \psi_i(\tau_k), && k = 1, \ldots, q, \\
\psi_i(0) &= \psi_i(T), && i = 1, \ldots, n.
\end{aligned}
$$

By Lemma 15.1 $\psi_i(t) \leq 0$, that is, $y(t) \leq z(t)$ in $[0, T]$. Define the sequences $\{\alpha_\nu(t)\}$, $\{\beta_\nu(t)\}$:

$$
\begin{aligned}
\alpha_0 &= \alpha, && \beta_0 = \beta, \\
\alpha_{\nu+1} &= Q(\alpha_\nu, \beta_\nu) && \beta_{\nu+1} = Q(\beta_\nu, \alpha_\nu) \quad (\nu = 0, 1, 2, \ldots).
\end{aligned}
$$

By property II we have that

$$
\alpha_0 \leq \alpha_1 \leq \cdots \leq \alpha_\nu \leq \beta_\nu \leq \cdots \leq \beta_1 \leq \beta_0
$$

in $[0, T]$. Then by standard arguments we conclude that

$$
\lim_{\nu \to \infty} \alpha_\nu(t) = \rho(t), \qquad \lim_{\nu \to \infty} \beta_\nu(t) = r(t)
$$

uniformly and monotonically in $[0, T]$, whence it follows that

$$\rho_i' = f_i(t, \rho_i, [\rho]_{p_i}, [r]_{q_i}), \quad t \neq \tau_k,$$
$$\Delta \rho_i = I_{ki}(\rho_i, [\rho]_{p_i}, [r]_{q_i}), \quad t = \tau_k, k = 1, \ldots, q,$$
$$\rho_i(0) = \rho_i(T), \quad i = 1, \ldots, n;$$
$$r_i' = f_i(t, r_i, [r]_{p_i}, [\rho]_{q_i}), \quad t \neq \tau_k,$$
$$\Delta r_i = I_{ki}(r_i, [r]_{p_i}, [\rho]_{q_i}), \quad t = \tau_k, k = 1, \ldots, q,$$
$$r_i(0) = r_i(T), \quad i = 1, \ldots, n.$$

We shall prove that ρ, r are coupled minimal and maximal quasisolutions of the periodic value problem (15.1) in $D[\alpha, \beta]$. Let y, z be coupled quasisolutions of the periodic value problem (15.1) in $D[\alpha, \beta]$. Suppose that for some $\nu \in \mathbb{N}$ we have $\alpha_\nu(t) \leq y(t), z(t) \leq \beta_\nu(t)$ in $[0, T]$. Set $\psi = \alpha_{\nu+1,i} - y_i$. Then from condition 4 and the mixed quasimonotonicity of f and I_k it follows that

$$\psi' = f_i(t, \alpha_{\nu,i}, [\alpha_\nu]_{p_i}, [\beta_\nu]_{q_i}) - M_i(\alpha_{\nu+1,i} - \alpha_{\nu,i})$$
$$\quad - f_i(t, y_i, [y]_{p_i}, [z]_{q_i})$$
$$\leq f_i(t, \alpha_{\nu,i}, [y]_{p_i}, [z]_{q_i}) - M_i(\alpha_{\nu+1,i} - \alpha_{\nu,i})$$
$$\quad - f_i(t, y_i, [y]_{p_i}, [z]_{q_i})$$
$$\leq -M_i(\alpha_{\nu,i} - y_i) - M_i(\alpha_{\nu+1,i} - \alpha_{\nu,i})$$
$$= -M_i \psi, \quad t \neq \tau_k,$$
$$\Delta \psi(\tau_k) \leq -L_{ki} \psi(\tau_k), \quad k = 1, \ldots, q,$$
$$\psi(0) = \psi(T), \quad i = 1, \ldots, n.$$

By Lemma 15.1 we conclude that $\alpha_{\nu+1,i} \leq y_i$ in $[0, T]$.

By similar arguments we conclude that $\alpha_{\nu+1} \leq y, z \leq \beta_{\nu+1}$ in $[0, T]$ and since $\alpha_0 = \alpha \leq y, z \leq \beta = \beta_0$ in $[0, T]$, then by induction it follows that $\alpha_\nu \leq y, z \leq \beta_\nu$ in $[0, T]$ for each $\nu \in \mathbb{N}$. Consequently, $\rho \leq y, z \leq r$, that is, ρ, r are coupled minimal and maximal quasisolutions of (15.1) in $D[\alpha, \beta]$. Since each solution x of (15.1) which belongs to $D[\alpha, \beta]$ can be regarded as coupled quasisolutions x, x of (15.1), then we also have that $\rho \leq x \leq r$ in $[0, T]$. $\square$

Remark 15.1   If $q_i = 0$, $i = 1 \ldots, n$ then ρ, r are minimal and maximal solutions of the periodic value problem (15.1) contained in $D[\alpha, \beta]$.

16. Numerical-analytical method for finding periodic solutions

In this section we shall justify the application of the numerical-analytical method from [68] for finding the periodic solutions of the impulsive periodic differential equation

$$\frac{\mathrm{d}x}{\mathrm{d}t} = f(t,x), \quad t \neq \tau_k,$$

$$\Delta x = I_k(x), \qquad t = \tau_k \tag{16.1}$$

where $t \in \mathbb{R}$, $k \in \mathbb{Z}$, $x \in \Omega \subset \mathbb{R}^n$.

Introduce the following conditions (H16).

H16.1 Equation (16.1) is T-periodic, that is, there exists a $q \in \mathbb{N}$ such that

$$\tau_{k+q} = \tau_k + T, \qquad f(t+T, x) = f(t, x), \qquad I_{k+q}(x) = I_k(x)$$

for $t \in \mathbb{R}$, $k \in \mathbb{Z}$, and $x \in \Omega$.

H16.2 The function $f : \mathbb{R} \times \Omega \to \mathbb{R}^n$ is continuous in the sets $(\tau_{k-1}, \tau_k] \times \Omega$ $(k \in \mathbb{Z})$ and for each $x \in \Omega$ and $k \in \mathbb{Z}$ there exists the finite limit of $f(t, y)$ as $(t, y) \to (\tau_k, x)$, $t > \tau_k$.

H16.3 The functions $I_k : \Omega \times \mathbb{R}^n$ $(k \in \mathbb{Z})$ are continuous in Ω.

H16.4 There exist constants $K_1 > 0$, $K_2 > 0$ and $M > 0$ such that

$$|f(t, x) - f(t, y)| \le K_1 |x - y|,$$

$$|I_k(x) - I_k(y)| \le K_2 |x - y|,$$

$$|f(t, x)| \le M,$$

$$|I_k(x)| \le M$$

for $x, y \in \Omega$, $t \in \mathbb{R}$, $k \in \mathbb{Z}$.

H16.5 There exists a non-empty compact set D contained in Ω together with its $(\frac{1}{2}MT + MQ)$-neighbourhood, where

$$Q = \sup\left\{ \left(1 - \frac{t}{T}\right) i[0, t) + \frac{t}{T} i[t, T) : t \in [0, T] \right\}. \tag{16.2}$$

H16.6 The following inequalities hold:

$$-1 + \frac{K_1 T}{3} + K_2 Q < K_1 K_2 \left(\frac{TQ}{3} - S\right) < 1$$

where

$$S = \sup\left\{ \left(1 - \frac{t}{T}\right) \sum_{0 \le \tau_k < t} \alpha(\tau_k) + \frac{t}{T} \sum_{t \le \tau_k < T} \alpha(\tau_k) : t \in [0, T] \right\} \tag{16.3}$$

and

$$\alpha(t) = 2t\left(1 - \frac{t}{T}\right) \qquad (t \in [0, T]).$$

Remark 16.1 The numbers Q and S defined above depend on the concrete position of the points τ_k of the interval $[0, T]$. In the general case for Q and S the estimates

$$Q \le q, \qquad S \le \sum_{0 \le \tau_k < T} \alpha(\tau_k) \le \frac{1}{2} qT$$

are valid, which follow from the inequalities $\alpha(t) \le \frac{1}{2} T$ and

$$\left(1 - \frac{t}{T}\right) \sum_{0 \le \tau_k < t} \gamma_k + \frac{t}{T} \sum_{t \le \tau_k < T} \gamma_k \le \sum_{0 \le \tau_k < T} \gamma_k \qquad (16.4)$$

where $t \in [0, T]$ and $\gamma_k \ge 0$.

We shall also note that in the proofs of the theorems below we shall use the following relations which are valid for $t \in [0, T]$:

$$\left(1 - \frac{t}{T}\right) \int_0^t ds + \frac{t}{T} \int_t^T ds = \alpha(t), \qquad (16.5)$$

$$\left(1 - \frac{t}{T}\right) \int_0^t \alpha(s) \, ds + \frac{t}{T} \int_t^T \alpha(s) \, ds = \frac{1}{3} \alpha^2(t) + \frac{1}{6} T\alpha(t),$$

$$\le \frac{1}{3} T\alpha(t). \qquad (16.6)$$

For fixed $x_0 \in D$ in the interval $[0, T]$ we recurrently define the sequence of functions $\{x_m(t, x_0)\}$:

$$x_0(t, x_0) \equiv x_0,$$

$$x_{m+1}(t, x_0) = x_0 + \left(1 - \frac{t}{T}\right) \int_0^t f(s, x_m(s, x_0)) \, ds - \frac{t}{T} \int_t^T f(s, x_m(s, x_0)) \, ds$$

$$+ \left(1 - \frac{t}{T}\right) \sum_{0 \le \tau_k < t} I_k(x_m(\tau_k, x_0)) - \frac{t}{T} \sum_{t \le \tau_k < T} I_k(x_m(\tau_k, x_0)), \qquad (16.7)$$

$$m = 0, 1, 2, \ldots.$$

We shall find sufficient conditions under which the sequence $x_m(t, x_0)$ tends uniformly to the T-periodic solution $\varphi(t)$ of (16.1) for which $\varphi(0) = x_0$.

Theorem 16.1 *Let conditions (H16) hold and let $x_0 \in D$. Then:*

1. *the functions $x_m(t, x_0)$ satisfy the relations*

$$x_m(0, x_0) = x_m(T, x_0) = x_0, \qquad (16.8)$$

$$x_m(t, x_0) \in \Omega \qquad (t \in [0, T], m = 0, 1, 2, \ldots); \qquad (16.9)$$

2. *the sequence $x_m(t, x_0)$ is uniformly convergent in the interval $[0, T]$ and its limit $x(t, x_0)$ satisfies the relations:*

$$x(0, x_0) = x(T, x_0) = x_0, \qquad (16.10)$$

$$x(t, x_0) \in \Omega \qquad (t \in [0, T]), \tag{16.11}$$

$$x(t, x_0) = x_0 + \left(1 - \frac{t}{T}\right) \int_0^t f(s, x(s, x_0))\, ds - \frac{t}{T} \int_t^T f(s, x(s, x_0))\, ds$$

$$+ \left(1 - \frac{t}{T}\right) \sum_{0 \leq \tau_k < t} I_k(x(\tau_k, x_0)) - \frac{t}{T} \sum_{t \leq \tau_k < T} I_k(x(\tau_k, x_0)); \tag{16.12}$$

3. if $\varphi(t)$ is a T-periodic solution of (16.1) for which $\varphi(0) = \varphi(T) = x_0$ and $\varphi(t) \in \Omega$, $t \in [0, T]$, then

$$\varphi(t) \equiv x(t, x_0) \qquad (t \in [0, T]).$$

Proof   1. The validity of (16.8) is obvious. From (16.7), H16.4, (16.5) and (16.2) we obtain inductively that

$$|x_m(t, x_0) - x_0| \leq M\alpha(t) + MQ \leq \frac{1}{2}MT + MQ \tag{16.13}$$

for $t \in [0, T]$ and $m = 0, 1, 2, \ldots$. This, together with H16.5, implies (16.9).

2. In order to prove the uniform convergence of the sequence $x_m(t, x_0)$ we shall estimate the difference $x_j(t, x_0) - x_{j-1}(t, x_0)$. From (16.13) it immediately follows that

$$|x_1(t, x_0) - x_0(t, x_0)| \leq M\alpha(t) + MQ \leq \frac{1}{2}MT + Mq.$$

Let us set $a_1 = M$, $b_1 = MQ$ and suppose that for $j = 1, 2, \ldots, m$ estimates of the following form are valid:

$$|x_j(t, x_0) - x_{j-1}(t, x_0)| \leq a_j\alpha(t) + b_j \leq \frac{1}{2}a_jT + b_j. \tag{16.14}$$

Then for $j = m + 1,$, taking into account (16.7), H16.4, (16.2), (16.3), (16.5) and (16.6), we obtain

$$|x_{m+1}(t, x_0) - x_m(t, x_0)|$$

$$\leq \left(1 - \frac{t}{T}\right) \int_0^t K_1(a_m\alpha(s) + b_m)\, ds + \frac{t}{T} \int_t^T K_1(a_m\alpha(s) + b_m)\, ds$$

$$+ \left(1 - \frac{t}{T}\right) \sum_{0 \leq \tau_k < t} K_2(a_m\alpha(\tau_k) + b_m) + \frac{t}{T} \sum_{t \leq \tau_k < T} K_2(a_m\alpha(\tau_k) + b_m)$$

$$\leq \frac{1}{3}K_1 a_m T\alpha(t) + K_1 b_m\alpha(t) + K_2 a_m S + K_2 b_m Q.$$

Consequently, an estimate of the form (16.14) is valid for each $j \in \mathbb{N}$ and the numbers a_j, b_j are determined successively from the relations

$$a_{j+1} = \frac{1}{3}K_1 T a_j + K_1 b_j,$$

$$b_{j+1} = K_2 S a_j + K_2 Q b_j, \tag{16.15}$$

and the initial values $a_1 = M$, $b_1 = MQ$. Denote $c_j = \text{col}\,(a_j, b_j)$. Then from (16.15), we obtain that

$$c_{j+r} = G^r c_j \qquad (j \in \mathbb{N}, r \in \mathbb{N}) \tag{16.16}$$

where

$$G = \begin{bmatrix} \frac{1}{3}K_1 T & K_1 \\ K_2 S & K_2 Q \end{bmatrix}.$$

From (16.14) and (16.16) it follows that the sequence $x_m(t, x_0)$ is uniformly convergent in the interval $[0, T]$ if the eigenvalues λ_1, λ_2 of the matrix G satisfy the inequality $|\lambda_i| < 1$ $(i = 1, 2)$. As is known [23], the roots of the equation $F(\lambda) \equiv \lambda^2 + a\lambda + b = 0$ $(a, b \in \mathbb{R})$ lie inside the unit circle $|\lambda| < 1$ of the complex plane if the following inequalities hold:

$$-1 + |a| < b < 1. \tag{16.17}$$

The characteristic polynomial of the matrix G is

$$F(\lambda) = \lambda^2 - (\frac{1}{3}K_1 T + K_2 Q)\lambda + K_1 K_2 (\frac{1}{3}TQ - S)$$

and for it the condition (16.17) takes the form

$$-1 + \frac{1}{3}K_1 T + K_2 Q < K_1 K_2 (\frac{1}{3}TQ - S) < 1.$$

But by H16.6 this condition is met. Therefore, the sequence $x_m(t, x_0)$ is uniformly convergent in the interval $[0, T]$; let $x(t, x_0)$ be its limit. Passing to the limit in (16.8), (16.9) and (16.7), we conclude that relations (16.10), (16.11) and (16.12) are satisfied.

3. Since $\varphi(t)$ is a T-periodic solution of (16.1) and $\varphi(0) = x_0$ then $\varphi(t)$ satisfies the relations

$$\varphi(t) = x_0 + \int_0^t f(s, \varphi(s))\, ds + \sum_{0 \le \tau_k < t} I_k(\varphi(\tau_k)) \qquad (t \in [0, T]) \tag{16.18}$$

and

$$\int_0^T f(s, \varphi(s))\, ds + \sum_{0 \le \tau_k < T} I_k(\varphi(\tau_k)) = 0. \tag{16.19}$$

From (16.18) and (16.19) it follows that $\varphi(t)$ and $x(t, x_0)$ satisfy the equation

$$x(t) = x_0 + \left(1 - \frac{t}{T}\right) \int_0^t f(s, x(s))\, ds - \frac{t}{T} \int_t^T f(s, x(s))\, ds$$
$$+ \left(1 - \frac{t}{T}\right) \sum_{0 \le \tau_k < t} I_k(x(\tau_k)) - \frac{t}{T} \sum_{t \le \tau_k < T} I_k(x(\tau_k)). \tag{16.20}$$

We shall prove that equation (16.20) cannot have two distinct solutions, whence it follows that $\varphi(t) \equiv x(t, x_0)$, $t \in [0, T]$. Suppose that this is not true and let $x(t)$ and

$y(t)$ be two distinct solutions of (16.20). Then for the function $r(t) = |x(t) - y(t)|$ for $t \in [0, T]$ we obtain the estimate

$$r(t) \le \left(1 - \frac{t}{T}\right) \int_0^t K_1 r(s) \, ds + \frac{t}{T} \int_t^T K_1 r(s) \, ds$$

$$+ \left(1 - \frac{t}{T}\right) \sum_{0 \le \tau_k < t} K_2 r(\tau_k) + \frac{t}{T} \sum_{t \le \tau_k < T} K_2 r(\tau_k). \tag{16.21}$$

From (16.21) we obtain inductively that

$$r(t) \le \tilde{A}_m \alpha(t) + \tilde{B}_m \qquad (m \in \mathbb{N}, t \in [0, T]) \tag{16.22}$$

where the constants $\tilde{A}_m$ and $\tilde{B}_m$ satisfy the recurrence relations

$$\tilde{A}_{m+1} = \frac{1}{3} K_1 T \tilde{A}_m + K_1 \tilde{B}_m,$$

$$\tilde{B}_{m+1} = K_2 S \tilde{A}_m + K_2 Q \tilde{B}_m,$$

and $\tilde{A}_0 = 0$, $\tilde{B}_0 = r_0 = \max_{t \in [0,T]} r(t) \ge 0$.

Since the moduli of the eigenvalues of the matrix G do not exceed 1, then $\tilde{A}_m \to 0$ and $\tilde{B}_m \to 0$ as $m \to \infty$. Then from (16.22) it follows that $r(t) \equiv 0$, that is, $x(t) \equiv y(t)$, $t \in [0, T]$. $\square$

Remark 16.2   From what was proved in Theorem 16.1 we can draw an additional conclusion about the rate of convergence of the sequence of functions $x_m(t, x_0)$ towards the solution $\varphi(t)$. From the dependence (16.15) it follows that the numbers a_k, b_k have the form

$$a_k = A_1 \lambda_1^{k-1} + A_2 \lambda_2^{k-1},$$

$$b_k = B_1 \lambda_1^{k-1} + B_2 \lambda_2^{k-1}, \tag{16.23}$$

where λ_1, λ_2 are the roots of the characteristic equation $F(\lambda) = 0$ and the constants A_1, A_2, B_1, B_2 are determined uniquely from λ_1, λ_2, a_1, a_2, b_1, b_2. Then from (16.23) there follows the existence of a constant $d > 0$ which depends on A_1, A_2, B_1, B_2 such that

$$\frac{1}{2} a_k T + b_k \le d \rho^{k-1} \qquad (k \in \mathbb{N}) \tag{16.24}$$

where $\rho = \max(|\lambda_1|, |\lambda_2|)$.

Taking into account (16.14) and (16.24) we obtain the estimate

$$|x_{m+r}(t, x_0) - x_m(t, x_0)| \le \sum_{k=m+1}^{m+r} |x_k(t, x_0) - x_{k-1}(t, x_0)|$$

$$\le \sum_{k=m+1}^{m+r} \left(\frac{1}{2} a_k T + b_k\right)$$

$$\le \frac{d \rho^m}{1 - \rho},$$

whence passing to the limit as $r \to \infty$ we find

$$|x_m(t, x_0) - \varphi(t)| \le \frac{d\rho^m}{1 - \rho}. \tag{16.25}$$

Let us discuss the question of the existence of periodic solutions of equation (16.1). By Theorem 16.1 the finding of a T-periodic solution of (16.1) is reduced to the computation of the functions $x_m(t, x_0)$ if it is known that such a solution does exist, and if the point x_0 through which it passes at $t = 0$ is also known. If we know the functions $x_m(t, x_0)$, the question of the existence of T-periodic solutions of (16.1) can be solved in the following way.

Introduce the function $\Delta : D \to \mathbb{R}^n$, $x_0 \to \Delta(x_0)$ by the formula

$$\Delta(x_0) = \int_0^T f(s, x(s, x_0))\, ds + \sum_{0 \le \tau_k < T} I_k(x(\tau_k, x_0))$$

where $x(t, x_0) = \lim_{m \to \infty} x_m(t, x_0)$ and $x_m(t, x_0)$ are determined according to (16.7). Since $x(t, x_0)$ satisfies the relation

$$x(t, x_0) = x_0 + \int_0^t f(s, x(s, x_0))\, ds - \frac{t}{T} \int_t^T f(s, x(s, x_0))\, ds$$

$$+ \sum_{0 \le \tau_k < t} I_k(x(\tau_k, x_0)) - \frac{t}{T} \sum_{0 \le \tau_k < T} I_k(x(\tau_k, x_0)),$$

then the equality $\Delta(x_0) = 0$ implies that $x(t, x_0)$ is a solution of (16.1). In this case from (16.8) it follows that the T-periodic extension of $x(t, x_0)$ is a T-periodic solution of (16.1).

Thus the question of the existence of T-periodic solutions of (16.1) is related to the question of the existence of zeros of the function $\Delta(x_0)$. However, it is practically impossible to find the function $\Delta(x_0)$. That is why the following problem arises: how can one deduce the existence of zeros of the function $\Delta(x_0)$ from the investigation of the function

$$\Delta_m(x_0) = \int_0^T f(s, x_m(s, x_0))\, ds + \sum_{0 \le \tau_k < T} I_k(x_m(\tau_k, x_0))?$$

This problem can be solved by the following theorem.

Theorem 16.2 *Let the following conditions hold.*

1. *Conditions (H16) are met.*
2. *For some $m \in \mathbb{N}$ the mapping $\Delta_m : D \to \mathbb{R}^n$ has an isolated singular point x^0 $(\Delta(x^0) = 0)$.*
3. *The index of the singular point x^0 is different from zero.*
4. *There exists a closed convex domain $F \subset D$ with a unique singular point x^0 such that on the boundary ∂F the following inequality holds*

$$\inf_{x \in \partial F} |\Delta_m(x)| > \frac{d\rho^m}{1 - \rho}(k_1 T + K_2 q), \tag{16.26}$$

where the constants d and ρ are the same as in estimate (16.25).

Then equation (16.1) has a T-periodic solution $x = \varphi(t)$ for which $\varphi(0) \in D$.

Proof   The index of the isolated singular point x^0 of the continuous mapping $\Delta_m(x_0)$ is equal to the characteristic of the vector field induced by this mapping on a sphere S_ϵ with sufficiently small radius and centre at x^0. Since there is only one singular point x^0 in F and F is homeomorphic to the unit ball in $\mathbb{R}^n$, then the characteristic of the vector field $\Delta_m(x)$ on the sphere S_ϵ is equal to the characteristic of this field on ∂F. The vector fields $\Delta_m(x)$ and $\Delta(x)$ are homotopic on ∂F. The last assertion follows from the fact that the family of vector fields continuous on ∂F, continuously depending on $\theta \in [0,1]$,

$$V(\theta, x_0) = \Delta_m(x_0) + \theta(\Delta(x_0) - \Delta_m(x_0))$$

connecting the fields $V(0, x_0) = \Delta_m(x_0)$ and $V(1, x_0) = \Delta(x_0)$ is nowhere vanishing. Indeed, from the estimate (16.25) it follows that

$$|\Delta(x_0) - \Delta_m(x_0)| \leq \int_0^T K_1 |x(s, x_0) - x_m(s, x_0)| \, ds$$

$$+ \sum_{0 \leq \tau_k < T} K_2 |x(\tau_k, x_0) - x_m(\tau_k, x_0)|$$

and in view of (16.25), we find that for $x_0 \in \partial F$

$$|V(\theta, x_0)| \geq |\Delta_m(x_0)| - |\Delta_m(x_0) - \Delta(x_0)| > 0.$$

Since the characteristics of vector fields homotopic on a compact set are equal [51], the characteristic of the field $\Delta(x)$ on ∂F is equal to the index of the singular point x^0 on the field $\Delta_m(x)$, hence it is different from zero. By Theorem 5-12 of [5] this is sufficient for the vector field $\Delta(x)$ to have a singular point $x_0 \in F$, that is, $\Delta(x_0) = 0$, which means that equation (16.1) has a T-periodic solution $x = \varphi(t)$ and $\varphi(0) = x_0 \in F \subset D$. □

Note that if the function $\Delta_m(x)$ is continuously differentiable in a neighbourhood of the point x^0 and $\det \frac{\partial \Delta_m}{\partial x}(x^0) \neq 0$ then the index of the field Δ_m at the point x^0 is different from zero.

Consider the T-periodic equation

$$\frac{dx}{dt} = \epsilon f(t, x), \quad t \neq \tau_k,$$
$$\Delta x = \epsilon I_k(x), \quad t = \tau_k$$

(16.27)

where $\epsilon \in (0, \bar{\epsilon})$ is a small parameter. The domain F on whose boundary ∂F inequality (16.26) is valid can be a ball of sufficiently small radius and centre at the isolated singular point. Therefore the following theorem is valid.

Theorem 16.3   *Let conditions (H16) hold, the averaged equation*

$$\frac{dy}{dt} = \epsilon f_0(y) = \frac{\epsilon}{T}\left(\int_0^T f(t,y)\,dt + \sum_{k=1}^q I_k(y)\right)$$

have an isolated equilibrium point $y = y_0$, $f_0(y_0) = 0$, and the index of the mapping $f_0(y)$ at the point y_0 be different from zero.
    Then there exists $\epsilon_0 \in (0,\bar{\epsilon})$ such that for $\epsilon \in (0,\epsilon_0]$ equation (16.27) has a T-periodic solution $x = \varphi(t,\epsilon)$ and $\lim_{\epsilon\to 0}\varphi(t,\epsilon) = y_0$.

17. Method of bilateral approximations for finding periodic solutions

We shall apply the method of bilateral approximations [54, 59] for finding the T-periodic solutions of equation (16.1). Let the functions $g(t,x,y)$ and $J_k(x,y)$ be such that

$$g(t,x,x) = f(t,x), \qquad J_k(x,x) = I_k(x) \qquad (t \in \mathbb{R}, k \in \mathbb{Z}, x \in \Omega), \qquad (17.1)$$

and instead of (16.1) consider the equation

$$\frac{dx}{dt} = g(t,x,x), \quad t \neq \tau_k,$$
$$\Delta x = J_k(x,x), \quad t = \tau_k. \qquad (17.2)$$

Assume that the following conditions (H17) hold.

H17.1 $\tau_0 = 0$, $\tau_k < \tau_{k+1}$ $(k \in \mathbb{Z})$ and there exists $q \in \mathbb{N}$ such that $\tau_{k+q} = \tau_k + T$ $(k \in \mathbb{Z})$.
H17.2 The function $g : \mathbb{R} \times \Omega \times \Omega \to \mathbb{R}^n$ is continuous in the sets $(\tau_k, \tau_{k+1}] \times \Omega \times \Omega$ $(k \in \mathbb{Z})$ and for any $x, y \in \Omega$ and $k \in \mathbb{Z}$ there exists the finite limit of $g(t,u,v)$ as $(t,u,v) \to (\tau_k, x, y)$, $t > \tau_k$.
H17.3 The functions $J_k : \Omega \times \Omega \to \mathbb{R}^n$ $(k = 1,\dots,q)$ are continuous in $\Omega \times \Omega$.
H17.4 $g(t+T,x,y) = g(t,x,y)$ and $J_{k+q}(x,y) = J_k(x,y)$ for $x,y \in \Omega$, $t \in \mathbb{R}$, $k \in \mathbb{Z}$.
H17.5 There exist $M, \mu, L_k, l_k \in \mathbb{R}^n$ such that the following inequalities hold

$$\mu \leq g(t,x,y) \leq M, \qquad l_k \leq J_k(x,y) \leq L_k, \qquad (17.3)$$

$$g(t,x,y) \leq g(t,u,v), \qquad J_k(x,y) \leq J_k(u,v) \qquad (17.4)$$

for $t \in \mathbb{R}$; $k \in \mathbb{Z}$; $x,y,u,v \in \Omega$, $x \leq u$, $v \leq y$, where $x \leq u$ means that $x_i \leq u_i$ $(i = 1,\dots,n)$.
H17.6 $\Omega = \{x \in \mathbb{R}^n : a \leq x \leq b\}$ and

$$b - a > \frac{1}{2}T(M - \mu) + 2\sum_{k=1}^q \max(|L_k|, |l_k|) = 2\epsilon$$

where $|x| = \mathrm{col}\,(|x_1|,\dots,|x_n|)$, $\max(x,y) = \mathrm{col}\,(\max(x_1,y_1),\dots,\max(x_n,y_n))$.

Remark 17.1   We shall note that if N is a non-negative $n \times n$ matrix and $-N \leq \frac{\partial f}{\partial x}(t, x) \leq N$, then the function

$$g(t, x, y) = \frac{1}{2}(f(t, x) + Nx) + \frac{1}{2}(f(t, y) - Ny)$$

satisfies conditions (17.1) and (17.4).

If the function $h(t, x)$ is such that

$$h(t, x) - h(t, u) \leq f(t, x) - f(t, u) \leq -h(t, x) + h(t, u)$$

for $x \leq u$ then the function

$$g(t, x, y) = \frac{1}{2}(f(t, x) + h(t, x)) + \frac{1}{2}(f(t, y) - h(t, y))$$

also satisfies conditions (17.1) and (17.4).

Let $x_0 \in D = \{x \in \mathbb{R}^n : a + \epsilon \leq x \leq b - \epsilon\}$ and define successively the sequences $\{u_m(t, x_0)\}$ and $\{v_m(t, x_0)\}$ of T-periodic functions given in the interval $[0, T]$ by the formulae

$$u_0(t, x_0) = x_0 - \frac{1}{2}(M - \mu)\alpha(t) + \left(1 - \frac{t}{T}\right) \sum_{0 \leq \tau_k < t} l_k - \frac{t}{T} \sum_{t \leq \tau_k < T} L_k, \quad (17.5)$$

$$v_0(t, x_0) = x_0 + \frac{1}{2}(M - \mu)\alpha(t) + \left(1 - \frac{t}{T}\right) \sum_{0 \leq \tau_k < t} L_k - \frac{t}{T} \sum_{t \leq \tau_k < T} l_k, \quad (17.6)$$

$$u_{m+1}(t, x_0) = x_0 + \left(1 - \frac{t}{T}\right) \int_0^t g(s, u_m(s, x_0), v_m(s, x_0))\, ds$$

$$- \frac{t}{T} \int_t^T g(s, v_m(s, x_0), u_m(s, x_0))\, ds$$

$$+ \left(1 - \frac{t}{T}\right) \sum_{0 \leq \tau_k < t} J_k(u_m(\tau_k, x_0), v_m(\tau_k, x_0))$$

$$- \frac{t}{T} \sum_{t \leq \tau_k < T} J_k(v_m(\tau_k, x_0), u_m(\tau_k, x_0)), \quad (17.7)$$

$$v_{m+1}(t, x_0) = x_0 + \left(1 - \frac{t}{T}\right) \int_0^t g(s, v_m(s, x_0), u_m(s, x_0))\, ds$$

$$- \frac{t}{T} \int_t^T g(s, u_m(s, x_0), v_m(s, x_0))\, ds$$

$$+ \left(1 - \frac{t}{T}\right) \sum_{0 \leq \tau_k < t} J_k(v_m(\tau_k, x_0), u_m(\tau_k, x_0))$$

$$- \frac{t}{T} \sum_{t \leq \tau_k < T} J_k(u_m(\tau_k, x_0), v_m(\tau_k, x_0)), \quad (17.8)$$

where $\alpha(t) = 2t(1 - t/T)$ for $t \in [0, T]$.

We shall find sufficient conditions under which the sequences $u_m(t, x_0)$ and $v_m(t, x_0)$ two-sided and monotonically tend to the T-periodic solution $\tilde{x}(t, x_0)$ of equation (17.2) for which $\tilde{x}(0, x_0) = x_0$.

Theorem 17.1   *Let conditions (H17) hold and let $x_0 \in D$. Then:*

1. *the functions $u_m(t, x_0)$, $v_m(t, x_0)$ satisfy the relations:*

$$u_m(0, x_0) = u_m(T, x_0) = v_m(0, x_0) = v_m(T, x_0) = x_0, \qquad (17.9)$$

$$u_0(t, x_0) \leq u_1(t, x_0) \leq \cdots \leq u_m(t, x_0), \qquad (17.10)$$

$$v_0(t, x_0) \geq v_1(t, x_0) \geq \cdots \geq v_m(t, x_0), \qquad (17.11)$$

$$a \leq u_m(t, x_0) \leq v_m(t, x_0) \leq b \qquad (17.12)$$

for $t \in [0, T]$, $m = 0, 1, 2, \ldots$;

2. *the sequences $\{u_m(t, x_0)\}$, $\{v_m(t, x_0)\}$ are uniformly convergent in the interval $[0, T]$ and their limits $u(t, x_0)$, $v(t, x_0)$ satisfy the relations*

$$u(0, x_0) = u(T, x_0) = v(0, x_0) = v(T, x_0) = x_0, \qquad (17.13)$$

$$u_m(t, x_0) \leq u(t, x_0) \leq v(t, x_0) \leq v_m(t, x_0) \qquad (t \in [0, T], m = 0, 1, 2, \ldots), \qquad (17.14)$$

$$\begin{aligned}
u(t, x_0) = x_0 &+ \left(1 - \frac{t}{T}\right) \int_0^t g(s, u(s, x_0), v(s, x_0))\, ds \\
&- \frac{t}{T} \int_t^T g(s, v(s, x_0), u(s, x_0))\, ds \\
&+ \left(1 - \frac{t}{T}\right) \sum_{0 \leq \tau_k < t} J_k(u(\tau_k, x_0), v(\tau_k, x_0)) \\
&- \frac{t}{T} \sum_{t \leq \tau_k < T} J_k(v(\tau_k, x_0), u(\tau_k, x_0)),
\end{aligned} \qquad (17.15)$$

$$\begin{aligned}
v(t, x_0) = x_0 &+ \left(1 - \frac{t}{T}\right) \int_0^t g(s, v(s, x_0), u(s, x_0))\, ds \\
&- \frac{t}{T} \int_t^T g(s, u(s, x_0), v(s, x_0))\, ds \\
&+ \left(1 - \frac{t}{T}\right) \sum_{0 \leq \tau_k < t} J_k(v(\tau_k, x_0), u(\tau_k, x_0)) \\
&- \frac{t}{T} \sum_{t \leq \tau_k < T} J_k(u(\tau_k, x_0), v(\tau_k, x_0)).
\end{aligned} \qquad (17.16)$$

Proof   1. The validity of (17.9) is obvious. From (17.5), (17.6), H17.6 and property (16.4) it follows that

$$a \le u_0(t, x_0) \le v_0(t, x_0) \le b. \tag{17.17}$$

Taking into account (17.3) and (17.8), we obtain that

$$v_1(t, x_0) \le x_0 + \left(1 - \frac{t}{T}\right) \int_0^t M \, ds - \frac{t}{T} \int_t^T \mu \, ds$$

$$+ \left(1 - \frac{t}{T}\right) \sum_{0 \le \tau_k < t} L_k - \frac{t}{T} \sum_{t \le \tau_k < T} l_k$$

$$= v_0(t, x_0)$$

for $t \in [0, T]$. Analogously, $u_0(t, x_0) \le u_1(t, x_0)$ for $t \in [0, T]$.

From (17.7), (17.8), (17.17) and condition (17.4) it follows that

$$v_1(t, x_0) - u_1(t, x_0)$$

$$= \left(1 - \frac{t}{T}\right) \int_0^t [g(s, v_0(s, x_0), u_0(s, x_0)) - g(s, u_0(s, x_0), v_0(s, x_0))] \, ds$$

$$+ \frac{t}{T} \int_t^T [g(s, v_0(s, x_0), u_0(s, x_0)) - g(s, u_0(s, x_0), v_0(s, x_0))] \, ds$$

$$+ \left(1 - \frac{t}{T}\right) \sum_{0 \le \tau_k < t} [J_k(v_0(\tau_k, x_0), u_0(\tau_k, x_0)) - J_k(u_0(\tau_k, x_0), v_0(\tau_k, x_0))]$$

$$+ \frac{t}{T} \sum_{t \le \tau_k < T} [J_k(v_0(\tau_k, x_0), u_0(\tau_k, x_0)) - J_k(u_0(\tau_k, x_0), v_0(\tau_k, x_0))]$$

$$\ge 0.$$

Thus for $t \in [0, T]$ we have

$$a \le u_0(t, x_0) \le u_1(t, x_0) \le v_1(t, x_0) \le v_0(t, x_0) \le b. \tag{17.18}$$

By induction on m by virtue of (17.18) it is proved that for $m = 0, 1, 2, \ldots$ and $t \in [0, T]$

$$u_{m+1}(t, x_0) \le u_m(t, x_0) \le v_m(t, x_0) \le v_{m+1}(t, x_0).$$

2. Consider the space $\mathcal{B} = PC([0, T], \mathbb{R}^n)$ of piecewise continuous functions $x : [0, T] \to \mathbb{R}^n$ which have points of discontinuity $\tau_k \in [0, T]$ and are continuous from the left in $(0, T]$. Let the norm of $x \in \mathcal{B}$ be $\|x\|_\mathcal{B} = \sup_{t \in [0, T]} \|x(t)\|$, where $\|y\| = \max_{1 \le i \le n} |y_i|$ for $y \in \mathbb{R}^n$.

From conditions H17.2 and H17.3 it follows that the functions $u_m(t, x_0)$ and $v_m(t, x_0)$ belong to $\mathcal{B}$. Since the sequences $\{u_m(t, x_0)\}$ and $\{v_m(t, x_0)\}$ are uniformly bounded and quasiequicontinuous, then by Lemma 2.4 they have convergent subsequences. But from the monotonicity of the sequences $\{u_m(t, x_0)\}$ and $\{v_m(t, x_0)\}$ it follows that $u_m(t, x_0)$ and $v_m(t, x_0)$ are convergent in B which means that there exist functions $u(t, x_0)$ and $v(t, x_0)$ of B for which

$$\lim_{m \to \infty} u_m(t, x_0) = u(t, x_0), \qquad \lim_{m \to \infty} v_m(t, x_0) = v(t, x_0) \qquad (17.19)$$

uniformly with respect to $t \in [0, T]$.

Relations (17.9)–(17.12) and (17.19) imply immediately the validity of relations (17.13)–(17.16). $\square$

Consider the equations

$$x(t) = x_0 + \left(1 - \frac{t}{T}\right) \int_0^t g(s, x(s), x(s))\, ds - \frac{t}{T} \int_t^T g(s, x(s), x(s))\, ds$$
$$+ \left(1 - \frac{t}{T}\right) \sum_{0 \le \tau_k < t} J_k(x(\tau_k), x(\tau_k)) - \frac{t}{T} \sum_{t \le \tau_k < T} J_k(x(\tau_k), x(\tau_k)). \qquad (17.20)$$

Theorem 17.2 *Let conditions (H17) hold and let $x_0 \in D$. Then equation (17.20) has a T-periodic solution $x^*(t)$ and the following relations hold:*

$$x^*(0) = x^*(T) = x_0,$$

$$u_m(t, x_0) \le x^*(t) \le v_m(t, x_0) \qquad (t \in [0, T], m = 0, 1, 2, \ldots), \qquad (17.21)$$

$$u(t, x_0) \le x^*(t) \le v(t, x_0) \qquad (t \in [0, T]). \qquad (17.22)$$

Proof Consider the set $\tilde{B}$ of functions $x \in B = PC([0, T], \mathbb{R}^n)$ which satisfy the conditions

$$x(0) = x(T) = x_0, \qquad a \le x(t) \le b \qquad (t \in [0, T]).$$

Define the operator $\mathcal{F} : \tilde{B} \to B$ by the formula

$$\mathcal{F}[x](t) = x_0 + \left(1 - \frac{t}{T}\right) \int_0^t g(s, x(s), x(s))\, ds - \frac{t}{T} \int_t^T g(s, x(s), x(s))\, ds$$
$$+ \left(1 - \frac{t}{T}\right) \sum_{0 \le \tau_k < t} J_k(x(\tau_k), x(\tau_k)) - \frac{t}{T} \sum_{t \le \tau_k < T} J_k(x(\tau_k), x(\tau_k)).$$

The following assertions are valid.

I. The set $\tilde{B}$ is bounded, convex and closed in B.

II. $\mathcal{F}$ maps $\tilde{B}$ into itself. Indeed, if $x \in \tilde{B}$, then

$$\mathcal{F}[x](t) \le x_0 + \left(1 - \frac{t}{T}\right) \int_0^t M\, ds - \frac{t}{T} \int_t^T \mu\, ds$$
$$+ \left(1 - \frac{t}{T}\right) \sum_{0 \le \tau_k < t} L_k - \frac{t}{T} \sum_{t \le \tau_k < T} l_k \qquad (17.23)$$

$$= v_0(t, x_0) \le b.$$

Analogously,
$$\mathcal{F}[x](t) \geq u_0(t, x_0) \geq a \qquad (17.24)$$
and since $\mathcal{F}[x](0) = \mathcal{F}[x](T) = x_0$ and $\mathcal{F}[x] \in \mathcal{B}$, then $\mathcal{F}[x] \in \tilde{\mathcal{B}}$.

III. The set $\mathcal{F}\tilde{\mathcal{B}}$ is relatively compact in $\mathcal{B}$.

For the proof of this assertion we apply Lemma 2.4 taking into account that $\mathcal{F}\tilde{\mathcal{B}}$ is uniformly bounded and quasiequicontinuous. We shall mention only that the quasiequicontinuity of $\mathcal{F}\tilde{\mathcal{B}}$ follows from H17.2, H17.3 and the equality

$$\mathcal{F}[x](t_2) - \mathcal{F}[x](t_1) = \int_{t_1}^{t_2} g(s, x(s), x(s))\, ds$$

$$+ \frac{t_1 - t_2}{T}\left(\int_0^T g(s, x(s), x(s))\, ds + \sum_{k=1}^q J_k(x(\tau_k), x(\tau_k)) \right)$$

for $x \in \tilde{\mathcal{B}}$ and $t_1, t_2 \in (\tau_{k-1}, \tau_k]$, $t_1 < t_2$ $(k = 1, \ldots, q)$.

Hence by the Schauder–Tychonoff theorem the operator $\mathcal{F}$ has a fixed point $x^* \in \tilde{\mathcal{B}}$, that is, there exists a T-periodic function $x^*(t)$ satisfying (17.20). From (17.23) and (17.24) there follows the estimate

$$u_0(t, x_0) \leq x^*(t) \leq v_0(t, x_0)$$

from which by induction on $m = 0, 1, 2, \ldots$ we obtain that

$$u_m(t, x_0) \leq x^*(t) \leq v_m(t, x_0) \qquad (t \in [0, T], m = 0, 1, 2, \ldots). \qquad (17.25)$$

In (17.25) we pass to the limit and obtain (17.22). $\square$

Theorem 17.3   *Let the following conditions be fulfilled:*

1. *conditions (H17) hold and $x_0 \in D$;*
2. *the functions g and J_k satisfy the estimates*

$$g(t, x, y) - g(t, y, x) \leq K(x - y),$$
$$J_k(x, y) - J_k(y, x) \leq C(x - y) \qquad (17.26)$$

   where $a \leq y \leq x \leq b$ and K and C are non-negative $n \times n$ matrices;
3. *the moduli of the eigenvalues of the matrix*

$$P = \begin{bmatrix} \frac{1}{3}TK & K \\ SC & QC \end{bmatrix}$$

   are less than 1.

Then equation (17.20) has a unique T-periodic solution $\tilde{x}(t, x_0)$ for which $\tilde{x}(0, x_0) = x_0$ and

$$\tilde{x}(t, x_0) = u(t, x_0) = v(t, x_0) \qquad (t \in [0, T]).$$

Proof   For $m = 0$ the following estimate is valid:

$$v_0(t, x_0) - u_0(t, x_0) = \alpha(t)(M - \mu)$$

$$+ \left(1 - \frac{t}{T}\right) \sum_{0 \le \tau_k < t} (L_k - l_k) + \frac{t}{T} \sum_{t \le \tau_k < T} (L_k - l_k)$$

$$\le \alpha(t)(M - \mu) + \sum_{k=1}^{q} (L_k - l_k)$$

$$= \alpha(t)a_0 + b_0.$$

For $m = j$ let us have that

$$v_j(t, x_0) - u_j(t, x_0) \le \alpha(t)a_j + b_j.$$

Then by (17.7), (17.8), (17.26), (17.2)–(17.6) we obtain

$$v_{j+1}(t, x_0) - u_{j+1}(t, x_0)$$

$$\le \left(1 - \frac{t}{T}\right) \int_0^t K(\alpha(s)a_j + b_j) \, ds + \frac{t}{T} \int_t^T K(\alpha(s)a_j + b_j) \, ds$$

$$+ \left(1 - \frac{t}{T}\right) \sum_{0 \le \tau_k < t} C(\alpha(\tau_k)a_j + b_j) + \frac{t}{T} \sum_{t \le \tau_k < T} C(\alpha(\tau_k)a_j + b_j)$$

$$\le \alpha(t)[\tfrac{1}{3}TKa_j + Kb_j] + [SCa_j + QCb_j].$$

Consequently, for $m = 0, 1, 2, \dots$ the following estimates are valid:

$$v_m(t, x_0) - u_m(t, x_0) \le \alpha(t)a_m + b_m \qquad (t \in [0, T])$$

where

$$a_{m+1} = \frac{1}{3}TKa_m + Kb_m, \quad a_0 = M - \mu,$$

$$b_{m+1} = SCa_m + QCb_m, \quad b_0 = \sum_{k=1}^{q} (L_k - l_k). \tag{17.27}$$

From (17.27) in view of condition 3 of the theorem it follows that $a_m \to 0$ and $b_m \to 0$ as $m \to \infty$, that is,

$$\lim_{m \to \infty} [v_m(t, x_0) - u_m(t, x_0)] = 0 \qquad \text{(uniformly with respect to } t \in [0, T]\text{)}.$$

Then $u(t, x_0) = v(t, x_0)$ and by Theorem 17.2

$$\tilde{x}(t, x_0) = x^*(t) = u(t, x_0) = v(t, x_0) \qquad (t \in [0, T]). \qquad \square$$

Under the conditions of Theorem 17.3 we shall consider the question of the existence of a T-periodic solution of equation (17.2). Introduce the mapping $\Delta : D \to \mathbb{R}^n$, $x_0 \to \Delta(x_0)$:

$$\Delta(x_0) = \int_0^T g(s, \tilde{x}(s, x_0), \tilde{x}(s, x_0)) \, ds + \sum_{k=1}^{q} J_k(\tilde{x}(\tau_k, x_0), \tilde{x}(\tau_k, x_0))$$

where $\tilde{x}(t, x_0)$ is the T-periodic solution of equation (17.20) for which $\tilde{x}(0, x_0) = x_0$. Since

$$\tilde{x}(t, x_0) = x_0 + \int_0^t g(s, \tilde{x}(s, x_0), \tilde{x}(s, x_0)) \, ds$$

$$+ \sum_{0 \le \tau_k < t} J_k(\tilde{x}(\tau_k, x_0), \tilde{x}(\tau_k, x_0)) - \frac{t}{T}\Delta(x_0),$$

then $\tilde{x}(t, x_0)$ is a T-periodic solution of (17.2) if and only if $\Delta(x_0) = 0$.

From condition (17.4) and (17.21) it follows that

$$\Delta_m(x_0) \le \Delta(x_0) \le \Delta^m(x_0) \qquad (x_0 \in D) \qquad (17.28)$$

where

$$\Delta_m(x_0) = \int_0^T g(s, u_m(s, x_0), v_m(s, x_0)) \, ds + \sum_{k=1}^q J_k(u_m(\tau_k, x_0), v_m(\tau_k, x_0)),$$

$$\Delta^m(x_0) = \int_0^T g(s, v_m(s, x_0), u_m(s, x_0)) \, ds + \sum_{k=1}^q J_k(v_m(\tau_k, x_0), u_m(\tau_k, x_0)).$$

The inequalities (17.28) imply the following theorem.

Theorem 17.4   Let the conditions of Theorem 17.3 hold and, for some integer $m \ge 0$, $\Delta_m(x_0) > 0$ (or $\Delta^m(x_0) < 0$). Then equation (17.2) has no T-periodic solution $x(t)$ for which $x(0) = x_0$.

The following theorem is also valid.

Theorem 17.5   Let the following conditions be fulfilled:

1. the conditions of Theorem 17.3 hold;
2. for some integer $m \ge 0$ the mapping $\Delta_m(x)$ has an isolated singular point x^0 $(\Delta_m(x^0) = 0)$;
3. the index of the singular point x^0 is different from zero;
4. there exists a closed convex domain $F \subset D$ with a unique singular point x^0 such that on its boundary ∂F the inequality

$$\left\| \left(\frac{1}{3}T^2 K + \sum_{k=1}^q \alpha(\tau_k)C \right) a_m + (TK + qC)b_m \right\| \le \inf_{x \in \partial F} \|\Delta_m(x)\| \qquad (17.29)$$

holds, where a_m and b_m are defined by formulae (17.27).

Then equation (17.2) has a T-periodic solution $x(t)$ for which $x(0) \in F$.

Proof   Following the proof of Theorem 17.2, it suffices to prove that

$$\|\Delta_m(x)\| > \|\Delta(x) - \Delta_m(x)\| \qquad (x \in \partial F)$$

which follows from (17.28), (17.29) and the estimate

$$0 \le \Delta(x) - \Delta_m(x) \le \Delta^m(x) - \Delta_m(x)$$

$$\le \int_0^T K[v_m(s, x_0) - u_m(s, x_0)]\, ds$$

$$+ \sum_{k=1}^{q} C[v_m(\tau_k, x_0) - u_m(\tau_k, x_0)]$$

$$\le \int_0^T K[\alpha(s)a_m + b_m]\, ds$$

$$+ \sum_{k=1}^{q} C[\alpha(\tau_k)a_m + b_m]$$

$$= \Big[\frac{1}{3}T^2 K + \sum_{k=1}^{q} \alpha(\tau_k)C\Big]a_m + [TK + qC]b_m. \qquad \square$$

18. Projection-iterative method for finding periodic solutions

In this section, following [30], we shall describe some results on the application of the projection-iterative method for finding the periodic solutions of the periodic impulsive differential equation

$$\frac{dx}{dt} = f(t, x), \quad t \ne \tau_k,$$

$$\Delta x = I_k(x), \quad t = \tau_k. \tag{18.1}$$

where $t \in \mathbb{R}$, $k \in \mathbb{Z}$, $x \in \Omega \subset \mathbb{R}^n$.

First we shall introduce some notation and auxiliary assertions. Without loss of generality we assume that equation (18.1) is periodic with period $T = 2\pi$, that is, that the following condition holds.

H18.1   There exists $q \in \mathbb{N}$ such that

$$\tau_{k+q} = \tau_k + 2\pi, \qquad f(t + 2\pi, x) = f(t, x), \qquad I_{k+q}(x) = I_k(x)$$

for $t \in \mathbb{R}$, $k \in \mathbb{Z}$, $x \in \Omega$.

Let $|x|$ be the Euclidean norm of the vector $x \in \mathbb{R}^n$. Denote by $\mathcal{B}$ the space of all 2π-periodic functions $x \in PC(\mathbb{R}, \mathbb{R}^n)$ with norm $\|x\| = \sup_{t \in [0, 2\pi]} |x(t)|$. With each function $x \in \mathcal{B}$ we associate its Fourier series

$$x(t) \sim \frac{a_0}{2} + \sum_{k=1}^{\infty} (a_k \cos kt + b_k \sin kt)$$

where

$$a_k = \frac{1}{\pi} \int_0^{2\pi} x(t) \cos kt \, dt \qquad (k = 0, 1, 2, \ldots),$$

$$b_k = \frac{1}{\pi} \int_0^{2\pi} x(t) \sin kt \, dt \qquad (k = 1, 2, \ldots),$$

and, for fixed $m \in \mathbb{N}$, introduce the operators

$$P_0 x(t) = \frac{1}{2\pi} \int_0^{2\pi} x(t) \, dt,$$

$$P_m x(t) = \sum_{k=1}^{m} (a_k \cos kt + b_k \sin kt), \qquad (18.2)$$

$$Q_m x(t) = \sum_{k=m+1}^{\infty} (a_k \cos kt + b_k \sin kt).$$

We shall need the following auxiliary assertion [30].

Lemma 18.1   For the functions $x \in \mathcal{B}$ the following estimates are valid:

$$\left\| \int_0^t P_m x(s) \, ds \right\| \leq 2\sqrt{2} \tau(m) \|x\|, \qquad (18.3)$$

$$\left\| \int_0^t Q_m x(s) \, ds \right\| \leq 2\sqrt{2} \sigma(m) \|x\|, \qquad (18.4)$$

where

$$\tau^2(m) = \sup_{t \in [0, 2\pi]} \sum_{k=1}^{m} \frac{1}{k^2} \sin^2 \frac{kt}{2},$$

$$\sigma^2(m) = \sup_{t \in [0, 2\pi]} \sum_{k=m+1}^{\infty} \frac{1}{k^2} \sin^2 \frac{kt}{2}.$$

Remark 18.1   It can be proved that [30]

$$\sigma^2(m) \leq \sum_{k=m+1}^{\infty} \frac{1}{k^2} \leq \frac{\pi^2}{6} \qquad (m = 1, 2, \ldots)$$

and

$$\tau^2(m) \leq \sup_{t \in [0, 2\pi]} \sum_{k=1}^{\infty} \frac{1}{k^2} \sin^2 \frac{kt}{2} \leq \frac{\pi^2}{8} \qquad (m = 1, 2, \ldots).$$

Introduce the following conditions.

H18.2 The function $f : \mathbb{R} \times \Omega \to \mathbb{R}^n$ is continuous in the sets $(\tau_{k-1}, \tau_k] \times \Omega$ $(k \in \mathbb{Z})$ and for any $x \in \Omega$ and $k \in \mathbb{Z}$ there exists the finite limit of $f(t, y)$ as $(t, y) \to (\tau_k, x)$, $t > \tau_k$.

H18.3 The functions $I_k : \Omega \to \mathbb{R}^n$ $(k \in \mathbb{Z})$ are continuous in Ω.

H18.4 There exist constants $K_1 > 0$, $K_2 > 0$ and $M > 0$ such that

$$|f(t,x) - f(t,y)| \le K_1 |x - y|,$$

$$|I_k(x) - I_k(y)| \le K_2 |x - y|,$$

$$|f(t,x)| \le M,$$

$$|I_k(x)| \le M$$

for $x, y \in \Omega$, $t \in \mathbb{R}$, $k \in \mathbb{Z}$.

H18.5 There exists a non-empty compact set D contained in Ω together with its ϵ-neighbourhood, where

$$\epsilon = 2(\sqrt{2}\sigma(m) + \sqrt{2}\tau(m) + q)M. \tag{18.5}$$

H18.6 The following inequalities hold:

$$2\sqrt{2}\sigma(m)K_1 < 1, \tag{18.6}$$

$$q_m = \frac{2\sqrt{2}\sigma(m)K_1 + 2qK_2}{1 - 2\sqrt{2}\tau(m)K_1} < 1. \tag{18.7}$$

Let equation (18.1) have a 2π-periodic solution $x = \varphi(t)$ and let the point $x_0 = \varphi(0) \in D$ be known. Then the successive approximations to this solution can be obtained, defining recurrently in the interval $[0, 2\pi]$ the sequence $\{x_\nu(t, x_0)\}$ of functions by the formulae:

$$x_0(t, x_0) \equiv x_0$$

$$x_\nu(t, x_0) = x_0 + \int_0^t P_m f(s, x_\nu(s, x_0)) \, ds + \int_0^t Q_m f(s, x_{\nu-1}(s, x_0)) \, ds$$

$$+ \sum_{0 \le \tau_k < t} I_k(x_{\nu-1}(\tau_k, x_0)) - \frac{t}{2\pi} \sum_{0 \le \tau_k < 2\pi} I_k(x_{\nu-1}(\tau_k, x_0)), \tag{18.8}$$

where $\nu \in \mathbb{N}$, $t \in [0, 2\pi]$.

More precisely, the following assertion is valid.

Theorem 18.1 *Let conditions H18.1–H18.6 hold and let $x_0 \in D$. Then:*

1. the functions $x_\nu(t, x_0)$ satisfy the relations

$$x_\nu(0, x_0) = x_\nu(2\pi, x_0) = x_0, \tag{18.9}$$

$$x_\nu(t, x_0) \in \Omega \qquad (t \in [0, 2\pi]); \tag{18.10}$$

2. the sequence $\{x_\nu(t, x_0)\}$ is uniformly convergent in the interval $[0, 2\pi]$ and its limit $x(t, x_0)$ satisfies the relations:

$$x(0, x_0) = x(2\pi, x_0) = x_0, \tag{18.11}$$

$$x(t, x_0) \in \Omega \qquad (t \in [0, 2\pi]), \tag{18.12}$$

$$x(t, x_0) = x_0 + \int_0^t f(s, x(s, x_0)) \, ds + \sum_{0 \leq \tau_k < t} I_k(x(\tau_k, x_0))$$

$$- \frac{t}{2\pi} \left[\int_0^{2\pi} f(s, x(s, x_0)) \, ds + \sum_{0 \leq \tau_k < 2\pi} I_k(x(\tau_k, x_0)) \right]; \tag{18.13}$$

3. if $\varphi(t)$ is a 2π-periodic solution of (18.1) for which $\varphi(0) = \varphi(2\pi) = x_0$ and $\varphi(t) \in \Omega$, $t \in [0, 2\pi]$, then

$$\varphi(t) \equiv x(t, x_0) \qquad (t \in [0, 2\pi]).$$

Moreover, the following estimate is valid:

$$|x_\nu(t, x_0) - \varphi(t)| \leq \frac{\epsilon}{1 - q_m} q_m^\nu \qquad (t \in [0, 2\pi], \nu \in \mathbb{N}). \tag{18.14}$$

Proof   Denote by $\tilde{B}$ the set of all functions $x \in B$ for which

$$x(0) = x(2\pi) = x_0 \qquad \text{and} \qquad x(t) \in \Omega \qquad (t \in [0, 2\pi]).$$

Fix an arbitrary $z \in \tilde{B}$ and on the set $\tilde{B}$ define the operator T_z by the formula

$$T_z[x](t) = x_0 + \int_0^t P_m f(s, x(s)) \, ds + \int_0^t Q_m f(s, z(s)) \, ds$$

$$+ \sum_{0 \leq \tau_k < t} I_k(z(\tau_k)) - \frac{t}{2\pi} \sum_{0 \leq \tau_k < 2\pi} I_k(z(\tau_k)). \tag{18.15}$$

Since $\int_0^{2\pi} P_m f(s, y(s)) \, ds = 0$, $\int_0^{2\pi} Q_m f(s, y(s)) \, ds = 0$ for each $y \in \tilde{B}$, then from (18.15) it follows that

$$T_z[x](0) = T_z[x](2\pi) = x_0 \tag{18.16}$$

for $x, z \in \tilde{B}$. Moreover, from H18.4, (18.15) and Lemma 18.1 it follows that

$$\|T_z[x] - x_0\| \leq 2(\sqrt{2}\tau(m) + \sqrt{2}\sigma(m) + q)M = \epsilon \tag{18.17}$$

for $x, z \in \tilde{B}$. Consequently. the operator T_z, for fixed $z \in \tilde{B}$, maps the set $\tilde{B}$ into itself.

Let $x, y \in \tilde{B}$. Then from H18.4, (18.3) and (18.15) there follows the estimate

$$\|T_z[x] - T_z[y]\| \leq \left\| \int_0^t P_m[f(s, x(s)) - f(s, y(s))] \, ds \right\|$$

$$\leq 2\sqrt{2}\tau(m)K_1\|x - y\|$$

which means that the operator $T_z : \tilde{B} \to \tilde{B}$ is a contraction because of (18.6). By Banach's fixed point theorem there exists a unique $x \in \tilde{B}$ for which $T_z[x] = x$. This means that to any fixed function $z \in \tilde{B}$ there corresponds a unique function $x \in \tilde{B}$ for which $x(t) \equiv T_z[x](t)$, $t \in [0, 2\pi]$. Hence the sequence $\{x_\nu(t, x_0)\}$ is determined successively and uniquely from the equations $x_\nu = T_{x_{\nu-1}}[x_\nu]$, so from relations (18.16) and (18.17) it follows by induction that the functions $x_\nu(t, x_0)$ satisfy (18.9) and (18.10).

2. From H18.4, (18.3), (18.4) and (18.8) it follows that

$$\|x_{\nu+1} - x_\nu\| \leq \left\| \int_0^t P_m[f(s, x_{\nu+1}(s, x_0)) - f(s, x_\nu(s, x_0))] \, ds \right\|$$

$$+ \left\| \int_0^t Q_m[f(s, x_\nu(s, x_0)) - f(s, x_{\nu-1}(s, x_0))] \, ds \right\|$$

$$+ 2 \sum_{0 \leq \tau_k < 2\pi} |I_k(x_\nu(\tau_k, x_0)) - I_k(x_{\nu-1}(\tau_k, x_0))|$$

$$\leq 2\sqrt{2}\tau(m)K_1\|x_{\nu+1} - x_\nu\| + (2\sqrt{2}\sigma(m)K_1 + 2qK_2)\|x_\nu - x_{\nu-1}\|$$

from which we find

$$\|x_{\nu+1} - x_\nu\| \leq q_m\|x_\nu - x_{\nu-1}\|. \tag{18.18}$$

As a consequence, from (18.18) we obtain the estimate

$$\|x_{\nu+r} - x_\nu\| \leq \frac{q_m^\nu}{1 - q_m}\|x_1 - x_0\| \qquad (\nu, r \in \mathbb{N}) \tag{18.19}$$

and from (18.7) we conclude that the sequence x_ν is convergent in $\tilde{B}$. Since $\tilde{B}$ is a closed set there exists a function $x(t, x_0) \in \tilde{B}$ such that

$$\lim_{\nu \to \infty} x_\nu(t, x_0) = x(t, x_0) \qquad \text{(uniformly on } t \in [0, 2\pi]\text{)}.$$

Passing to the limit in (18.8), (18.9) and (18.10) we find that the function satisfies relations (18.11), (18.12) and the equation

$$x(t, x_0) = x_0 + \int_0^t P_m f(s, x(s, x_0)) \, ds + \int_0^t Q_m f(s, x(s, x_0)) \, ds$$

$$+ \sum_{0 \leq \tau_k < t} I_k(x(\tau_k, x_0)) - \frac{t}{2\pi} \sum_{0 \leq \tau_k < 2\pi} I_k(x(\tau_k, x_0)). \tag{18.20}$$

But this equation is equivalent to (18.13), in view of (18.2).

3. Since $\varphi(t)$ is a 2π-periodic solution of (18.1) for which $\varphi(0) = x_0$, then $\varphi(t)$ satisfies the equation

$$\varphi(t) = x_0 + \int_0^t f(s, \varphi(s))\, ds + \sum_{0 \le \tau_k < t} I_k(\varphi(\tau_k))$$

$$- \frac{t}{2\pi}\left[\int_0^{2\pi} f(s, \varphi(s))\, ds + \sum_{0 \le \tau_k < 2\pi} I_k(\varphi(\tau_k))\right].$$

Taking into account (18.2) we conclude that $\varphi(t)$ also satisfies equation (18.20). But by what was proved in item 1, equation (18.20) has a unique solution in $\tilde{B}$. Consequently, $x(t, x_0) \equiv \varphi(t)$, $t \in [0, 2\pi]$. Taking into account that $\|x_1 - x_0\| \le \epsilon$ and passing to the limit as $r \to \infty$ in (18.19), we obtain (18.14). □

The question of the existence of a 2π-periodic solution $x(t, x_0)$ of (18.1) for which $x(0, x_0) = x_0$ is solved as in Sections 16 and 17 by the investigation of the question about the zeros of the functions

$$\Delta(x_0) = \int_0^{2\pi} f(s, x(s, x_0))\, ds + \sum_{0 \le \tau_k < 2\pi} I_k(x(\tau_k, x_0)),$$

$$\Delta_\nu(x_0) = \int_0^{2\pi} f(s, x_\nu(s, x_0))\, ds + \sum_{0 \le \tau_k < 2\pi} I_k(x_\nu(\tau_k, x_0))$$

where $x_0 \in D$, $x(t, x_0) = \lim_{\nu \to \infty} x_\nu(t, x_0)$ and $x_\nu(t, x_0)$ are defined by (18.8).

The following theorem, which we give without proof, is valid.

Theorem 18.2   *Let the following conditions hold.*

1. *Conditions H18.1–H18.6 are met.*
2. *For some $\nu \in \mathbb{N}$ the mapping $\Delta_\nu : D \to \mathbb{R}^n$ has an isolated singular point x^0 $(\Delta_\nu(x^0) = 0)$.*
3. *The index of the singular point x^0 is different from zero.*
4. *There exists a closed convex domain $F \subset D$ with a unique singular point x^0 such that on the boundary ∂F the inequality*

$$\lim_{x \in \partial F} |\Delta_\nu(x)| > \frac{\epsilon q_m^\nu}{1 - q_m}(2\pi K_1 + q K_2)$$

is valid, where the constants ϵ and q_m are determined in (18.5) and (18.6).

Then equation (18.1) has a 2π-periodic solution $\varphi(t)$ for which $\varphi(0) \in D$. □

19. Method of boundary functions

In this section we shall justify an asymptotic method for finding a periodic solution of a linear singularly perturbed system of differential equations with fast and slow variables. To this end an appropriate modification to the method of boundary functions due to A. B. Vasil'eva [86] has been made.

Consider the periodic value problem

$$\epsilon\frac{dx}{dt} = A_1(t)x + B_1(t)y + f_1(t), \quad t \neq \tau_k,$$
$$\frac{dy}{dt} = A_2(t)x + B_2(t)y + f_2(t), \quad t \neq \tau_k; \tag{19.1}$$

$$\Delta x(\tau_k) = P^{(k)}x(\tau_k) + a_k,$$
$$\Delta y(\tau_k) = S^{(k)}y(\tau_k) + b_k; \tag{19.2}$$

$$x(0) = x(T), \qquad y(0) = y(T) \tag{19.3}$$

where $t \in \mathbb{R}$, $k \in \mathbb{Z}$, $x \in \mathbb{R}^m$, $y \in \mathbb{R}^n$ and $\epsilon > 0$ is a small parameter.

Introduce the following conditions (H19).

H19.1 System (19.1), (19.2) is T-periodic, that is, there exists a $q \in \mathbb{N}$ such that for $t \in \mathbb{R}$ and $k \in \mathbb{Z}$ we have

$$A_i(t+T) = A_i(t), \qquad B_i(t+T) = B_i(t), \qquad f_i(t+T) = f_i(t), \qquad (i = 1,2)$$

$$\tau_{k+q} = \tau_k + T, \qquad P^{(k+q)} = P^{(k)}, \qquad S^{(k+q)} = S^{(k)},$$

$$a_{k+q} = a_k, \qquad b_{k+q} = b_k.$$

H19.2 There exists $r \in \mathbb{N}$ such that

$$A_1(\,\cdot\,) \in PC^{r+1}(\mathbb{R}, \mathbb{R}^{m\times m}), \quad B_1(\,\cdot\,) \in PC^{r+1}(\mathbb{R}, \mathbb{R}^{m\times n}),$$
$$A_2(\,\cdot\,) \in PC^{r+1}(\mathbb{R}, \mathbb{R}^{n\times m}), \quad B_2(\,\cdot\,) \in PC^{r+1}(\mathbb{R}, \mathbb{R}^{n\times n}).$$

H19.3 $A_1(t) = \text{diag}\,(A_{11}(t), A_{22}(t))$ where $A_{11} \in \mathbb{R}^{m_1\times m_1}$, $A_{22} \in \mathbb{R}^{m_2\times m_2}$, $m_1 + m_2 = m$ and there exists an $\alpha > 0$ such that

$$\Re\lambda_i(A_{11}(t)) \leq -2\alpha < 0, \qquad \Re\lambda_j(A_{22}(t)) \geq 2\alpha > 0 \tag{19.4}$$

for $t \in [0,T]$, $i = 1,\ldots,m_1$; $j = 1,\ldots,m_2$.

H19.4 The matrices $P^{(k)}$ have the following block structure

$$P^{(k)} = \begin{bmatrix} P_{11}^{(k)} & P_{12}^{(k)} \\ P_{21}^{(k)} & P_{22}^{(k)} \end{bmatrix}$$

and the matrices $E_n + S^{(k)}$, $E_{m_1} + P_{11}^{(k)}$, $E_{m_2} + P_{22}^{(k)}$ are non-singular.

H19.5 $f_1(\,\cdot\,) \in PC(\mathbb{R}, \mathbb{R}^m)$, $f_2(\,\cdot\,) \in PC(\mathbb{R}, \mathbb{R}^n)$.

H19.6 The linear equation

$$\frac{d\xi}{dt} = [B_2(t) - A_2(t)A_1^{-1}(t)B_1(t)]\xi, \quad t \neq \tau_k,$$
$$\Delta\xi = S^{(k)}\xi, \qquad\qquad\qquad t = \tau_k, \tag{19.5}$$

has a unique T-periodic solution $\xi \equiv 0$.

Without loss of generality we assume that

$$\tau_0 < 0 < \tau_1 < \cdots < \tau_q < T < \tau_{q+1}.$$

Introduce the notation $z = (x, y) \in \mathbb{R}^m \times \mathbb{R}^n$, $x = (x_1, x_2) \in \mathbb{R}^{m_1} \times \mathbb{R}^{m_2}$.

Under these assumptions we shall prove that for ϵ small enough the periodic value problem (19.1)–(19.3) has a solution $z(t, \epsilon) = (x(t, \epsilon), y(t, \epsilon))$. We shall seek this solution in the form

$$z(t, \epsilon) = \bar{z}(t, \epsilon) + \Pi^{(k)}z(t_k, \epsilon) + Q^{(k)}z(s_k, \epsilon) \tag{19.6}$$

for $\tau_k < t \leq \tau_{k+1}$, where $\bar{z}(t, \epsilon)$ and the boundary functions

$$\Pi^{(k)}z(t_k, \epsilon) \qquad \text{and} \qquad Q^{(k)}z(s_k, \epsilon)$$

have the representations

$$\bar{z}(t, \epsilon) = \sum_{i=0}^{\infty} \epsilon^i \bar{z}_i(t) \qquad (t \in \mathbb{R}),$$
$$\Pi^{(k)}z(t_k, \epsilon) = \sum_{i=0}^{\infty} \epsilon^i \Pi_i^{(k)}z(t_k) \quad (k \in \mathbb{Z}), \tag{19.7}$$
$$Q^{(k)}z(s_k, \epsilon) = \sum_{i=0}^{\infty} \epsilon^i Q_i^{(k)}z(s_k) \quad (k \in \mathbb{Z})$$

where

$$t_k = \frac{t - \tau_k}{\epsilon}, \qquad s_k = \frac{t - \tau_{k+1}}{\epsilon}, \qquad \text{for } t \in (\tau_k, \tau_{k+1}].$$

Remark 19.1 Due to the T-periodicity it suffices to determine just q boundary functions $\Pi^{(k)}z$, $Q^{(k)}z$ $(k = 1 \ldots, q)$.

The local influence of the boundary functions in the approximation of the solution is taken into account by the following additional conditions:

$$\Pi_i^{(k)}z(+\infty) = 0, \qquad Q_i^{(k)}z(-\infty) = 0 \qquad (k \in \mathbb{Z}, i = 0, 1, 2\ldots). \tag{19.8}$$

Later we shall repeatedly use the so-called reduced problem

$$0 = A_1(t)x + B_1(t)y + f_1(t),$$
$$\frac{dy}{dt} = A_2(t)x + B_2(t)y + f_2(t), \quad t \neq \tau_k,$$
$$\Delta y(\tau_k) = S^{(k)}y(\tau_k) + b_k, \tag{19.9}$$
$$y(0) = y(T)$$

which is obtained from the periodic boundary value problem (19.1)–(19.3) by setting $\epsilon = 0$.

Problem (19.9) can be represented in the form

$$x = f_0(t, y(t)),$$
$$\frac{dy}{dt} = [B_2(t) - A_2(t)A_1^{-1}(t)B_1(t)]y + F_0(t), \quad t \neq \tau_k,$$
$$\Delta y(\tau_k) = S^{(k)}y(\tau_k) + b_k, \tag{19.10}$$
$$y(0) = y(T)$$

where

$$f_0(t, y) = -A_1^{-1}(t)B_1(t)y - A_1^{-1}(t)f_1(t)$$
$$F_0(t) = f_2(t) - A_2(t)A_1^{-1}(t)f_1(t).$$

By condition H19.6 and Theorem 4.1, problem (19.10) has a unique solution $(\bar{x}(t), \bar{y}(t))$ which is T-periodic and

$$\bar{x}(t) = f_0(t, \bar{y}(t)),$$
$$\bar{y}(t) = W(t, 0)\bar{y}(0) + \int_0^t W(t, s)F_0(s)\, ds + \sum_{0 < \tau_k < t} W(t, \tau_k^+)b_k,$$
$$\bar{y}(0) = [E_n - W(T, 0)]^{-1}\left[\int_0^T W(T, s)F_0(s)\, ds + \sum_{k=1}^q W(T, \tau_k^+)b_k\right]$$

where $W(t, s)$ is the Cauchy matrix for equation (19.5).

First we explain the justification of the algorithm for the determination of the functions $\bar{z}_i(t)$, $\Pi_i^{(k)}z(t_k)$, $Q_i^{(k)}z(s_k)$ entering the representation (19.6) of the solution $z(t, \epsilon)$.

We substitute (19.6) into (19.1) and obtain, for $k = 0, \ldots, q$ and $t \in (\tau_k, \tau_{k+1}]$,

$$\epsilon\frac{d\bar{x}}{dt} + \frac{d\Pi^{(k)}x}{dt_k} + \frac{dQ^{(k)}x}{ds_k} = f_1(t)$$
$$+ A_1(t)\bar{x}(t, \epsilon) + A_1(\tau_k + \epsilon t_k)\Pi^{(k)}x(t_k, \epsilon)$$
$$+ A_1(\tau_{k+1} + \epsilon s_k)Q^{(k)}x(s_k, \epsilon)$$
$$+ B_1(t)\bar{y}(t, \epsilon) + B_1(\tau_k + \epsilon t_k)\Pi^{(k)}y(t_k, \epsilon)$$
$$+ B_1(\tau_{k+1} + \epsilon s_k)Q^{(k)}y(s_k, \epsilon), \tag{19.11}$$

$$\frac{d\bar{y}}{dt} + \frac{1}{\epsilon}\frac{d\Pi^{(k)}y}{dt_k} + \frac{1}{\epsilon}\frac{dQ^{(k)}y}{ds_k} = f_2(t)$$
$$+ A_2(t)\bar{x}(t, \epsilon) + A_2(\tau_k + \epsilon t_k)\Pi^{(k)}x(t_k, \epsilon)$$
$$+ A_2(\tau_{k+1} + \epsilon s_k)Q^{(k)}x(s_k, \epsilon)$$

$$+ B_2(t)\bar{y}(t, \epsilon) + B_2(\tau_k + \epsilon t_k)\Pi^{(k)}y(t_k, \epsilon)$$

$$+ B_2(\tau_{k+1} + \epsilon s_k)Q^{(k)}y(s_k, \epsilon). \tag{19.12}$$

Substituting (19.6) into (19.2) we obtain

$$\bar{x}(\tau_k^+, \epsilon) + \Pi^{(k)}x(0, \epsilon) + Q^{(k)}x\left(\frac{\tau_k - \tau_{k+1}}{\epsilon}, \epsilon\right)$$

$$= (E_m + P^{(k)})$$

$$\times \left[\bar{x}(\tau_k, \epsilon) + \Pi^{(k-1)}x\left(\frac{\tau_k - \tau_{k-1}}{\epsilon}, \epsilon\right) + Q^{(k-1)}x(0, \epsilon)\right] + a_k, \tag{19.13}$$

$$\bar{y}(\tau_k^+, \epsilon) + \Pi^{(k)}y(0, \epsilon) + Q^{(k)}y\left(\frac{\tau_k - \tau_{k+1}}{\epsilon}, \epsilon\right)$$

$$= (E_n + S^{(k)})$$

$$\times \left[\bar{y}(\tau_k, \epsilon) + \Pi^{(k-1)}y\left(\frac{\tau_k - \tau_{k-1}}{\epsilon}, \epsilon\right) + Q^{(k-1)}y(0, \epsilon)\right] + b_k, \tag{19.14}$$

and substitution into (19.3) yields

$$\bar{z}(0, \epsilon) + \Pi^{(0)}z\left(-\frac{\tau_0}{\epsilon}, \epsilon\right) + Q^{(0)}z\left(-\frac{\tau_1}{\epsilon}, \epsilon\right)$$

$$= \bar{z}(T, \epsilon) + \Pi^{(q)}z\left(\frac{T - \tau_q}{\epsilon}, \epsilon\right) + Q^{(q)}z\left(\frac{T - \tau_{q+1}}{\epsilon}, \epsilon\right). \tag{19.15}$$

In equations (19.11) and (19.12) we replace $\bar{z}$, $\Pi^{(k)}z$ and $Q^{(k)}z$ by the expressions (19.7) and represent the matrices

$$A_j(\tau_k + \epsilon t_k), \qquad A_j(\tau_{k+1} + \epsilon s_k), \qquad B_j(\tau_k + \epsilon t_k), \qquad B_j(\tau_{k+1} + \epsilon s_k) \qquad (j = 1, 2)$$

as power series with respect to ϵ. We equate the coefficients at each power ϵ^i (those depending on t, t_k, s_k are equated separately) and obtain a system of equations for the coefficients of (19.7). In equations (19.13) and (19.14) we equate the coefficients at ϵ^i neglecting the terms $\Pi^{(k-1)}z\left(\frac{\tau_k - \tau_{k-1}}{\epsilon}, \epsilon\right)$ and $Q^{(k)}z\left(\frac{\tau_k - \tau_{k+1}}{\epsilon}, \epsilon\right)$. From the T-periodicity it follows that necessarily the following should hold:

$$\Pi^{(0)}z\left(-\frac{\tau_0}{\epsilon}, \epsilon\right) = \Pi^{(q)}z\left(\frac{T - \tau_q}{\epsilon}, \epsilon\right),$$

$$Q^{(0)}z\left(-\frac{\tau_1}{\epsilon}, \epsilon\right) = Q^{(q)}z\left(\frac{T - \tau_{q+1}}{\epsilon}, \epsilon\right).$$

Consequently, relation (19.15) takes the form

$$\bar{x}(0, \epsilon) = \bar{x}(T, \epsilon), \qquad \bar{y}(0, \epsilon) = \bar{y}(T, \epsilon). \tag{19.16}$$

In this way for the determination of the zero approximations from (19.11), (19.12), (19.13), (19.14) and (19.16) we obtain the relations

$$
\begin{aligned}
0 &= A_1(t)\bar{x}_0 + B_1(t)\bar{y}_0 + f_1(t), \\
\bar{y}_0' &= A_2(t)\bar{x}_0 + B_2(t)\bar{y}_0 + f_2(t), \qquad t \neq \tau_k, \\
\Delta \bar{y}_0(\tau_k) &= S^{(k)}\bar{y}_0(\tau_k) + b_k + [(E_n + S^{(k)})Q_0^{(k-1)}y(0) - \Pi_0^{(k)}y(0)], \\
\bar{y}(0) &= \bar{y}(T);
\end{aligned}
\tag{19.17}
$$

$$
\frac{d\Pi_0^{(k)}y(t_k)}{dt_k} = 0, \qquad \frac{dQ_0^{(k)}y(s_k)}{ds_k} = 0;
\tag{19.18}
$$

$$
\frac{d\Pi_0^{(k)}x(t_k)}{dt_k} = A_1(\tau_k)\Pi_0^{(k)}x(t_k) + B_1(\tau_k)\Pi_0^{(k)}y(t_k),
$$

$$
\frac{dQ_0^{(k)}x(s_k)}{ds_k} = A_1(\tau_{k+1})Q_0^{(k)}x(s_k) + B_1(\tau_{k+1})Q_0^{(k)}y(s_k);
\tag{19.19}
$$

$$
\bar{x}_0(\tau_k^+) + \Pi_0^{(k)}x(0) = (E_m + P^{(k)})[\bar{x}_0(\tau_k) + Q_0^{(k-1)}x(0)] + a_k.
\tag{19.20}
$$

From (19.18), in view of condition (19.8), it follows that

$$
\Pi_0^{(k)}y(t_k) \equiv 0, \qquad Q_0^{(k)}y(s_k) \equiv 0.
\tag{19.21}
$$

Thus from (19.17) it follows that the functions $\bar{x}_0(t)$, $\bar{y}_0(t)$ satisfy system (19.10) which has a unique solution. Hence

$$
\bar{x}_0 = \bar{x}(t), \qquad \bar{y}_0 = \bar{y}(t).
$$

Taking into account (19.19), (19.21) and the block diagonal structure of $A_1(t)$, we obtain

$$
\frac{d\Pi_{0,1}^{(k)}x(t_k)}{dt_k} = A_{11}(\tau_k)\Pi_{0,1}^{(k)}x(t_k),
\tag{19.22}
$$

$$
\frac{d\Pi_{0,2}^{(k)}x(t_k)}{dt_k} = A_{22}(\tau_k)\Pi_{0,2}^{(k)}x(t_k),
\tag{19.23}
$$

$$
\frac{dQ_{0,1}^{(k)}x(s_k)}{ds_k} = A_{11}(\tau_{k+1})Q_{0,1}^{(k)}x(s_k),
\tag{19.24}
$$

$$
\frac{dQ_{0,2}^{(k)}x(s_k)}{ds_k} = A_{22}(\tau_{k+1})Q_{0,2}^{(k)}x(s_k)
\tag{19.25}
$$

where $\Pi_i^{(k)}x = (\Pi_{i,1}^{(k)}x, \Pi_{i,2}^{(k)}x)$ and $Q_i^{(k)}x = (Q_{i,1}^{(k)}x, Q_{i,2}^{(k)}x)$ belong to $\mathbb{R}^{m_1} \times \mathbb{R}^{m_2}$ ($i = 0, 1, 2 \ldots$).

From (19.23), (19.24) and condition (19.8) it follows that necessarily

$$
\Pi_{0,2}^{(k)}x(0) = 0, \qquad Q_{0,1}^{(k)}x(0) = 0
$$

which implies that

$$\Pi_{0,2}^{(k)}x(t_k) \equiv 0, \qquad Q_{0,1}^{(k)}x(s_k) \equiv 0. \tag{19.26}$$

Taking into account (19.20) and (19.26) we find that the initial values $Q_{0,2}^{(k)}x(0)$ and $\Pi_{0,1}^{(k)}x(0)$ can be determined successively from the formulae

$$Q_{0,2}^{(k)}x(0) = (E_{m_2} + P_{22}^{(k+1)})^{-1}$$
$$\times [\bar{x}_{0,2}(\tau_{k+1}^+) + P_{21}^{(k+1)}\bar{x}_{0,1}(\tau_{k+1}) - a_{k+1,2}] - \bar{x}_{0,2}(\tau_{k+1}), \tag{19.27}$$

$$\Pi_{0,1}^{(k)}x(0) = (E_{m_1} + P_{11}^{(k)})\bar{x}_{0,1}(\tau_k)$$
$$+ P_{12}^{(k)}[\bar{x}_{0,2}(\tau_k) + Q_{0,2}^{(k-1)}x(0)] - \bar{x}_{0,1}(\tau_k^+) + a_{k,1}. \tag{19.28}$$

Based on (19.22), (19.25), (19.27) and (19.28), we determine

$$\Pi_{0,1}^{(k)}x(t_k) = e^{A_{11}(\tau_k)t_k}\Pi_{0,1}^{(k)}x(0) \qquad (t_k \geq 0),$$

$$Q_{0,2}^{(k)}x(s_k) = e^{A_{22}(\tau_{k+1})s_k}Q_{0,2}^{(k)}x(0) \qquad (s_k \leq 0).$$

From conditions (19.4) and Lemma 3.2 of [86] it follows that there exist constants $c > 0$ and $\epsilon_0 > 0$ such that for any $\epsilon \in (0, \epsilon_0]$

$$|\Pi_{0,1}^{(k)}x(t_k)| \leq ce^{-\alpha t_k} \qquad (t_k \geq 0),$$

$$|Q_{0,2}^{(k)}x(s_k)| \leq ce^{\alpha s_k} \qquad (s_k \leq 0).$$

Moreover, for each $k \in \mathbb{Z}$ the following relations are valid:

$$\Pi_0^{(k+q)}z(\tau) = \Pi_0^{(k)}z(\tau), \qquad Q_0^{(k+q)}z(\sigma) = Q_0^{(k)}z(\sigma).$$

In the same way for the determination of the ith approximations we obtain the relations

$$\bar{x}_{i-1}' = A_1(t)\bar{x}_i + B_1(t)\bar{y}_i,$$
$$\bar{y}_i' = A_2(t)\bar{x}_i + B_2(t)\bar{y}_i, \qquad t \neq \tau_k,$$
$$\Delta\bar{y}_i(\tau_k) = S^{(k)}\bar{y}_i(\tau_k) + (E_n + S^{(k)})Q_i^{(k-1)}y(0) - \Pi_i^{(k)}y(0), \tag{19.29}$$
$$\bar{y}_i(0) = \bar{y}_i(T);$$

$$\frac{d\Pi_i^{(k)}x(t_k)}{dt_k} = A_1(\tau_k)\Pi_i^{(k)}x(t_k) + B_1(\tau_k)\Pi_i^{(k)}y(t_k) + T_i^{(k)}(t_k), \tag{19.30}$$

$$\frac{d\Pi_i^{(k)}y(t_k)}{dt_k} = R_i^{(k)}(t_k), \tag{19.31}$$

$$\frac{dQ_i^{(k)}x(s_k)}{ds_k} = A_1(\tau_{k+1})Q_i^{(k)}x(s_k) + B_1(\tau_{k+1})Q_i^{(k)}y(s_k) + G_i^{(k)}(s_k), \tag{19.32}$$

$$\frac{dQ_i^{(k)}y(s_k)}{ds_k} = H_i^{(k)}(s_k), \tag{19.33}$$

$$\bar{x}_i(\tau_k^+) + \Pi_i^{(k)} x(0) = (E_m + P^{(k)})[\bar{x}_i(\tau_k) + Q_i^{(k-1)} x(0)] \qquad (19.34)$$

where

$$T_i^{(k)}(t_k) = \sum_{\nu=1}^{i} \frac{t_k^\nu}{\nu!} \left[\frac{d^\nu A_1}{dt^\nu}(\tau_k) \Pi_{i-\nu}^{(k)} x(t_k) + \frac{d^\nu B_1}{dt^\nu}(\tau_k) \Pi_{i-\nu}^{(k)} y(t_k) \right],$$

$$R_i^{(k)}(t_k) = \sum_{\nu=0}^{i-1} \frac{t_k^\nu}{\nu!} \left[\frac{d^\nu A_2}{dt^\nu}(\tau_k) \Pi_{i-1-\nu}^{(k)} x(t_k) + \frac{d^\nu B_2}{dt^\nu}(\tau_k) \Pi_{i-1-\nu}^{(k)} y(t_k) \right],$$

$$\qquad (19.35)$$

$$G_i^{(k)}(s_k) = \sum_{\nu=1}^{i} \frac{s_k^\nu}{\nu!} \left[\frac{d^\nu A_1}{dt^\nu}(\tau_{k+1}) Q_{i-\nu}^{(k)} x(s_k) + \frac{d^\nu B_1}{dt^\nu}(\tau_{k+1}) Q_{i-\nu}^{(k)} y(s_k) \right],$$

$$H_i^{(k)}(s_k) = \sum_{\nu=0}^{i-1} \frac{s_k^\nu}{\nu!} \left[\frac{d^\nu A_2}{dt^\nu}(\tau_{k+1}) Q_{i-1-\nu}^{(k)} x(s_k) + \frac{d^\nu B_2}{dt^\nu}(\tau_{k+1}) Q_{i-1-\nu}^{(k)} y(s_k) \right].$$

From condition (19.8) it follows that the initial data for equations (19.31) and (19.33) necessarily have the form

$$\Pi_i^{(k)} y(0) = -\int_0^\infty R_i^{(k)}(\tau)\,d\tau, \qquad Q_i^{(k)} y(0) = \int_{-\infty}^0 H_i^{(k)}(\tau)\,d\tau. \qquad (19.36)$$

Then the solutions of (19.31) and (19.33) are

$$\Pi_i^{(k)} y(t_k) = -\int_{t_k}^\infty R_i^{(k)}(\tau)\,d\tau, \qquad Q_i^{(k)} y(s_k) = \int_{-\infty}^{s_k} H_i^{(k)}(\tau)\,d\tau. \qquad (19.37)$$

System (19.29) for the determination of $\bar{x}_i$, $\bar{y}_i$ is of the form (19.10). Therefore taking into account (19.36) we conclude that (19.29) has a unique T-periodic solution $\bar{x}_i(t)$, $\bar{y}_i(t)$ which is determined successively from the formulae

$$\bar{y}_i(t) = W(t,0)\bar{y}_i(0) + \int_0^t W(t,s)F_i(s)\,ds + \sum_{0 < \tau_k < t} W(t,\tau_k^+)\gamma_i^{(k)},$$

$$\qquad (19.38)$$

$$\bar{x}_i(t) = -A_1^{-1}(t)B_1(t)\bar{y}_i(t) + A_1^{-1}(t)\bar{x}'_{i-1}(t),$$

where

$$\bar{y}_i(0) = [E_n - W(T,0)]^{-1} \left[\int_0^T W(T,s)F_i(s)\,ds + \sum_{k=1}^q W(T,\tau_k^+)\gamma_i^{(k)} \right],$$

$$F_i(t) = A_2(t)A_1^{-1}(t)\bar{x}'_{i-1}(t),$$

$$\gamma_i^{(k)} = [E_n + S^{(k)}]\int_{-\infty}^0 H_i^{(k-1)}(\tau)\,d\tau + \int_0^\infty R_i^{(k)}(\tau)\,d\tau.$$

Taking into account condition H19.3, we write equations (19.30) and (19.32) in the form

$$\frac{d\Pi_{i,1}^{(k)} x(t_k)}{dt_k} = A_{11}(\tau_k)\Pi_{i,1}^{(k)} x(t_k) + [B_1(\tau_k)\Pi_i^{(k)} y(t_k) + T_i^{(k)}(t_k)]_1, \qquad (19.39)$$

$$\frac{d\Pi_{i,2}^{(k)}x(t_k)}{dt_k} = A_{22}(\tau_k)\Pi_{i,2}^{(k)}x(t_k) + [B_1(\tau_k)\Pi_i^{(k)}y(t_k) + T_i^{(k)}(t_k)]_2, \qquad (19.40)$$

$$\frac{dQ_{i,1}^{(k)}x(s_k)}{ds_k} = A_{11}(\tau_{k+1})Q_{i,1}^{(k)}x(s_k) + [B_1(\tau_{k+1})Q_i^{(k)}y(s_k) + G_i^{(k)}(s_k)]_1, \qquad (19.41)$$

$$\frac{dQ_{i,2}^{(k)}x(s_k)}{ds_k} = A_{22}(\tau_{k+1})Q_{i,2}^{(k)}x(s_k) + [B_1(\tau_{k+1})Q_i^{(k)}y(s_k) + G_i^{(k)}(s_k)]_2. \qquad (19.42)$$

Taking into account (19.4) and condition (19.8), we conclude that the initial data for equations (19.40) and (19.41) have the form

$$\Pi_{i,2}^{(k)}x(0) = -\int_0^\infty e^{-A_{22}(\tau_k)s}[B_1(\tau_k)\Pi_i^{(k)}y(s) + T_i^{(k)}(s)]_2 \, ds, \qquad (19.43)$$

$$Q_{i,1}^{(k)}x(0) = \int_{-\infty}^0 e^{-A_{11}(\tau_{k+1})s}[B_1(\tau_{k+1})Q_i^{(k)}y(s) + G_i^{(k)}(s)]_1 \, ds. \qquad (19.44)$$

With these initial data the solutions of (19.40) and (19.41) are

$$\Pi_{i,2}^{(k)}x(t_k) = -\int_{\tau_k}^\infty e^{A_{22}(\tau_k)(t_k-s)}[B_1(\tau_k)\Pi_i^{(k)}y(s) + T_i^{(k)}(s)]_2 \, ds, \qquad (19.45)$$

$$Q_{i,1}^{(k)}x(s_k) = \int_{-\infty}^{s_k} e^{A_{11}(\tau_{k+1})(s_k-s)}[B_1(\tau_{k+1})Q_i^{(k)}y(s) + G_i^{(k)}(s)]_1 \, ds. \qquad (19.46)$$

Now the initial data for equations (19.39) and (19.42) are determined taking into account (19.34), (19.43) and (19.44). We conclude that

$$Q_{i,2}^{(k)}x(0) = (E_{m_2} + P_{22}^{(k+1)})^{-1}[\bar{x}_{i,2}(\tau_{k+1}^+) + \Pi_{i,2}^{(k+1)}x(0)$$
$$- P_{21}^{(k+1)}(\bar{x}_{i,1}(\tau_{k+1}) + Q_{i,1}^{(k)}x(0))] - \bar{x}_{i,2}(\tau_{k+1}),$$

$$\Pi_{i,1}^{(k)}x(0) = (E_{m_1} + P_{11}^{(k)})[\bar{x}_{i,1}(\tau_k) + Q_{i,1}^{(k-1)}x(0)]$$
$$+ P_{12}^{(k)}[\bar{x}_{i,2}(\tau_k) + Q_{i,2}^{(k-1)}x(0)] - \bar{x}_{i,1}(\tau_k^+).$$

With these initial data the solutions of (19.39) and (19.42) are

$$\Pi_{i,1}^{(k)}x(t_k) = e^{A_{11}(\tau_k)t_k}\Pi_{i,1}^{(k)}x(0)$$
$$+ \int_0^{t_k} e^{A_{11}(\tau_k)(t_k-s)}[B_1(\tau_k)\Pi_i^{(k)}y(s) + T_i^{(k)}(s)]_1 \, ds,$$

$$Q_{i,2}^{(k)}x(s_k) = e^{A_{22}(\tau_{k+1})s_k}Q_{i,2}^{(k)}x(0)$$
$$+ \int_0^{s_k} e^{A_{22}(\tau_{k+1})(s_k-s)}[B_1(\tau_{k+1})Q_i^{(k)}y(s) + G_i^{(k)}(s)]_2 \, ds.$$

In this way the coefficients of the expansions (19.7) are determined uniquely one after the other. Moreover, the periodicity conditions are obviously met:

$$\Pi_i^{(k+q)}z(\tau) = \Pi_i^{(k)}z(\tau), \qquad Q_i^{(k+q)}z(\sigma) = Q_i^{(k)}z(\sigma) \qquad (k \in \mathbb{Z})$$

for $\tau \geq 0$, $\sigma \leq 0$ and $i = 0, 1, 2, \ldots$.

From (19.35) it follows that analogous conditions of periodicity with respect to k are satisfied by the functions $T_i^{(k)}(\tau)$, $R_i^{(k)}(\tau)$, $G_i^{(k)}(\sigma)$, $H_i^{(k)}(\sigma)$ for $\tau \geq 0$, $\sigma \leq 0$, $i \in \mathbb{N}$. Moreover, for the boundary functions $\Pi_i^{(k)} z(t_k)$ and $Q_i^{(k)} z(\sigma_k)$, for $k \in \mathbb{Z}$ and $i = 0, 1, 2, \ldots$, the following estimates are valid:

$$\begin{aligned}
|\Pi_i^{(k)} z(t_k)| &\leq c e^{-\alpha t_k} \quad (t_k \geq 0), \\
|Q_i^{(k)} z(s_k)| &\leq c e^{\alpha s_k} \quad (s_k \leq 0)
\end{aligned} \tag{19.47}$$

where the constant $c > 0$ does not depend on $k \in \mathbb{Z}$ and $\epsilon \in (0, \epsilon_0]$.

From (19.47) there follows the convergence of all improper integrals entering formulae (19.36)–(19.38), (19.43)–(19.46).

Remark 19.2 Henceforth we shall repeatedly shrink the interval $(0, \epsilon_0]$ of variation of the values of the parameter ϵ for which some assertion or inequality is valid. Each time, the right endpoint of this interval will again be denoted by ϵ_0.

The objective of the subsequent arguments will be to prove that the periodic value problem (19.1)–(19.3) has a unique solution when $\epsilon > 0$ is small enough. Moreover, in the metric of the space $PC(\mathbb{R}, \mathbb{R}^{m+n})$ we shall estimate the difference between this solution and the rth approximation

$$z_r(t, \epsilon) = \sum_{i=0}^{r} \epsilon^i [\bar{z}_i(t) + \Pi_i^{(k)} z(t_k) + Q_i^{(k)} z(s_k)], \qquad \tau_k < t \leq \tau_{k+1}. \tag{19.48}$$

To this end we shall first consider the following auxiliary T-periodic equation

$$\begin{aligned}
\epsilon \xi' &= A_1(t)\xi + \psi(t), \quad t \neq \tau_k, \\
\Delta \xi(\tau_k) &= P^{(k)} \xi(\tau_k) + c_k, \quad k \in \mathbb{Z}
\end{aligned} \tag{19.49}$$

under the assumption that the following condition holds.

H19.7 The matrices $A_1(\cdot) \in PC(\mathbb{R}, \mathbb{R}^{m \times m})$ and $P^{(k)} \in \mathbb{R}^{m \times m}$ satisfy conditions H19.3 and H19.4, and the function $\psi(\cdot) \in PC(\mathbb{R}, \mathbb{R}^m)$ and the vectors $c_k \in \mathbb{R}^m$ are such that

$$\psi(t + T) = \psi(t), \qquad c_{k+q} = c_k \qquad (t \in \mathbb{R}, k \in \mathbb{Z}).$$

Equation (19.49) is decomposed into two equations

$$\begin{aligned}
\epsilon \xi_1' &= A_{11}(t)\xi_1 + \psi_1(t), \qquad\qquad t \neq \tau_k, \\
\Delta \xi_1(\tau_k) &= P_{11}^{(k)} \xi_1(\tau_k) + P_{12}^{(k)} \xi_2(\tau_k) + c_{k,1},
\end{aligned} \tag{19.50}$$

and

$$\begin{aligned}
\epsilon \xi_2' &= A_{22}(t)\xi_2 + \psi_2(t), \qquad\qquad t \neq \tau_k, \\
\Delta \xi_2(\tau_k) &= P_{22}^{(k)} \xi_2(\tau_k) + P_{21}^{(k)} \xi_1(\tau_k) + c_{k,2}.
\end{aligned} \tag{19.51}$$

Denote by $Y_i(t, s, \epsilon)$ $(i = 1, 2)$ the Cauchy matrix of the homogeneous linear equation

$$\epsilon \eta_i' = A_{ii}(t)\eta_i.$$

In [27] it is proved that under condition H7 the matrices $Y_i(t, s, \epsilon)$ $(i = 1, 2)$ satisfy the estimates

$$|Y_1(t, s, \epsilon)| \leq K_0 e^{-\alpha(t-s)/\epsilon} \quad (0 \leq s \leq t \leq T),$$

$$|Y_2(t, s, \epsilon)| \leq K_0 e^{\alpha(t-s)/\epsilon} \quad (0 \leq t \leq s \leq T),$$

$$(19.52)$$

where the constant $K_0 > 0$ does not depend on the parameter $\epsilon \in (0, \epsilon_0]$, and ϵ_0 is chosen small enough.

Let $U_i(t, s, \epsilon)$ $(i = 1, 2)$ be the Cauchy matrix for the linear impulsive equation

$$\epsilon u_i' = A_{ii}(t)u_i, \quad t \neq \tau_k,$$

$$\Delta u_i(\tau_k) = P_{ii}^{(k)} u_i(\tau_k).$$

$$(19.53)$$

Taking into account (19.52) and the formula (2.13) for the Cauchy matrix of a linear homogeneous impulsive equation we conclude that there exist constants $\epsilon_0 > 0$ and $K > 0$ such that for $\epsilon \in (0, \epsilon_0]$ the following inequalities are valid:

$$|U_1(t, s, \epsilon)| \leq K e^{-\alpha(t-s)/\epsilon} \quad (0 \leq s \leq t \leq T),$$

$$|U_2(t, s, \epsilon)| \leq K e^{\alpha(t-s)/\epsilon} \quad (0 \leq t \leq s \leq T).$$

$$(19.54)$$

From estimates (19.54) it follows that for sufficiently small values of ϵ equation (19.53) has a unique T-periodic solution $u_i \equiv 0$ $(i = 1, 2)$.

Let $d_k \in \mathbb{R}^m$ and $d_{k+q} = d_k$ $(k \in \mathbb{Z})$.

Lemma 19.1   *Let condition H19.7 hold. Then there exist constants $\epsilon_0 > 0$ and $K_1 > 0$ such that for $\epsilon \in (0, \epsilon_0]$ the equation*

$$\epsilon \xi_1' = A_{11}(t)\xi_1 + \psi_1(t), \quad t \neq \tau_k,$$

$$\Delta \xi_1(\tau_k) = P_{11}^{(k)} \xi_1(\tau_k) + d_{k,1}$$

$$(19.55)$$

has a unique T-periodic solution $\xi_1(t, \epsilon) \in PC(\mathbb{R}, \mathbb{R}^{m_1})$ and the following estimate is valid

$$\|\xi_1(\,\cdot\,, \epsilon)\| \leq K_1 \max(\|\psi_1\|, \sup_k |d_{k,1}|). \tag{19.56}$$

Lemma 19.2   *Let condition H19.7 hold. Then there exist constants $\epsilon_0 > 0$ and $K_2 > 0$ such that for $\epsilon \in (0, \epsilon_0]$ the equation*

$$\epsilon \xi_2' = A_{22}(t)\xi_2 + \psi_2(t), \quad t \neq \tau_k,$$

$$\Delta \xi_2(\tau_k) = P_{22}^{(k)} \xi_2(\tau_k) + d_{k,2}$$

$$(19.57)$$

has a unique T-periodic solution $\xi_2(t, \epsilon) \in PC(\mathbb{R}, \mathbb{R}^{m_2})$ and the following estimate is valid

$$\|\xi_2(\,\cdot\,, \epsilon)\| \leq K_2 \max(\|\psi_2\|, \sup_k |d_{k,2}|). \tag{19.58}$$

Proof of Lemmas 19.1 and 19.2   The existence and uniqueness of the solutions $\xi_i(t,\epsilon)$, $i = 1,2$, are based on Theorem 4.1. Estimates (19.56) and (19.58) are obtained taking into account (19.54) and the explicit form of these solutions:

$$\xi_1(t,\epsilon) = U_1(t,0,\epsilon)\xi_1(0,\epsilon)$$

$$+ \frac{1}{\epsilon}\int_0^t U_1(t,s,\epsilon)\psi_1(s)\,\mathrm{d}s + \sum_{0<\tau_k<t} U_1(t,\tau_k^+,\epsilon)d_{k,1},$$

$$\xi_1(0,\epsilon) = [E_{m_1} - U_1(T,0,\epsilon)]^{-1}$$

$$\times \left[\frac{1}{\epsilon}\int_0^T U_1(T,s,\epsilon)\psi_1(s)\,\mathrm{d}s + \sum_{k=1}^q U_1(T,\tau_k^+,\epsilon)d_{k,1}\right]; \tag{19.59}$$

$$\xi_2(t,\epsilon) = U_2(t,T,\epsilon)\xi_2(T,\epsilon)$$

$$+ \frac{1}{\epsilon}\int_T^t U_2(t,s,\epsilon)\psi_2(s)\,\mathrm{d}s - \sum_{t\leq\tau_k<T} U_2(t,\tau_k^+,\epsilon)d_{k,2},$$

$$\xi_2(T,\epsilon) = [E_{m_2} - U_2(0,T,\epsilon)]^{-1}$$

$$\times \left[\frac{1}{\epsilon}\int_T^0 U_2(0,s,\epsilon)\psi_2(s)\,\mathrm{d}s - \sum_{k=1}^q U_2(0,\tau_k^+,\epsilon)d_{k,2}\right]. \tag{19.60}$$

$\square$

Lemma 19.3   *Let condition H19.7 hold. Then there exist constants $\epsilon_0 > 0$ and $K_3 > 0$ such that for $\epsilon \in (0,\epsilon_0]$ equation (19.49) has a unique T-periodic solution $\xi(t,\epsilon) \in PC(\mathbb{R},\mathbb{R}^m)$ and the following estimate is valid*

$$\|\xi(\,\cdot\,,\epsilon)\| \leq K_3 \max(\|\psi\|, \sup_k |c_k|). \tag{19.61}$$

Proof   Let the sequences $\{f_k\} \subset \mathbb{R}^{m_1}$ and $\{h_k\} \subset \mathbb{R}^{m_2}$ be such that $f_{k+q} = f_k$, $h_{k+q} = h_k$ ($k \in \mathbb{Z}$). Denote by $\xi_1(t,\epsilon,f_1,\dots,f_q)$ the T-periodic solution of equation (19.55) in which

$$d_{k,1} = f_k + c_{k,1}, \tag{19.62}$$

and by $\xi_2(t,\epsilon,h_1,\dots,h_q)$ the T-periodic solution of equation (19.57) in which

$$d_{k,2} = h_k + c_{k,2}. \tag{19.63}$$

By Lemmas 19.1 and 19.2 these solutions exist and are unique for ϵ small enough.

In order to obtain the T-periodic solution of system (19.50), (19.51) we must choose f_k, h_k so as to satisfy the relations

$$P_{21}^{(k)}\xi_1(\tau_k,\epsilon,f_1,\dots,f_q) = h_k \qquad (k = 1,\dots,q),$$

$$P_{12}^{(k)}\xi_2(\tau_j,\epsilon,h_1,\dots,h_q) = f_j \qquad (j = 1,\dots,q). \tag{19.64}$$

Taking into account (19.59), (19.60), (19.62) and (19.63), we conclude that (19.64) is a linear algebraic system with respect to f_k, h_k, with determinant tending to 1 as $\epsilon \to 0$. Consequently, there exists an $\epsilon_0 > 0$ such that for $\epsilon \in (0, \epsilon_0]$ system (19.64) has a unique solution.

Estimate (19.61) follows from estimates (19.56), (19.58) and relations (19.62)–(19.64). $\square$

Theorem 19.1   *Let conditions (H19) hold. Then there exist constants $\epsilon_* > 0$ and $M > 0$ such that for $\epsilon \in (0, \epsilon_*]$ the periodic value problem (19.1)–(19.3) has a unique solution $z(t, \epsilon)$ satisfying the estimate*

$$\|z(\,\cdot\,, \epsilon) - z_r(\,\cdot\,, \epsilon)\| \le M\epsilon^{r+1}. \tag{19.65}$$

Proof   Carrying out the change of variables

$$u = x - x_r(t, \epsilon), \qquad v = y - y_r(t, \epsilon) \tag{19.66}$$

we transform problem (19.1)–(19.3) into the problem

$$
\begin{aligned}
\epsilon u' &= A_1(t)u + B_1(t)v + g_1(t, \epsilon), && t \ne \tau_k, \\
v' &= [B_2(t) - A_2(t)A_1^{-1}(t)B_1(t)]v \\
&\quad + D(t)[A_1(t)u + B_1(t)v] + g_2(t, \epsilon), && t \ne \tau_k, \\
\Delta u(\tau_k) &= P^{(k)}u(\tau_k) + \alpha_k(\epsilon), \\
\Delta v(\tau_k) &= S^{(k)}v(\tau_k) + \beta_k(\epsilon),
\end{aligned}
\tag{19.67}
$$

$$u(0) = u(T), \qquad v(0) = v(T),$$

where

$$D(t) = A_2(t)A_1^{-1}(t),$$

$$g_1(t, \epsilon) = A_1(t)x_r(t, \epsilon) + B_1(t)y_r(t, \epsilon) - \epsilon\frac{dx_r}{dt} + f_1(t),$$

$$g_2(t, \epsilon) = A_2(t)x_r(t, \epsilon) + B_2(t)y_r(t, \epsilon) - \frac{dy_r}{dt} + f_2(t),$$

$$\alpha_k(\epsilon) = \sum_{i=0}^{r} \epsilon^i \left[(E_m + P^{(k)})\Pi_i^{(k-1)} x\left(\frac{\tau_k - \tau_{k-1}}{\epsilon}, \epsilon\right) - Q_i^{(k)} x\left(\frac{\tau_k - \tau_{k+1}}{\epsilon}, \epsilon\right) \right],$$

$$\beta_k(\epsilon) = \sum_{i=0}^{r} \epsilon^i \left[(E_n + S^{(k)})\Pi_i^{(k-1)} y\left(\frac{\tau_k - \tau_{k-1}}{\epsilon}, \epsilon\right) - Q_i^{(k)} y\left(\frac{\tau_k - \tau_{k+1}}{\epsilon}, \epsilon\right) \right].$$

(19.68)

From relations (19.17)–(19.20), (19.29)–(19.35), (19.48) and (19.68) we conclude that there exist constants $\epsilon_0 > 0$ and $c_0 > 0$ such that for $\epsilon \in (0, \epsilon_0]$ the following estimates are valid:

$$|D(t)| \leq c_0,$$

$$|g_1(t,\epsilon)| \leq c_0\epsilon^{r+1},$$

$$|g_2(t,\epsilon)| \leq c_0\epsilon^r (e^{-\alpha(t-\tau_k)/\epsilon} + e^{\alpha(t-\tau_{k+1})/\epsilon}), \tag{19.69}$$

$$|\alpha_k(\epsilon)| \leq c_0\epsilon^{r+1},$$

$$|\beta_k(\epsilon)| \leq c_0\epsilon^{r+1}$$

for $k \in \mathbb{Z}$ and $\tau_k < t \leq \tau_{k+1}$.

Let $\rho > 0$ be fixed and introduce the set

$$T_\rho = \{w \in PC(\mathbb{R}, \mathbb{R}^n) : \|w\| \leq \rho, w(t+T) = w(t), t \in \mathbb{R}\}.$$

From Lemma 19.3 it follows that the equation

$$\epsilon h' = A_1(t)h + B_1(t)w + g_1(t,\epsilon), \quad t \neq \tau_k,$$
$$\Delta h(\tau_k) = P^{(k)}h(\tau_k) + \alpha_k(\epsilon), \tag{19.70}$$

has a unique T-periodic solution $h(t) = h(t,w,\epsilon) \in PC(\mathbb{R}, \mathbb{R}^m)$. From estimates (19.69) and (19.61) it follows that there exist constants $L_0 > 0$ and $L_1 > 0$ such that

$$\|h(\,\cdot\,,w,\epsilon)\| \leq L_0\|w\| + L_1\epsilon^{r+1},$$
$$\|h(\,\cdot\,,w_1,\epsilon) - h(\,\cdot\,,w_2,\epsilon)\| \leq L_0\|w_1 - w_2\| \tag{19.71}$$

for $\epsilon \in (0,\epsilon_0]$ and $w_1, w_2 \in T_\rho$.

Define the operator $\phi_\epsilon : T_\rho \to PC(\mathbb{R}, \mathbb{R}^n)$ as follows: with each $w \in T_\rho$ associate $\tilde{v} = \phi_\epsilon w$, where $\tilde{v} = \tilde{v}(t,\epsilon)$ is the solution of the periodic value problem

$$\tilde{v}' = [B_2(t) - A_2(t)A_1^{-1}(t)B_1(t)]\tilde{v}$$
$$\qquad + D(t)[A_1(t)h(t) + B_1(t)w(t)] + g_2(t,\epsilon), \quad t \neq \tau_k,$$
$$\Delta\tilde{v}(\tau_k) = S^{(k)}\tilde{v}(\tau_k) + \beta_k(\epsilon), \tag{19.72}$$
$$\tilde{v}(0) = \tilde{v}(T).$$

The solution $\phi_\epsilon w$ of (19.72) is unique and can be represented in the form

$$\phi_\epsilon w(t) = \int_0^t W(t,s)D(s)[A_1(s)h(s) + B_1(s)w(s)]\,ds + G(t,h,\epsilon), \tag{19.73}$$

where $W(t,s)$ is the Cauchy matrix for equation (19.5) and

$$G(t,h,\epsilon) = W(t,0)[E_n - W(T,0)]^{-1}$$
$$\times \left\{ \int_0^T W(T,s)[D(s)(A_1(s)h(s) + B_1(s)w(s)) + g_2(s,\epsilon)]\,ds \right.$$
$$\left. + \sum_{k=1}^q W(T,\tau_k^+)\beta_k(\epsilon) \right\}$$
$$+ \int_0^t W(t,s)g_2(s,\epsilon)\,ds + \sum_{0<\tau_k<t} W(t,\tau_k^+)\beta_k(\epsilon).$$

From (19.73), taking into account that $h(t) = h(t, w, \epsilon)$ is a solution of (19.70), we obtain

$$\phi_\epsilon w(t) = \epsilon \int_0^t W(t, s)D(s)h'(s) \, ds - \int_0^t W(t, s)D(s)g_1(s, \epsilon) \, ds + G(t, h, \epsilon). \quad (19.74)$$

After integration by parts we obtain that

$$\epsilon \int_0^t W(t, s)D(s)h'(s) \, ds$$

$$= \epsilon D(t)h(t) - \epsilon W(t, 0)D(0)h(0) - \epsilon \int_0^t \frac{\partial}{\partial s}[W(t, s)D(s)]h(s) \, ds \quad (19.75)$$

$$+ \epsilon \sum_{0 < \tau_k < t} [W(t, \tau_k)D(\tau_k)h(\tau_k) - W(t, \tau_k^+)D(\tau_k^+)h(\tau_k^+)].$$

From (19.75) and the boundedness of W and $\frac{\partial}{\partial s}(WD)$ for $0 \le s \le t \le T$ it follows that there exists a constant $c_1 > 0$ such that

$$\left| \epsilon \int_0^t W(t, s)D(s)h'(s) \, ds \right| \le \epsilon c_1 \|h\| \quad (19.76)$$

for $t \in [0, T]$ and $\epsilon \in (0, \epsilon_0]$. Taking into account (19.69) we derive the estimates

$$\left| \int_0^t W(t, s)D(s)g_1(s, \epsilon) \, ds \right| \le c_2 \epsilon^{r+1}, \quad (19.77)$$

$$|G(t, h, \epsilon)| \le c_3 \epsilon \|h\| + c_4 \epsilon^{r+1}$$

for $t \in [0, T]$ and $\epsilon \in (0, \epsilon_0]$, where $c_i > 0$, $(i = 1, 2, 3, 4)$ do not depend on ϵ. From (19.74), (19.76), (19.77), (19.59) and (19.60) it follows that

$$\|\phi_\epsilon w\| \le c \epsilon \|w\| + c \epsilon^{r+1}, \quad (19.78)$$

$$\|\phi_\epsilon w_1 - \phi_\epsilon w_2\| \le c \epsilon \|w_1 - w_2\| \quad (19.79)$$

for $\epsilon \in (0, \epsilon_0]$, $w, w_1, w_2 \in T_\rho$, where $c > 0$ does not depend on ϵ.

Choose $\epsilon_* \in (0, \epsilon_0]$ so that

$$c\epsilon_*(\rho + \epsilon_*^r) < \rho, \qquad c\epsilon_* < 1. \quad (19.80)$$

From (19.78)–(19.80) it follows that for $\epsilon \in (0, \epsilon_*]$ the operator ϕ_ϵ maps T_ρ into itself and is a contraction. Denote by $v(t, \epsilon)$ the unique fixed point of ϕ_ϵ and let $u(t, \epsilon) = h(t, v(t, \epsilon), \epsilon)$. Then $(u(t, \epsilon), v(t, \epsilon))$ is the unique solution of the periodic value problem (19.67) for $\epsilon \in (0, \epsilon_*]$. From (19.78) and (19.71) there follows the estimate

$$|v(t, \epsilon)| \le M\epsilon^{r+1}, \qquad |u(t, \epsilon)| \le M\epsilon^{r+1} \quad (19.81)$$

for $t \in [0, T]$, $\epsilon \in (0, \epsilon_*]$, where $M > 0$ does not depend on ϵ.

From (19.81) and (19.66) there immediately follows (19.65). $\square$

Example 19.1 (Industrial robot)   The linearized equations of motion of an industrial robot with m degrees of freedom with an electric or hydraulic drive at every
joint have the form

$$\ddot{q} + A(t)\dot{q} + B(t)q = M(t) + f(t),$$

$$T\dot{M} + M = Ku \tag{19.82}$$

where $q \in \mathbb{R}^m$ is the phase vector, while $M \in \mathbb{R}^m$ is the vector of the exterior forces
or applied moments.

The second equation of system (19.82) describes the dynamics of the drive mech-
anisms. Here $u \in \mathbb{R}^m$ is the governing vector, $K = \mathrm{diag}\,(K_1,\ldots,K_m)$, and $T = \mathrm{diag}\,(T_1,\ldots,T_m)$, where the time constants T_i are small (of the order 10^{-2} s) and
are approximately equal to one another. Note that, in the case when the robot per-
forms cyclic operations, the vector of current perturbations $f(t) \in \mathbb{R}^m$ and the $m \times m$
matrices $A(t)$ and $B(t)$ are periodic functions of t.

System (19.82) can be written in the form

$$\epsilon \frac{dx}{dt} = A_1(t)x + Ku(t),$$

$$\frac{dy}{dt} = B_1(t)x + B_2(t)y + g_2(t) \tag{19.83}$$

where we have denoted

$$\epsilon = \min_{1 \leq i \leq m} T_i, \qquad \bar{T}_i = \frac{T_i}{\epsilon}, \qquad \bar{T} = \mathrm{diag}\,(\bar{T}_1,\ldots,\bar{T}_m),$$

$$x = \bar{T}M, \qquad A_1(t) = -\bar{T}^{-1}, \qquad y = \begin{bmatrix} q \\ \dot{q} \end{bmatrix}, \qquad B_1(t) = \begin{bmatrix} 0 \\ \bar{T}^{-1} \end{bmatrix},$$

$$B_2(t) = \begin{bmatrix} 0 & E_m \\ -B(t) & -A(t) \end{bmatrix}, \qquad g_2(t) = \begin{bmatrix} 0 \\ f(t) \end{bmatrix}.$$

The quality criterion

$$I = \epsilon(a|x(\tau_1)) + \int_0^\omega (u(t)|W(t)u(t))\,dt + \int_0^\omega (g_2(t)|y(t))\,dt \tag{19.84}$$

is associated with system (19.83), where ω is the period of system (19.83), $a \in \mathbb{R}^m$ is
a constant vector, while $\tau_1 \in (0,\omega)$ is a fixed moment.

An essential fact about the problem of minimizing the functional (19.84) with
respect to the periodic trajectories of system (19.83) is that the quality criterion
contains an extra-integral term $\epsilon(a|x(\tau_1))$: the average value of the drive forces or
moments for $t = \tau_1$. The problem of minimizing the functional (19.84) with respect
to the periodic trajectories of system (19.83) is a particular case of the following more
general problem.

Consider a controllable system described by the linear equations

$$\epsilon\frac{dx}{dt} = A_1(t)x + A_2(t)y + G_1(t)u + g_1(t),$$

$$\frac{dy}{dt} = B_1(t)x + B_2(t)y + G_2(t)u + g_2(t) \tag{19.85}$$

where $x \in \mathbb{R}^m$, $y \in \mathbb{R}^m$ is the state vector, $u \in \mathbb{R}^m$ is the control vector, $A_i(t)$, $B_i(t)$ and $G_i(t)$ $(i = 1, 2)$ are ω-periodic $m \times m$ matrices, $g_i(t)$ $(i = 1, 2)$ are ω-periodic m-dimensional vector-valued functions, while $\epsilon > 0$ is a small parameter.

For system (19.85) consider the problem of minimizing the functional

$$I = \sum_{i=1}^{p} \epsilon(a_i|x(\tau_i)) + \sum_{i=1}^{p}(b_i|y(\tau_i))$$

$$+ \int_0^\omega (u(t)|W(t)u(t))\,dt + \gamma_1 \int_0^\omega (g_1(t)|x(t))\,dt \tag{19.86}$$

$$+ \gamma_2 \int_0^\omega (g_2(t)|y(t))\,dt$$

in the class of functions that are piecewise continuous in $[0, \omega]$ and have discontinuities of the first kind at the points τ_i $(i = 1, \ldots, p)$. In (19.86), $a_i \in \mathbb{R}^m$, $b_i \in \mathbb{R}^m$, τ_i $(0 \leq \tau_1 < \cdots < \tau_p < \omega)$ are fixed moments, $W(t)$ is a positive definite symmetric ω-periodic $m \times m$ matrix, while γ_1 and γ_2 are positive real numbers.

The conjugate system for this problem has the form

$$\frac{d\psi}{dt} = -A_1^*(t)\frac{\psi}{\epsilon} - B_1^*(t)\eta + \gamma_1 g_1(t), \qquad t \neq \tau_i,$$

$$\frac{d\eta}{dt} = -A_2^*(t)\frac{\psi}{\epsilon} - B_2^*(t)\eta + \gamma_2 g_2(t), \qquad t \neq \tau_i, \tag{19.87}$$

$$\Delta\psi = \epsilon a_i, \qquad \Delta\eta = b_i, \qquad t = \tau_i$$

where $\Delta\psi = \psi(\tau_i^+) - \psi(\tau_i^-)$, $\Delta\eta = \eta(\tau_i^+) - \eta(\tau_i^-)$.

The maximum principle of Pontryagin et al. [64] implies that, if $\psi = \psi_\omega(t, \epsilon)$, $\eta = \eta_\omega(t, \epsilon)$ is an ω-periodic solution of system (19.87), the required control $\hat{u}_\epsilon(t)$ has the representation

$$\hat{u}_\epsilon(t) = \frac{1}{2}W^{-1}(t)[G_1^*(t)\psi_\omega(t, \epsilon) + G_2^*(t)\eta_\omega(t, \epsilon)].$$

Therefore, to solve the problem which we have set, it suffices to find an ω-periodic solution of system (19.87).

Let $i \in \mathbb{Z}$, $i = kp + r$, $1 \leq r < p$ and set $\tau_i = \tau_r + k\omega$, $a_i = a_r$, $b_i = b_r$. Then the instants τ_i and the vectors a_i and b_i satisfy the relations

$$\tau_{i+p} = \tau_i + \omega, \qquad a_{i+p} = a_i, \qquad b_{i+p} = b_i \qquad (i \in \mathbb{Z}).$$

Assume that the following conditions hold.

H19.8 The ω-periodic matrices $A_i(t)$, $B_i(t)$ $(i = 1, 2)$ have continuous derivatives up to order $n + 1$, inclusive, for $t \in \mathbb{R}$.

H19.9 The eigenvalues $\lambda_j(t)$ of the matrix $A_1(t)$ satisfy the condition $\Re\lambda_j(t) \geq \delta > 0$ for $t \in \mathbb{R}$.

H19.10 The functions $g_i(\,\cdot\,) \in PC^1(\mathbb{R}, \mathbb{R}^m)$ are ω-periodic.

H19.11 The linear homogeneous system

$$\frac{d\varphi}{dt} = [A_2^*(t)(A_1^*(t))^{-1}B_1^*(t) - B_2^*(t)]\varphi \qquad (19.88)$$

has no ω-periodic solutions other than the zero solution.

In (19.87) we set $\psi = \epsilon\xi$ and obtain the system

$$\epsilon\frac{d\xi}{dt} = \alpha_1(t)\xi + \beta_1(t)\eta + f_1(t), \qquad t \neq \tau_i,$$

$$\frac{d\eta}{dt} = \alpha_2(t)\xi + \beta_2(t)\eta + f_2(t), \qquad t \neq \tau_i, \qquad (19.89)$$

$$\Delta\xi = a_i, \qquad \Delta\eta = b_i, \qquad t = \tau_i$$

where $\alpha_j(t) = -A_j^*(t)$, $\beta_j(t) = -B_j^*(t)$, $f_j(t) = \gamma_j g_j(t)$ $(j = 1, 2)$.

By Theorem 19.1 there exists an $\epsilon_0 > 0$ such that for $\epsilon \in (0, \epsilon_0]$ system (19.89) (and system (19.87)) has a unique ω-periodic solution $\xi = \xi_\omega(t, \epsilon)$, $\eta = \eta_\omega(t, \epsilon)$ ($\psi = \epsilon\xi_\omega(t, \epsilon)$, $\eta = \eta_\omega(t, \epsilon)$). If, moreover, we suppose that the zero solution of system (19.88) is exponentially stable, then it can be proved [83] that the ω-periodic solution of system (19.89) (and of system (19.87)) is also exponentially stable.

Notes and comments for Chapter VI

For a result similar to Theorem 15.2 see [55]. The contents of Section 16 were adapted from A. M. Samoilenko and N. A. Perestyuk [73], Section 21. For other results see S. G. Hristova and D. D. Bainov [43] and R. N. Butris [20].

Section 17 contains the results of P. S. Simeonov and D. D. Bainov [85], while the results of Section 18 are due to D. D. Bainov and S. D. Milusheva [14].

Theorem 19.1 is due to M. A. Arolska-Hekimova and G. H. Sarafova [9]. Periodic singularly perturbed impulsive equations were considered also by M. A. Arolska-Hekimova [7] and by M. A. Hekimova and D. D. Bainov [37, 38]. Example 19.1 was taken from [38].

Finally we denote the following.

Almost periodic impulsive differential equations were considered by M. U. Ahmetov and N. A. Perestyuk [1, 3] and A. M. Samoilenko and N. A. Perestyuk [72, 73].

Periodic impulsive equations with deviating argument were discussed by M. A. Arolska and D. D. Bainov [8], D. D. Bainov and V. Chr. Covachev [11], R. N. Butris [20] and by S. G. Hristova and D. D. Bainov [41–43].

Periodic impulsive equations in a Banach (or in a Hilbert) space were investigated by D. D. Bainov, S. I. Kostadinov and A. D. Myshkis [13], V. I. Guţu [31, 32] and Yu. V. Rogovchenko and S. I. Trofimchuk [65].

References

1. Ahmetov, M. U. and Perestyuk, N. A., Periodic and almost periodic solutions of systems with impulse effect,*Math. Phys.*, **34** (1983), 3–8 (in Russian).
2. Ahmetov, M. U. and Perestyuk, N. A., Stability of periodic solutions of differential equations with impulse effect on surfaces,*Ukr. Math. J.*, **41** No. 12 (1989), 1596–1601 (in Russian).
3. Ahmetov, M. U. and Perestyuk, N. A., On the comparison method for differential equations with impulse effect, *Differential Equations*, **26**, No. 9 (1990), 1475–83 (in Russian).
4. Ahmetov, M. U. and Perestyuk, N. A., Periodic solutions of quasilinear impulsive systems in the critical case, *Ukr. Math. J.*, **43**, No. 3 (1991), 308–15 (in Russian).
5. Alexandrov, P. S., *Combinatorial Topology*, Gostekhizdat, Moscow–Leningrad, 1947 (in Russian).
6. Andronov, A. A., Witt, A. A. and Khaikin, S. E., *Oscillation Theory*, Nauka, Moscow, 1981 (in Russian).
7. Arolska-Hekimova, M. A., Existence and properties of periodic solutions of linear singularly perturbed systems of differential equations with impulse effect, *Serdica, Bulgariae Mathematicae Publicationes*, **13** (1987), 144–9 (in Russian).
8. Arolska, M. A. and Bainov, D. D., Periodic solutions of differential-difference systems of neutral type with impulse effect, *Annales Universitatis Scientiarum Budapestinensis de Rolando Eötvös Nominatae Sectio Mathematica*, **XXVI** (1983), 103–12 (in Russian).
9. Arolska-Hekimova, M. A. and Sarafova, G. H., Periodic solutions of linear singularly perturbed systems of differential equations with impulse effect, *Serdica, Bulgariae Mathematicae Publicationes* , **13** (1987), 63–75 (in Russian).
10. Babitskii, V. I. and Krupenin, V. L., *Oscillations in Strongly Nonlinear Systems*, Nauka, Moscow, 1985 (in Russian).
11. Bainov, D. D. and Covachev, V. Chr., Periodic solutions of impulsive systems with delay viewed as a small parameter (to appear).
12. Bainov, D. D., Hristova, S. G., Hu, S. and Lakshmikantham, V., Periodic boundary value problems for systems of first order impulsive equations, *Differential and Integral Equations*, **2**, No. 1 (1989), 37–43.
13. Bainov, D. D., Kostadinov, S. I. and Myshkis, A. D., Bounded and periodic solutions of differential equations with impulse effect in a Banach space, *Differential and Integral Equations*, **1**, No. 2 (1988), 223–30.

14. Bainov, D. D. and Milusheva, S. D., A projection-iterative method for solving the periodic problem for ordinary differential equations with impulse effect, *Kumamoto J. Math.*, **2** (1989), 40–8.

15. Bainov, D. D. and Simeonov, P. S., *Systems with Impulse Effect. Stability, Theory and Applications*, Ellis Horwood Series in Mathematics and its Applications, Ellis Horwood, Chichester, 1989.

16. Bainov, D. D. and Simeonov, P. S., *Integral Inequalities and Applications*, Kluwer Academic Publishers, Dordrecht, 1992.

17. Bautin, N. N., Theory of point transformations and dynamical theory of clocks, *Proc. International Conference on Non-linear Oscillations V*, Acad. Sci. Ukr., Kiev, Vol. II, 1963, 29–54 (in Russian).

18. Bellman, R., *Mathematical Methods in Medicine*, World Scientific Series in Modern Applied Mathematics, Volume 1, 1983.

19. Boichuk, A. A., Perestyuk, N. A. and Samoilenko, A. M., Periodic solutions of impulsive differential equations in the critical cases, *Differential Equations*, **XXVII**, No. 9 (1991), 1516–21 (in Russian).

20. Butris, R. N., On the periodic solutions of nonlinear differential-operator equations with impulse effect, *Ukr. Math. J.*, **43**, No. 9 (1991), 1260–4 (in Russian).

21. Coddington, E. A. and Levinson, N., *Theory of Ordinary Differential Equations*, McGraw-Hill, New York, 1955.

22. Coppel, W. A., *Dichotomies in Stability Theory*, Springer Verlag, Lecture Notes in Mathematics, **629**, Berlin, 1978.

23. Demidovich, B. P., *Lectures on Mathematical Stability Theory*, Nauka, Moscow, 1967 (in Russian).

24. Dishliev, A. B. and Bainov, D. D., Sufficient conditions for absence of 'beating' in systems of differential equations with impulses, *Appl. Anal*, **18** (1984), 67–73.

25. Dishliev, A. B. and Bainov, D. D., Conditions for the absence of the phenomenon 'beating' for systems of impulse differential equations, *Bull. Inst. Math. Acad. Sin.*, **13**, No. 2 (1985), 237–56.

26. Erbe, L. H. and Xinzhi Liu, Existence of periodic solutions of impulsive differential systems, *J. Appl. Math. and Stochastic Anal.*, **4** (1991), 137–46.

27. Flatto, L. and Levinson, N., Periodic solutions of singularly perturbed systems, *J. Rat. Mech. Anal.*, **4** (1955), 943–50.

28. Gopalsamy, K., Limit cycles in periodically perturbed population systems, *Bull. Math. Biol.*, **43**, No. 4 (1981), 463–85.

29. Grebenikov, E. A. and Ryabov, Yu. A., *Constructive Methods of Analysis of Nonlinear Systems*, Nauka, Moscow, 1979 (in Russian).

30. Grechko, V. I., On a projection-iterative method of finding periodic solutions of systems of ordinary differential equations, *Ukr. Math. J.*, **26**, No. 4 (1974), 534–9 (in Russian).

31. Guţu, V. I., Periodic solutions of linear differential equations with impulses in a Banach space, *Differential Equations and Mathematical Physics*, 1989, pp. 59–66 (in Russian).

32. Guţu, V. I., Periodic solutions of weakly nonlinear differential equations in a Hilbert space, *Differential Equations and Mathematical Phys-ics*, 1989, pp. 67–71 (in Russian).

33. Guţu, V. I., Period solutions of one- and two-dimensional equations with impulses, *Differential Equations*, **26**, No. 5 (1990), 904–6 (in Russian).

34. Halanay, A. and Wexler, D., *Qualitative Theory of Impulsive Systems*, Editura Academiei Republici Socialiste România, Bucharest, 1968.

35. Hale, J. K., *Oscillations in Nonlinear Systems*, McGraw-Hill, New York, 1963.

36. Hcu, C. S., Impulsive parametric excitation, ASME Paper N WA/APM 19 (1971), p. 8.

37. Hekimova, M. A. and Bainov, D. D., Periodic solutions of singularly-perturbed systems of differential equations with impulse effect, *Zeitschrift für Angewandte Mathematik und Physik*, **36** (1985), 520–37.

38. Hekimova, M. A. and Bainov, D. D., A method for asymptotic integration of a singularly perturbed system with impulses and its application in the theory of optimal control, *J. Math. Pures et Appl.*, **65**, No. 3 (1986), 307–21.

39. Hristova, S. G. and Bainov, D. D., Periodic solutions of quasilinear non-autonomous systems with impulses, *Math. Mech. in the Appl. Sci.*, **8** (1986), 247–55.

40. Hristova, S. G. and Bainov, D. D., Existence of periodic solutions of nonlinear systems of differential equations with impulse effect, *Journal of Mathematical Analysis and Applications*, **125**, No. 1 (1987), 192–202.

41. Hristova, S. G. and Bainov, D. D., The method of the small parameter in the theory of differential equations with impulse effect, *Riv. Mat. Univ. Parma*, (4) **13** (1987), 91–100.

42. Hristova, S. G. and Bainov, D. D., A projection-iterative method for finding periodic solutions of nonlinear systems of difference-differential equations with impulses, *J. Approx. Theory*, **49**, No. 4 (1987), 311–20.

43. Hristova, S. G. and Bainov, D. D., Numeric-analytic method for finding the periodic solutions of nonlinear differential-difference equations with impulses, *Computing*, **38** (1987), 363–8.

44. Hristova, S. G. and Bainov, D. D., Application of Lyapunov's functions to finding periodic solutions of systems of differential equations with impulses, *Bol. Soc. Paran. Mat.* (2ª série), **9**, No. 2 (1988), 151–63.

45. Hu, S. and Lakshmikantham, V., Periodic boundary value problems for second order impulsive differential systems, *Nonlinear Analysis, Theory, Methods and Applications*, **13**, No. 1 (1989), 75–85.

46. Hutson, V. C. L. and Pym, J. S., *Applications of Functional Analysis and Operator Theory*, Academic Press, London, 1980.

47. Kalitin, B. S., On the oscillations of the pendulum with shock impulse, *Differential Equations*, **V**, No. 7 (1969), 1267–74 (in Russian).

48. Kalitin, B. S., On the oscillations of the pendulum with shock impulse II, *Differential Equations*, **VI**, No. 12 (1970), 2174–81 (in Russian).

49. Kalitin, B. S., On the limiting cycles of pendulum systems with impulse distur-
    bance, *Differential Equations*, **VII**, No. 3 (1971), 540–2 (in Russian).

50. Kobrinskii, A. E. and Kobrinskii, A. A., *Vibroshock Systems*, Nauka, Moscow,
    1973 (in Russian).

51. Krasenosel'skii, M. A., *Monodromy Operator along the Trajectories of Differential
    Equations*, Nauka, Moscow, 1966 (in Russian).

52. Kruger-Thiemer, E., Formal theory of drug dosage regiments, I, *J. Theor. Biol.*,
    13 (1966).

53. Krylov, N. M. and Bogolyubov, N. N., *Introduction to Nonlinear Mechanics*,
    Publishing House of the Academy of Sciences of Ukr. SSR, Kiev, 1937 (in Rus-
    sian).

54. Kurpel', N. S., On bilateral approximations to periodic solutions of differen-
    tial equations, *Proc. International Conference on Non-linear Oscillations V*,
    Inst. Math. Acad. Sci. Ukr., Kiev, Vol. I, 1970, pp. 348–52 (in Russian).

55. Lakshmikantham, V., Bainov, D. D. and Simeonov, P. S., *Theory of Impulsive
    Differential Equations*, World Scientific Series in Modern Applied Mathematics,
    Vol. 6, Singapore, 1989.

56. Liz, E. and Nieto, J. J., Periodic solutions of dicontinuous impulsive differential
    systems, *Journal of Mathematical Analysis and Applications*, **161**, No. 2 (1991),
    388–94.

57. Mil'man, V. D. and Myshkis, A. D., On the stability of motion in the presence
    of impulses, *Sib. Math. J.*, **1**, No. 2 (1960), 233–7 (in Russian).

58. Mil'man, V. D. and Myshkis, A. D., Random impulses in linear dynamical sys-
    tems. In *Approximate Methods for Solving Differential Equations*, Publishing
    House of the Academy of Sciences of Ukr. SSR, Kiev, 1963, pp. 64–81 (in Rus-
    sian).

59. Mitropol'skii, Yu. A. and Martynyuk, D. I., *Periodic and Quasiperiodic Oscilla-
    tions of Systems with Delay*, Višča Škola, Kiev, 1979 (in Russian).

60. Myshkis, A. D. and Samoilenko, A. M., Systems with impulses at fixed moments
    of time, *Mat. Sb.*, **74**, No. 2 (1967), 202–8 (in Russian).

61. Pandit, S.G. and Deo, S. G., *Differential Systems Involving Impulses*, Lecture
    Notes 954, Springer Verlag, Berlin, 1982.

62. Pierson-Gorez, C., Impulsive differential equations with periodic and nonlinear
    boundary conditions (to appear).

63. Pliss, V. A., *Nonlocal Problems of Oscillation Theory*, Nauka, Moscow, 1964 (in
    Russian).

64. Pontryagin, L. S., Boltianskii, V. G., Gamkrelidze, R. V. and Miščenko, E. F.,
    Mathematical Theory of Optimal Control, Nauka, Moscow, 1969 (in Russian).

65. Rogovchenko, Yu. V. and Trofimchuk, S. I., Bounded and periodic solutions of
    weakly nonlinear impulsive evolutionary systems, *Ukr. Math. J.*, **39** No. 2 (1987),
    260–4 (in Russian).

66. Rogovchenko, Yu.V. and Trofimchuk, S.I., Periodic solutions of weakly nonlinear
    systems with impulse effect, *Ukr. Math. J.*, **41**, No. 5 (1989), 622–6 (in Russian).

67. Romanovskii, Yu. M., Stepanova, N. V. and Chernavskii, D. S., *Mathematical Biophysics*, Nauka, Moscow, 1984 (in Russian).

68. Samoilenko, A.M., Numerical-analytical method of investigation of periodic systems of ordinary differential equations, I, II, *Ukr. Math. J.*, **17**, No. 4 (1965), 82–93; **18**, No. 2 (1966), 50–9 (in Russian).

69. Samoilenko, A. M. and Elgondyev, K. K., *Investigation of Linear Differential Equations with Impulse Effect in* $\mathbb{R}^2$, Kiev, Inst. Math. Acad. Sci. Ukr. Preprint 89.59 (in Russian).

70. Samoilenko, A. M. and Perestyuk, N. A., Stability of the solutions of differential equations with impulse effect, *Differential Equations*, **13**, No 11 (1977), 1981–92 (in Russian).

71. Samoilenko, A. M. and Perestyuk, N. A., Periodic solutions of weakly nonlinear systems with impulse effect, *Differential Equations*, **XIV**, No. 6 (1978), 1034–45 (in Russian).

72. Samoilenko, A. M. and Perestyuk, N. A., Periodic and almost periodic solutions of differential equations with impulse effect, *Ukr. Math. J.*, **34**, No. 1 (1982), 66–73 (in Russian).

73. Samoilenko, A. M. and Perestyuk, N. A., *Differential Equations with Impulse Effect*, Višča Škola, Kiev, 1987 (in Russian).

74. Samoilenko, A. M., Perestyuk, N. A. and Trofimchuk, S. I., Generalized solutions of impulsive systems and the phenomenon 'beating', *Ukr. Math. J.*, **43**, No. 5 (1991), 657–63 (in Russian).

75. Simeonov, P. S., Existence, uniqueness and continuability of the solutions of systems with impulse effect, *Godishnik VUZ Appl. Math.*, **22**, No. 3 (1986), 69–78 (in Russian).

76. Simeonov, P. S., Continuous dependence of the solutions of systems with impulse effect, *Godishnik VUZ Appl. Math.*, **22**, No. 3 (1986), 57–67 (in Russian).

77. Simeonov, P. S. and Bainov, D. D., Differentiability of solutions of systems with impulse effect with respect to initial data and parameter, *Institute of Mathematics, Academia Sinica*, **15**, No 2 (1987), 251–69.

78. Simeonov, P. S. and Bainov, D. D., On an integral inequality for piecewise continuous functions, *J. Math. and Phys. Sci.*, **21**, No. 4 (1987), 315–23.

79. Simeonov, P. S. and Bainov, D. D., Perturbation theorems for systems with impulse effect, *Int. J. Systems Sci.*, **19**, No. 7 (1988), 1213–23.

80. Simeonov, P. S. and Bainov, D. D., Integral and differential inequalities for a class of piecewise-continuous functions, *Annales Polonici Mathematici*, **XLVIII** (1988), 207–16 (in Russian).

81. Simeonov, P. S. and Bainov, D. D., On the asymptotic equivalence of systems with impulse effect, *Journal of Mathematical Analysis and Applications* , **135**, No. 2 (1988), 591–610.

82. Simeonov, P. S. and Bainov, D. D., Exponential stability of the solutions of the initial-value problem for systems with impulse effect, *J. Computational and Appl. Math.*, **23** (1988), 353–65.

83. Simeonov, P. S. and Bainov, D. D., Stability of the solutions of singularly perturbed systems with impulse effect, *JMAA*, **136**, No. 2 (1988), 575–88.

84. Simeonov, P. S. and Bainov, D. D., Orbital stability of the periodic solutions of autonomous systems with impulse effect, *Int. J. Systems Sci.*, **19**, No. 12 (1988), 2561–85.

85. Simeonov, P. S. and Bainov, D. D., Application of the method of the two-sided approximations to the solutions of the periodic problem for impulsive differential equations, *Tamkang J. Math.*, **22**, No. 3 (1991), 275–84.

86. Vasil'eva, A. B. and Butuzov, V. F., *Asymptotic Expansions of the Solutions of Singularly-Perturbed Equations*, Nauka, Moscow, 1973 (in Russian).

87. Vatsala, A. S. and Yong Sun, Periodic boundary value problems of impulsive differential equations, *Applicable Analysis*, **44** (1992), 145–58.

88. Volterra, V., *Leçons sur la Théorie Mathématique de la Lutte pour la Vie*, Gauthier-Villars, Paris, 1931.

89. Weinberg, M. M. and Trenogin, V. A., *Theory of Branching of the Solutions of Nonlinear Equations*, Nauka, Moscow, 1969 (in Russian).

90. Yakubovich, V. A. and Starzinskii, V. M., *Linear Differential Equations with Periodic Coefficients and their Applications*, Nauka, Moscow, 1972 (in Russian).

91. Zavališčin, S. T., Sesekin, A. N. and Drozdenko, S. E., *Dynamical Systems with Impulse Structure*, Middle-Ural Publishing House, Sverdlovsk, 1983 (in Russian).

92. Zhang, B. G. and Gopalsamy, K., Global attractivity and oscillations in a periodic delay-logistic equation, *JMAA*, **150**, (1990), 274–83.

Author Index

Subject Index

Milton Keynes UK
Ingram Content Group UK Ltd.
UKHW040103071024
449327UK00019B/772